Building and Using Comparable Corpora

Serge Sharoff · Reinhard Rapp
Pierre Zweigenbaum · Pascale Fung
Editors

Building and Using Comparable Corpora

Springer

Editors
Serge Sharoff
Centre for Translation Studies
University of Leeds
Leeds
UK

Reinhard Rapp
University of Mainz
Mainz
Germany

Pierre Zweigenbaum
LIMSI-CNRS
Université de Paris-Sud
Orsay
France

Pascale Fung
Department of Science and Technology,
Electronic and Computer Engineering
The Hong Kong University
Hong Kong
People's Republic of China

ISBN 978-3-642-20127-1 ISBN 978-3-642-20128-8 (eBook)
DOI 10.1007/978-3-642-20128-8
Springer Heidelberg New York Dordrecht London

Library of Congress Control Number: 2013947575

Printed on acid-free paper

Springer is part of Springer Science+Business Media (www.springer.com)

Preface

This book came from the experience of a series of annual BUCC workshops. The first workshop of this kind was held in 2008 at LREC in Marrakech organised by Pierre Zweigenbaum, Éric Gaussier and Pascale Fung. Since then, the workshops changed the continents (Singapore in 2009, Malta in 2010, Portland, Oregon, 2011, Istanbul 2012); the organising committee included Reinhard Rapp, Serge Sharoff and Marko Tadic, but its main topic remained the same, focusing on the need to use comparable corpora as training data for linguistic research and NLP applications. The chapters for this volume were collected mostly from the best submissions to the workshops at the end of 2011 or through specific requests to the most prominent authors in this field. After completing the editorial process the collection of chapters is presented to your attention.

The volume starts with a chapter overviewing the state of the art. It discusses the rationale behind the use of comparable corpora, as well the issues involved in their collection, annotation and use. The rest of the volume consists of two parts. Part I is devoted to methods of compiling comparable corpora and measuring the degree of comparability between their documents. Part II is on applications which use comparable corpora in various contexts such as Machine Translation or computer-assisted human translation.

In Part I there are eight chapters.

"Mining Parallel Documents Using Low Bandwidth and High Precision CLIR from the Heterogeneous Web" by Shi and Fung proposes a method for mining parallel documents, which is based on the principles of cross-lingual information retrieval. The quality of resources obtained in this way is evaluated by using SMT.

"Automatic Comparable Web Corpora Collection and Bilingual Terminology Extraction for Specialized Dictionary Making" by Gurrutxaga et al. presents two tools, respectively, for compiling comparable corpora from the Web and for extracting bilingual terminology from them. The authors are specifically interested in under-resourced languages, Basque in their case, when the number of relevant webpages is relatively small. The paper describes the use of the standard tools (Boot- Cat, Kimatu, etc) and application of the context feature vectors for aligning monolingual term lists.

"Statistical Comparability: Methodological Caveats" by Köhler explores the issue of corpus comparability from the viewpoint of statistical testing. It shows how the notions of statistics of frequency distributions, such as homoscedasticity and skewness, can be applied to analyse comparable corpora, including the issues of their representativeness, homogeneity, as well as comparability.

"Methods for Collection and Evaluation of Comparable Documents" by Paramita et al., similar to "Automatic Comparable Web Corpora Collection and Bilingual Terminology Extraction for Specialized Dictionary Making", also focuses on the collection of comparable corpora from the Web for under-resourced languages. The authors investigate the use of the interwiki links in Wikipedia and retrieval of Twitter tweets by using URLs and topics as queries. They also propose methods to evaluate the retrieved documents using automatic classification of their comparability levels.

"Measuring the Distance Between Comparable Corpora Between Languages" by Sharoff explores methods for comparing corpora of unknown composition using keywords. First, he explores attempts at approximating the content of corpora collected from the Web using various methods, also in comparison to traditional corpora, such as the BNC. The procedure for estimating the corpus composition is based on selecting keywords, followed by clustering. This can apply to corpora within the same language, e.g., the BNC against ukWac as well as to corpora in different languages, e.g., webpages collected using the same procedure for English and Russian.

Li and Gaussier ("Exploiting Comparable Corpora for Lexicon Extraction: Measuring and Improving Corpus Quality") take care of an important property of comparable corpora: their degree of comparability. They propose a measure of comparability which is linked to the possibility of extracting word translations from comparable corpora. They show that this measure correlates with intuition on a range of artificial comparable corpora. They design a bilingual clustering method which increases this measure through a controlled extension of initial comparable corpora, and show that the bilingual lexicons they extract from these corpora are indeed improved by this process.

"Statistical Corpus and Language Comparison on Comparable Corpora" by Eckart and Quasthoff describes the construction of the Leipzig Corpus Collection which currently grows at a rate of 30 GB per month. It also gives an overview on a number of applications of this comparable data, and highlights some of its statistical properties such as n-gram frequencies, word co-occurrences and the distributions of word and sentence lengths. An integral part of the system is a web portal which gives an overview on the corpora and serves as a starting point for evaluating phenomena relating to corpus, genre and language comparison.

"Comparable Multilingual Patents as Large-Scale Parallel Corpora" by Lu and Tsou describes methods used for building a large-scale multilingual corpus of comparable patents for a range of languages, such as Chinese, Japanese, Korean and German. The chapter also discusses a procedure to extract parallel sentences from these patents to build an SMT system.

In Part II we also have eight chapters.

"Extracting Parallel Phrases from Comparable Data" by Hewavitharana and Vogel deals with the problem of discovering parallelness in comparable data at the sub-sentential level, i.e. to extract parallel phrases embedded in comparable sentences. They explore and quantitatively compare three different approaches: Using the standard Viterbi phrase alignment, using lexical features only without relying on the Viterbi path of word alignments, and using a maximum entropy classifier which is applied on large collections of phrase pair candidates. Their finding is that the second approach leads to the best results in terms of F-measure.

Similar to the first chapter of Part II, the "Exploiting Comparable Corpora", authored by Munteanu and Marcu, also deals with the extraction of parallel sub-sentential fragments from comparable corpora. However, they use a completely different approach which is based on signal filtering, whereby the signal is derived from word translation probabilities. Also, their evaluation procedure is task based. They showed that by adding the parallel data as extracted from the comparable corpora to the non-domain specific parallel training data of a statistical MT system, the translation quality improved.

Deléger et al. ("Paraphrase Detection in Monolingual Specialized/Lay Comparable Corpora") study same-language comparable corpora, where the dimension of comparability is a contrast in discourse type: texts intended for specialists of a domain (health) versus texts intended for lay people. They identify systematic variations in the expression of information in such comparable corpora. For this purpose, they test both a top-down approach, applying given variation patterns, and a bottom-up approach, discovering such patterns from the observation of data. The most common patterns evidence a preference for verb nominalisations and for relational adjectives in specialized language, as opposed to lay language.

Ji et al. ("Information Network Construction and Alignment from Automatically Acquired Comparable Corpora") describe an approach for acquiring cross-lingual comparable corpora which is based on concept extraction from videos. The corpora thus obtained are then used to identify translations of names using a weakly supervised and language-independent bootstrapping approach. The approach uses as seeds expressions that have the same forms in different languages, and—based on link comparison—iteratively mines more and more name translations.

Morin et al. ("Bilingual Terminology Mining from Language for Special Purposes Comparable Corpora") deal with small comparable corpora (250 kwords) in specialised domains, which reduces the discriminative power of the context vectors used in the standard approach of bilingual lexicon extraction. They propose two directions which they show improve bilingual lexicon extraction in this situation. First, to make them more discriminant, they boost the importance given to specific words which they consider as 'anchor points': transliterated words and neoclassical compounds. Second, they experiment with a small, in-domain parallel corpus from which they extract an additional bilingual lexicon which they use to extend the seed lexicon of the standard method.

Kageura and Abekawa ("The Place of Comparable Corpora in Providing Terminological Reference Information to Online Translators: A Strategic Framework") notice that the recent advances in the term extraction and alignment methods are not taken into translation practice. They are primarily interested in using comparable corpora to provide terminological resources, especially in the context of online collaborative translation. They advocate the use of comparable corpora for a *posteriori* enquiries after bilingual term candidates have been extracted.

"Old Needs, New Solutions: Comparable Corpora for Language Professionals" by Bernardini and Ferraresi is also concerned about the use of comparable corpora by professional translators. The authors investigate the context for using different types of corpora, including small *ad hoc* corpora (very small, but reliable), large web-derived reference corpora (with abundance of data, but little specialisation) and interactively constructed semi-automatic corpora, which occupy the middle ground and offer a positive trade-off between the effort needed to construct the corpora and their perceived usefulness.

"Exploiting the Incomparability of Comparable Corpora for Contrastive Linguistics and Translation Studies" by Neumann and Hansen-Schirra investigates the use of a comparable corpus of English and German, which includes both monolingually comparable texts and texts with their translations. The chapter provides insights from a feature matrix to reveal differences and commonalities between the original texts in two languages (English and German) as well as between originals and their translations in the same language.

Serge Sharoff
Reinhard Rapp
Pierre Zweigenbaum
Pascale Fung

Acknowledgments

In the process of preparing this volume we received a lot of help from the publishers, especially from Olga Chiarcos and Federica Corradi dell Acqua, as well as from the editors of related series, Ed Hovy and Nancy Ide. We really appreciate their advice. We are also grateful to our authors and reviewers, and for the support obtained from the 7th European Community Framework Programme via the projects TTC, Monotrans, HyghTra, and AutoWordNet.

Serge Sharoff
Reinhard Rapp
Pierre Zweigenbaum
Pascale Fung

Contents

Overviewing Important Aspects of the Last Twenty Years of Research in Comparable Corpora

Serge Sharoff, Reinhard Rapp and Pierre Zweigenbaum

"That is not said right," said the Caterpillar.
"Not quite right, I'm afraid," said Alice, timidly:
"some of the words have got altered."
Lewis Carroll, Alice's Adventures in Wonderland

1 Data-driven Turn

The beginning of the 1990s marked a radical turn in various NLP applications towards using large collections of texts. For translation-related studies this implied the use of parallel corpora, i.e. authentic translations. Probably the first research group to explore this approach was the one at the IBM Watson Centre [11]. However, the use of parallel data predates the appearance of the computer, as evidenced from the Rosetta Stone, which contained the same text in three languages, thus providing the vital clue to deciphering the Egyptian hieroglyphs by Jean-François Champollion in 1822 [12]. It is interesting that more modern computational methods are still used for solving somewhat similar tasks [48].

For producing statistically reliable results the corpora need to be large, while the usual sources of large parallel corpora are public organisations producing a large

S. Sharoff (✉)
University of Leeds, West Yorkshire, United Kingdom
e-mail: s.sharoff@leeds.ac.uk

R. Rapp
University of Mainz, Mainz, Germany
e-mail: reinhardrapp@gmx.de

P. Zweigenbaum
LIMSI, CNRS and ERTIM, INALCO, Paris, France
e-mail: pz@limsi.fr

S. Sharoff et al. (eds.), *Building and Using Comparable Corpora*,
DOI: 10.1007/978-3-642-20128-8_1, © Springer-Verlag Berlin Heidelberg 2013

de KDE ist ein Projekt zur Entwicklung Freier Software. Hauptprodukt ist die Software Compilation bestehend aus KDE Workspace (auf deutsch: KDE Arbeitsumgebung; früher: K Desktop Environment), einer Arbeitsumgebung, sowie vielen Zusatzprogrammen für den täglichen Gebrauch, den KDE Applications. KDE Workspace ist vorrangig für Computer gedacht, auf denen ein Unix-ähnliches Betriebssystem läuft, wie zum Beispiel BSD, Linux oder Solaris.

en KDE (pronounced /keɪdi:'i:/) is a free software project based around its flagship product, a cross-platform desktop environment designed to run on Linux, FreeBSD, Windows and Mac OS X systems. The goal of the project is to provide basic desktop functions and applications for daily needs as well as tools and documentation for developers to write stand-alone applications for the system.

Fig. 1 Wikipedia articles on KDE in German and English

amount of translations, which are available in the public domain (usually because of the status of such organisations). Examples of corpora frequently used in NLP research are the Canadian Hansards [38], European Parliament proceedings [49], or the United Nations documents [23]. Such repositories are often the main resource for testing new tools and methods in Statistical Machine Translation.

However, reliance only on existing parallel texts leads to serious limitations, since the domains and genres of texts from such institutional repositories often do not match well the targets of NLP applications, e.g., the accuracy of statistical machine translation crucially depends on a good match between the training corpus and the texts to be translated [5, 22]. Also many more texts are produced monolingually in each language than produced by professional translators. This is the reason why many researchers have switched to using comparable (=less parallel) resources to mine information about possible translations. The importance of this research strand was first recognised in the 1990s [29, 64].

2 Collecting Comparable Resources

2.1 *Degrees of Comparability*

It is important to note that the distinction between comparable (non-parallel) and parallel corpora is not a clear-cut line. Informally any collection of texts covering two different languages can be measured along the scale of 'fully parallel' to 'non-related' with several options in between.

Parallel texts

These are traditional parallel texts, which can be classified into:

- Texts which are true and accurate translations, such as the UN or EuroParl documents;

- Texts which are reasonable translations with minor language-specific variations, e.g., an example of search in the OpenOffice user manuals for *New York* might be replaced with *Beijing* in the Chinese version;

Strongly comparable texts

They are heavily edited translations or independent, but closely related texts reporting the same event or describing the same subject. This category includes:

- Texts coming from the same source with the same editorial control, but written in different languages, e.g. the BBC News in English and Romanian [58];
- Independently written texts concerning the same subject, e.g. Wikipedia articles linked via iwiki, see Fig. 1 from Wikipedia, or news items concerning exactly the same specific event from different news agencies, such as AFP, DPA and Reuters;
- In exclusively oral languages, multiple recordings of a shared story [51]; once transcribed and augmented with an English gloss, they provide a comparable corpus in which correspondences can be searched;
- In sign languages, another instance of languages which do not come in the form of written texts, translations or multiple narrations of a same story: [73] outline how the gradation of parallel to comparable corpora can apply to sign language corpora in one or multiple languages.

Weakly comparable texts

This category includes:

- Texts in the same narrow subject domain and genre, but describing different events, e.g., parliamentary debates on health care from the Bundestag, the House of Commons and the Russian Duma;
- Texts within the same broader domain and genre, but varying in subdomains and specific genres, e.g., a crawl of discussion forums in information technology might bring more technical discussions on Linux server administration in English vs more user-oriented discussions on AutoCAD drawing issues in French.

Unrelated texts

This category comprises the vast majority of Internet texts, which can still be used for comparative linguistic research. For example, one can use random snapshots of the Web for Chinese, English, German and Russian to deliver comparable language teaching materials for these languages [47, 74].

2.2 *Measuring Comparability*

There is an inevitable trade-off between the amount of noise and the amount of data along this scale: fewer texts are translated than produced monolingually, fewer events are covered by news agencies in exactly the same way in many languages than the number of monolingual stories in each of these languages. On the other hand, more parallel collections tend to be more useful for NLP applications, since more information can be extracted from greater parallelism in their content. In the 1990s and the beginning of the 2000s, work in computational translation studies (Statistical Machine Translation and Terminology Extraction) was mostly based on parallel corpora. Weakly-comparable and unrelated corpora have not been used a lot in computational research. Research presented in this volume (and in the events of the BUCC workshop series which preceded it) uses 'strongly-comparable' corpora.

In addition to these informal ways of assessing comparability, a more formal definition is based on measuring the distance between texts in their similarity space. This distance in the case of monolingual documents was first discussed by Adam Kilgarriff using the BNC as a benchmark: a *Known Similarity Corpus* was composed of documents known to be inherently similar within each category, while considerably different across the categories [46]. The distance between the documents in this approach can be measured by the degree of overlap between their keywords. There can be some difference in the way the keywords are extracted (top 500 words as used in [46], tf*idf, ll-score, etc.), as well as how the exact distance measure is defined (χ^2 in [46], cosine, Euclidean, etc.). Although, this suggestion from Kilgarriff was done within the same language, the idea can be extended further to measure corpus comparability by "translating" the documents from another language using either MT or simple dictionary mapping. Alternatively, instead of using more common words it is also possible to use Hapax Legomena (words occuring only once in each document) in order to identify potentially parallel documents [25, 59]. The advantage of this approach for closely related languages is that it makes it possible to by-pass the unreliable dictionaries and MT systems, while proper names and dates tend to be identical. If the aim is to investigate the relations between noisier collections (weakly comparable), it is possible to rely on classification of texts into topics and genres [75] under the assumption that the same labels are used for each language.

Irrespective of the approach to *measuring* text similarity, a benchmark for its *evaluation* is needed, which can be set in several ways:

- By using document-aligned parallel resources, such as Europarl or mining new parallel text collections;
- By using document-aligned comparable corpora, such as the dumps of the Wikipedia articles with information about their wiki categories and iwiki links between the languages;
- By collecting comparable resources using well-aligned keywords sent to a search engine, e.g., *autosave, configuring, debugger, user-friendly* for English versus *autoguardar, configurar, depurador, amigable* for Spanish [7].

Each of the approaches has its advantages and disadvantages. Evaluation using document-aligned parallel resources relies on texts which are known to be identical in terms of their topics, but such evaluation underestimates the degree of variation possible in comparable, originally produced texts. At the same time, a procedure for collecting comparable documents from the Web needs its own evaluation on the accuracy of the collection procedure. One way out of this loop is by using extrinsic evaluation, i.e., by judging how suitable a comparable corpus is for a given multilingual task, such as extraction of parallel sentences or terms [8] or, better, a more finalized task such as cross-language information retrieval [53].

2.3 Monolingual Comparable Corpora

Comparable corpora are usually built by selecting two different languages, specifying a set of dimensions (topic, genre, time period, etc.) and selecting texts in these two languages with similar values for these dimensions.

However, monolingual comparable corpora can also be built. In this case the language dimension is fixed and it is one of the other dimensions which varies, for instance the intended audience (domain specialists versus lay people [19, 24]), the time period (e.g., nineteenth century press vs contemporary press [71]) or the source (different news agencies reporting on events in the same time period [83]).

Finding word or term ‘translations’ across these new varying dimensions presents different questions: it is facilitated by the proximity of language and large number of shared words in the two parts of the corpus. For instance, word overlap allows [83] to pair documents and then sentences; word alignment is then used to identify paraphrase segments. In [19], the dimension of variation is the intended audience: morphosemantic relations are used to detect matching expressions in lay and specialized corpora, reflecting differences in the patterns used in these two discourse types. These morphosemantic relations are discovered in [18] through POS-tagged n-gram alignment, taking into account linguistically-motivated morphosemantic variations.

2.4 Mining Parallel and Comparable Corpora

The easiest way of mining parallel corpora is by directly re-using the output of translation work in the form of segment-aligned TMX files, such as coming from TAUS Data Association.[1] The problem is that the number of texts available in this form is limited. More parallel texts are directly accessible in the form of multilingual webpages, such as newspapers or corporate websites. Earlier attempts at collecting such documents were based on the possibility to map structural similarities between the links to such websites, e.g., http://europa.eu/index_bg.htm vs http://europa.eu/index_el.htm which differ in the language identifier within their URLs [13, 69].

More modern approaches add the possibility of enriching the link heuristics with information about the contents [6]. Discovery of such websites can also be automated [26, 81]. Another possibility for getting good-quality parallel data is to mine parallel

[1] http://www.tausdata.org/

RSS feeds [80]. Such approaches can help in finding sufficient amounts of parallel texts even for medium-density languages, such as Hungarian [82] or Ukrainian [4].

Moving further down the cline towards comparable corpora, similar techniques can be used for extracting parallel texts from comparable websites by their structural links [2] or their contents [58, 79].

With respect to the collection of comparable resources using topical crawlers, there has been an array of recent EU projects, all aimed at designing tools for utilising bilingual information in crawling [8, 9, 77].

3 Using Comparable Corpora

3.1 *Extraction of Bilingual Dictionaries*

This section aims at exemplifying the wealth of work in comparable corpora by looking in some detail at one particular subtopic: Extracting information on word translations automatically from corpora (often referred to as *bilingual lexicon extraction*), rather than compiling dictionaries in the traditional lexicographic way, is an established application of parallel and comparable corpora.

With their seminal papers [10, 11], Brown et al. showed that information on word translations (the so-called translation models) could be reliably and in high quality extracted from parallel corpora, which was confirmed by others (e.g. [32]). But parallel corpora were (and, although to a lesser degree, still are) a scarce resource, so some years later the idea came up whether it might also be possible to derive information on word translations from comparable corpora. Independently of each other, at ACL 1995 Fung [27] and Rapp [64] suggested two approaches on how this could be accomplished. Fung [27] utilized a context heterogeneity measure, thereby assuming that words with productive context in one language translate to words with productive context in another language, and words with rigid context translate into words with rigid context. In contrast, the underlying assumption in Rapp [64] is that words which are translations of each other show similar co-occurrence patterns across languages. For example, if the words *teacher* and *school* co-occur more often than chance in English, then the same can be expected for their translations in a corpus of another language.

The validity of this co-occurrence constraint is obvious for parallel corpora, but it also holds for non-parallel corpora. It can be observed that this constraint works best with parallel corpora, second-best with comparable corpora, and somewhat worse with unrelated corpora. Robustness is not a big issue in any of these cases. In contrast, when applying sentence alignment algorithms to parallel corpora, omissions, insertions, and transpositions of text segments can have critical negative effects. However, the co-occurrence constraint when applied to comparable corpora is much weaker than the word-order constraint as used with parallel corpora. This is why larger corpora and well-chosen statistical methods are needed.

It should be noted that the advantages of looking at comparable rather than parallel corpora are not only robustness and ease of acquisition, but also that usually fewer corpora are required. Let us assume, for example, that we are interested in extracting dictionaries covering all possible pairs involving 10 languages, which would be altogether 90 directed language pairs. As both parallel and comparable corpora can be used in both directions of a language pair, this effectively reduces to 45 pairs. To deal with these 45 pairs, in the comparable case we need 10 corpora, one for each language. But in the parallel case we may need up to 45 corpora, thereby assuming that each language pair is based on the translation of a different text. That is, in the comparable case the required number of corpora increases linearly with the number of languages considered, but in the parallel case it can increase quadratically. However, if we are lucky, the same text may have been translated into several or all languages of interest. This means that the number of parallel corpora required can be reduced significantly. This is one of the reasons why large multilingual corpora covering many languages, such as Europarl and JRC-Acquis are particularly useful.

The task of identifying word translations has become one of the most investigated applications of comparable corpora. Following Rapp [64], most work was done using vector space approaches based on a multitude of variations of the above co-occurrence constraint (which can be seen as an extension of Harris' distributional hypothesis [40] to the multilingual case). Among the pioneers, Tanaka and Iwasaki [78] pursued a matrix-based approach where the selection of a target word candidate is seen in analogy to word sense disambiguation. Fung and McKeown [30] used word relation matrices which, using dictionary information, are mapped across languages to find new translation pairs. The accuracy is reported to be around 30 %. Fung and Yee [28] introduce an Information Retrieval inspired vector space approach: Using an existing bilingual lexicon of seed words, the co-occurrence vector of a word to be considered is translated into the target language. Then, using standard vector similarity measures, the resulting target language vector is compared to the vectors of the words in the target language vocabulary. The target language word with the highest similarity is considered to be the correct translation.

Peters and Picchi [61, 62] apply such a method for cross-language information retrieval. Given a query term to be translated, they compute its characteristic context words, and then translate these using existing dictionaries. They then search for those passages in the target language where there is a significant presence of the translated context words. This way, for any query term of interest, they obtain a ranked list of documents containing equivalent terms in another language.

Rapp [65] further refines the vector space approach, thereby also taking word order into account. This leads to an accuracy of 72 % for a standard test word list commonly used in Psychology. In subsequent work, transitivity across languages is taken into account [67]. Hereby advantage is taken of the possibility that, if corpora of more than two languages are available, the translations from one language to another can be determined not only directly, but also indirectly via a pivot language. This way, the more languages are considered the more evidence for a particular translation assignment can be provided by mutual cross-validation.

A related but different concept, referred to as *bridge languages*, had been used before by Schafer and Yarowsky [72]. However, the emphasis here is on cognate similarity between closely related languages such as Czech and Serbian. That is, if a Czech to English dictionary is available, the English translations of Serbian words can be determined by computing their orthographically most similar Czech counterparts, and by looking up their translations.

In addition to the bridge language concept, they manage to avoid the need for a seed lexicon by successfully combining temporal occurrence similarity across dates in news corpora, cross-language context similarity, weighted Levenstein string edit distance, and relative frequency and burstiness similarity measures.

A similar multi-clue approach is also used by Koehn and Knight [50]. They utilize spelling similarity, the above mentioned co-occurrence constraint, a second-order co-occurrence constraint (e.g. *Wednesday* and *Thursday* have similar contexts, as do their translations in another language), and corpus frequency (which should correlate between translations). They report a 39 % accuracy on a test set consisting of the 1,000 most frequent English and German nouns.

The potential of the spelling similarity clue is also demonstrated by Gamallo Otero and Garcia [33]. By extracting translation equivalents with similar spelling from Portuguese and Spanish comparable corpora (Wikipedia), they were able to come up with 27,000 new pairs of lemmas and multiwords not found in existing dictionaries, with about 92 % accuracy.

An additional potentially interesting clue which can be seen as an extension of spelling similarity is described in Langlais et al. [52]. In the medical domain they use analogical learning to exploit the formal similarity of medical words in some languages (systematic compounding). Their system does not require corpora but is trained on an initial bilingual lexicon.

Chiao and Zweigenbaum [15] conduct co-occurrence based lexicon extraction in the medical domain and systematically test several weighting factors and similarity measures. They found that by introducing an additional reverse-translation filtering step the accuracy of their system could be improved from 50 to 70 %. This is further elaborated in Chiao et al. [14].

Also specializing on the medical domain, for bilingual lexicon extraction Dejean et al. [17] not only exploit a seed lexicon but also a readily available multilingual medical thesaurus. They could show that using hierarchical information contained in the thesaurus significantly improves results.

Gamallo Otero and Pichel Campos [34] extract bilingual pairs of lexico-syntactic patterns from a parallel corpus. Subsequently they construct context vectors for all source and target language words by recording their frequency of occurrence in these patterns. There is thus only one vector space for both languages, so that vectors can be readily compared. For the language pair English–Spanish they report an accuracy of 89 % for high-frequency words. The method is further refined by Gamallo Otero and Pichel Campos in [35].

Shezaf and Rappoport [76] describe an algorithm introducing so-called non-aligned signatures for improving noisy dictionaries. The algorithm is in effect similar to Fung and Yee [28] and Rapp [65], but (like [68]) rather than full co-occurrence vectors considers only salient context words (i.e. strong word associations).

As an application in contrastive linguistics, Defrancq [16] conducted a study for establishing cross-linguistic semantic relatedness between verbs in different languages based on monolingual corpora. A small number of verbs were semi-automatically investigated for their co-occurrences with particular interrogative elements, and then verbs were compared using Kullback-Leibler divergence.

Gaussier et al. [37], in an attempt to solve the problem of different word ambiguities in source and target language, use a geometric view and try to decompose the word vectors according to their senses. They investigate a number of methods, including canonical correlation analysis, multilingual probabilistic latent semantic analysis, thereby involving Fisher kernels. The best results with an improvement of 10 % are reported for a mixed method.

In contrast to the dominating vector space approaches based on word-co-occurrence data, Michelbacher et al. [21, 54] use linguistic relations like subcategorization, modification and coordination in a graph-based model. Also, other than most previous work, in their approach they distinguish between different parts of speech. Their basic approach is to use the SimRank algorithm to recursively compute node similarities. These are based on the similarity scores of neighboring nodes within a graph. Dorow et al. [21] proposed an extension towards cross-lingual semantic relatedness. It computes node-similarities between two graphs and allows for weighted graph edges.

Garera et al. [36] use a vector space model but consider dependency links rather than word co-occurrences. By doing so they obtain an improvement of 16 % for the language pair English-Spanish. They induce translation lexicons from comparable corpora based on multilingual dependency parses which takes long-range dependency into account. The system is shown to bring a 16 to 18 % improvement over a co-occurrence-based baseline. A similar approach is also pursued by Yu and Tsujii [84]. Their work is also based on the observation that a word and its translation share similar dependency relations, and they also obtain significant improvements.

There have also been a number of attempts to generate bilingual dictionaries from comparable corpora without the need of a seed lexicon. Diab and Finch [20] do so by using a computationally expensive bootstrapping approach which only requires very few seed translations. Otherwise their approach is related to Rapp [64], but they limit the co-occurrences they consider to those between the top 2,000 frequent tokens in the corpus and the top 150 frequent tokens, in four different collocation positions. Their method for searching new word translations is based on a gradient descent algorithm. They iteratively change the mapping of a given word until they reach a local minimum for the sum of squared differences between the association measure of all pairs of words in one language and the association measure of the pairs of translated words. Their reported accuracies are between 92.4 and 98.7 %, but for a pseudo translation task using two different corpora of the same language. So it might be a challenge to make the algorithm converge for non-related languages.

Haghighi et al. [39] approach the task of bilingual lexicon extraction by looking at word features such as co-occurrence counts and orthographic substrings, and then inducing translations using a generative model based on canonical correlation analysis, which explains the monolingual lexicons in terms of latent matchings. For a range of corpus types and languages they show that high-precision lexicons can be learned even without a seed lexicon.

Robitaille et al. [70] deal with bilingual dictionary construction for multi-word terms. For their list of seed terms they download taylor-made multilingual corpora from the Web. They then extract multi-word terms from these corpora, and use a compositional method to align them across languages. Coverage is increased using a bootstrapping method.

The following three publications replace seed lexica by Wikipedia interlanguage links, which are pointers between wikipedia articles in different languages that relate to the same headword. Hassan and Mihalcea [41] represent words using explicit semantic analysis, and then compute the semantic relatedness of these concept vectors across languages by exploiting the mappings from the Wikipedia interlanguage links. Rapp et al. [66] do something similar but replace explicit semantic analysis by a keyword extraction procedure used for representing documents, and then applying an alignment algorithm on the keyword lists. Both methods show a reasonably good performance and can be applied to other multilingual document collections as well if these are aligned at the document level. (Such alignments can be computed using algorithms for measuring document comparability, which, however, usually require a bilingual lexicon). Prochasson and Fung [63] also start from aligned Wikipedia (and other) documents. They conduct a supervised classification and then utilize context-vector similarity and a co-occurrence model between words of aligned documents in a machine learning approach.

Morin and Prochasson [56] present an effective way of extracting bilingual lexica. By utilizing structural properties of the documents they extract parallel sentences from the comparable corpora, and then extract the dictionaries from these. Hazem and Morin [42] treat the dictionary extraction task as a question answering problem and describe their respective system QAlign. In a previous paper Morin et al. [55] showed that the quality of the comparable corpus is very important for dictionary construction.

The problem that most methods for dictionary extraction from comparable corpora have difficulties with rare words had been discovered early, but was for the first time put in focus by Pekar et al. [60]. Their solution was to estimate missing co-occurrence values based on similar words of the same language. Note, however, that the more recent approaches utilizing aligned comparable corpora [41, 63, 66] serve the same purpose and are likely to produce better results.

Finally, let us mention that, as shown by Rapp and Zock [68] bilingual lexica can even be extracted from monolingual corpora just by computing the strongest associations of foreign words occurring in a corpus. The reason is that in the contexts of foreign words often their translations are mentioned. But of course this is only of practical value for languages which are often cited, such as English. However, these can serve as pivots, thus mediating translations between other language pairs. In this method co-occurrence information is solely required for the target language. For the source language, to identify what counts as a foreign word, only a vocabulary list is needed. Such a list can be extracted from a source language corpus, which relates the method to the comparable corpora topic.

3.2 Comparable Corpora for Statistical Machine Translation

Bilingual lexicons can be extracted with good success from parallel segments which have been extracted from comparable corpora. Given the limited availability of parallel corpora in many domains and for many language pairs, comparable corpora are often regarded as a potential source to help train Statistical Machine Translation (SMT) systems.

Most work in that area has been geared towards extracting parallel sub-parts of comparable corpora [44, 57, 85]. Using the collected parallel sub-parts helps train an SMT system and improve its performance over using a much larger out-of-domain parallel corpus. For instance, Abdul Rauf and Schwenk [1] obtained an increase of 2pt in BLEU score on Arabic to English translation, whereas Gahbiche-Braham et al. [31] increased by 6pt their BLEU score for Arabic to French translation.

In contrast, word translations directly extracted from comparable corpora currently have a too low precision to be useful for SMT. However they have been shown recently to improve the performance of a state-of-the-art cross-language information retrieval system [53], which indicates that further improvements in this line of research might pave the way to to applicability.

Another motivation for using comparable corpora in MT research can come from a cognitive perspective: Experience shows that persons who have learned a second language completely independently from their mother tongue can nevertheless translate between the languages. That is, human performance shows that there must be a way to bridge the gap between languages which does not rely on parallel data (in the context of human language learning with "parallel data" we could e.g. mean the use of mixed language in class). Using parallel data for MT is of course a nice shortcut and apparently much easier than understanding human language capabilities. But let us compare this approach to writing a chess program which simply enumerates very many possibilities of potential moves. This also tells us close to nothing about human reasoning. But language is not a domain as limited as chess. Therefore, in the long run it is likely that we will not get around understanding more about human language processing, and avoiding shortcuts by doing MT based on comparable corpora may well be a key to this.

4 Future Research Directions

The history of the BUCC workshops and the contributions to this volume identify several sources of interesting results. One comes from the fact that the Web is huge and it is getting easier to obtain reasonably similar texts for a range of languages.

The use of inter-wiki links in Wikipedia is a simple example of the growing space of similar texts. This presses the algorithms for more targeted detection of parallel and quasi-parallel segments in large collections, on the level of websites, documents, paragraphs and sentences. This leads to the possibility of using weakly comparable collections with the advantage of getting more closely related data for small domains (like wind energy) or less common language pairs (like German-Chinese).

Combination of features and supervision

Numerous types of information and functions on these types of information have been brought to bear to help identify matching words in comparable corpora: frequency of occurrence, co-occurrence counts, counts of lexico-syntactic patterns [34] or of dependency relations [3, 36], association measures, similarity measures, part-of-speech, cognates [72] and formal similarity [52], named entities and their relations [45], hierarchical information [17], co-presence in aligned comparable documents [63], to name but a few.

Most authors have contrasted these sources of information and tried to select those which worked best. Another path could be instead to try to combine them all together, pooling on the strengths of each type of information and function. This has been tried in only limited ways until now [43, 63]. Besides, most work has been performed in an unsupervised framework, whereas supervision is readily available in the standard setting through the availability of a partial bilingual dictionary. Supervision has proved very effective when used [3, 63]. Considering each type of information and each function on these types of information as features input to a supervised classifier might be a way to weight and combine them in an optimal way to identify word translations in comparable corpora, taking the best of each world.

References

1. Abdul Rauf, S., Schwenk, H.: Exploiting comparable corpora with TER and TERp. In: Proceedings of the 2nd Workshop on Building and Using Comparable Corpora: from Parallel to Non-parallel Corpora, pp. 46–54. Association for Computational Linguistics, Singapore (August 2009), http://www.aclweb.org/anthology/W/W09/W09-3109.pdf
2. Adafre, S., de Rijke, M.: Finding similar sentences across multiple languages in Wikipedia. In: Proceedings of the 11th Conference of the European Chapter of the Association for Computational Linguistics (EACL-2006), pp. 62–69. Trento (2006)
3. Andrade, D., Matsuzaki, T., Tsujii, J.: Learning the optimal use of dependency-parsing information for finding translations with comparable corpora. In: Proceedings of the 4th Workshop on Building and Using Comparable Corpora: Comparable Corpora and the Web, pp. 10–18. Association for Computational Linguistics, Portland (June 2011), http://www.aclweb.org/anthology/W11-1203
4. Babych, B., Hartley, A., Sharoff, S.: Translating from under-resourced languages: comparing direct transfer against pivot translation. In: Proceedings of the MT Summit XI, pp. 412–418. Copenhagen (2007), http://corpus.leeds.ac.uk/serge/publications/2007-mt-summit.pdf
5. Babych, B., Hartley, A., Sharoff, S., Mudraya, O.: Assisting translators in indirect lexical transfer. In: Proceedings of 45th ACL, pp. 739–746. Prague (2007), http://corpus.leeds.ac.uk/serge/publications/2007-ACL.pdf

6. Barbosa, L., Bangalore, S., Rangarajan Sridhar, V.K.: Crawling back and forth: using back and out links to locate bilingual sites. In: Proceedings of 5th International Joint Conference on Natural Language Processing, Chiang Mai (November 2011)
7. Baroni, M., Bernardini, S.: Bootcat: Bootstrapping corpora and terms from the Web. In: Proceedings of LREC2004. Lisbon (2004), http://sslmit.unibo.it/baroni/publications/lrec2004/bootcat_lrec_2004.pdf
8. Bel, N., Papavasiliou, V., Prokopidis, P., Toral, A., Arranz, V.: Mining and exploiting domain-specific corpora in the PANACEA platform. In: The 5th Workshop on Building and Using Comparable Corpora (2012)
9. Blancafort, H., Heid, U., Gornostay, T., Méchoulam, C., Daille, B., Sharoff, S.: User-centred views on terminology extraction tools: usage scenarios and integration into MT and CAT tools. In: Proceedings TRALOGY Conference "Translation Careers and Technologies: Convergence Points for the Future" (2011)
10. Brown, P., Pietra, S.D., Pietra, V.D., Mercer, R.: The mathematics of statistical machine translation: parameter estimation. Computat. Linguist. **19**(2), 263–312 (1993)
11. Brown, P.F., Cocke, J., Jelinek, F., Lafferty, J.D., Mercer, R.L., Roossin, P.S.: A statistical approach to machine translation. Computat. Linguist. **16**(2), 79–85 (1990)
12. Budge, E.A.T.W.: The Rosetta Stone. British Museum. London (1913)
13. Chen, J., Nie, J.: Parallel Web text mining for cross-language ir. In: Proceedings of RIAO, pp. 62–77 (2000)
14. Chiao, Y.C., Sta, J.D., Zweigenbaum, P.: A novel approach to improve word translations extraction from non-parallel, comparable corpora. In: Proceedings International Joint Conference on Natural Language Processing, Hainan (2004)
15. Chiao, Y.C., Zweigenbaum, P.: Looking for candidate translational equivalents in specialized, comparable corpora. In: COLING 2002 (2002)
16. Defrancq, B.: Establishing cross-linguistic semantic relatedness through monolingual corpora. Int. J. Corpus Linguist. **13**(4), 465–490 (2008)
17. Déjean, H., Gaussier, E., Sadat, F.: An approach based on multilingual thesauri and model combination for bilingual lexicon extraction. In: COLING 2002 (2002)
18. Deléger, L., Cartoni, B., Zweigenbaum, P.: Paraphrase detection in monolingual specialized/lay comparable corpora. In: Sharoff, S., Rapp, R., Fung, P., Zweigenbaum, P. (eds.) Building and Using Comparable Corpora. Springer, Dordrecht (2012)
19. Deléger, L., Zweigenbaum, P.: Extracting lay paraphrases of specialized expressions from monolingual comparable medical corpora. In: Fung, P., Zweigenbaum, P., Rapp, R. (eds.) Proceedings of the 2nd Workshop on Building and Using Comparable Corpora: from Parallel to Non-Parallel Corpora, pp. 2–10. Association for Computational Linguistics, Singapore (August 2009), http://aclweb.org/anthology/W/W09/W09-3102
20. Diab, M., Finch, S.: A statistical wordlevel translation model for comparable corpora. In: Proceedings of the Conference on Content-Based Multimedia Information Access (RIAO) (2000)
21. Dorow, B., Laws, F., Michelbacher, L., Scheible, C., Utt, J.: A graph-theoretic algorithm for automatic extension of translation lexicons. In: EACL 2009 Workshop on Geometrical Models of Natural Language Semantics (2009)
22. Eisele, A., Federmann, C., Saint-Amand, H., Jellinghaus, M., Herrmann, T., Chen, Y.: Using moses to integrate multiple rule-based machine translation engines into a hybrid system. In: Proceedings of the Third Workshop on Statistical Machine Translation at ACL2008, pp. 179–182 (2008)
23. Eisele, A., Chen, Y.: MultiUN: A multilingual corpus from United Nations documents. In: Proceedings of the Seventh conference on International Language Resources and Evaluation (LREC'10). Valletta, Malta (2010), http://www.euromatrixplus.net/multi-un/
24. Elhadad, N., Sutaria, K.: Mining a lexicon of technical terms and lay equivalents. In: Proceedings of the Workshop on BioNLP 2007: Biological, Translational, and Clinical Language Processing, pp. 49–56. Association for Computational Linguistics (2007)

25. Enright, J., Kondrak, G.: A fast method for parallel document identification. In: NAACL / Human Language Technologies, pp. 29–32. Rochester (2007)
26. Esplà-Gomis, M., Forcada, M.L.: Combining content-based and url-based heuristics to harvest aligned bitexts from multilingual sites with bitextor. Prague Bull. Math. Linguist. **93**, 77–86 (2010)
27. Fung, P.: Compiling bilingual lexicon entries from a non-parallel English-Chinese corpus. In: Proceedings of Third Annual Workshop on Very Large Corpora, pp. 173–183. Boston (1995)
28. Fung, P.: Extracting key terms from chinese and japanese texts. Int. J. Comput. Process. Orient. Lang. **12**(1), 99–121 (1998)
29. Fung, P.: A statistical view on bilingual lexicon extraction: from parallel corpora to non-parallel corpora. In: Machine Translation and the Information Soup, pp. 1–17. Springer, Berlin (1998), http://www.springerlink.com/content/pqkpwpw32f5r74ev/
30. Fung, P., McKeown, K.: Finding terminology translations from non-parallel corpora. In: Proceedings of the 5th Annual Workshop on Very Large Corpora, pp. 192–202 (1997)
31. Gahbiche-Braham, S., Bonneau-Maynard, H., Yvon, F.: Two ways to use a noisy parallel news corpus for improving statistical machine translation. In: Proceedings of the 4th Workshop on Building and Using Comparable Corpora: Comparable Corpora and the Web, pp. 44–51. Association for Computational Linguistics, Portland (June 2011), http://www.aclweb.org/anthology/W11-1207
32. Gale, W., Church, K.: A program for aligning sentences in bilingual corpora. Comput. linguist. **19**(1), 75–102 (1993)
33. Gamallo Otero, P., Garcia, M.: Extraction of bilingual cognates from wikipedia. In: Caseli, H., Villavicencio, A., Teixeira, A., Perdigo, F. (eds.) Computational Processing of the Portuguese Language. Lecture Notes in Artificial Intelligence, vol. 7243, pp. 63–72. Springer, Berlin (2012)
34. Gamallo Otero, P., Pichel Campos, J.R.: An approach to acquire word translations from non-parallel texts. In: EPIA, pp. 600–610 (2005)
35. Gamallo Otero, P., Pichel Campos, J.R.: Automatic generation of bilingual dictionaries using intermediary languages and comparable corpora. Comput. Linguist. Intell. Text Process. **6008**, 473–483 (2010)
36. Garera, N., Callison-Burch, C., Yarowsky, D.: Improving translation lexicon induction from monolingual corpora via dependency contexts and part-of-speech equivalences. In: CoNLL 09: Proceedings of the Thirteenth Conference on Computational Natural Language Learning, p. 129137. Morristown (2009)
37. Gaussier, E., Renders, J.M., Matveeva, I., Goutte, C., Djean, H.: A geometric view on bilingual lexicon extraction from comparable corpora. In: Proceedings of the 42nd Annual Meeting of the Association for Computational Linguistics, p. 526533. Barcelona (2004)
38. Germann, U.: Aligned Hansards of the 36th Parliament of Canada (2001), http://www.isi.edu/natural-language/download/hansard/
39. Haghighi, A., Liang, P., Berg-Kirkpatrick, T., Klein, D.: Learning bilingual lexicons from monolingual corpora. In: Proceedings of ACL-08: HLT, pp. 771–779. Columbus (2008)
40. Harris, Z.: Distributional structure. Word **10**(23), 146–162 (1954)
41. Hassan, S., Mihalcea, R.: Cross-lingual semantic relatedness using encyclopedic knowledge. In: EMNLP (2009)
42. Hazem, A., Morin, E.: Qalign: a new method for bilingual lexicon extraction from comparable corpora. Comput. Linguist. Intell. Text Process. **7182**, 83–96 (2012)
43. Hazem, A., Morin, E.: Extraction de lexiques bilingues partir de corpus comparables par combinaison de reprsentations contextuelles. In: Proceedings of the TALN 2013. ATALA, Les Sables d'Olonne (2013), in Press
44. Hewavitharana, S., Vogel, S.: Extracting parallel phrases from comparable data. In: Proceedings of the 4th Workshop on Building and Using Comparable Corpora: Comparable Corpora and the Web, pp. 61–68. Association for Computational Linguistics, Portland (June 2011), http://www.aclweb.org/anthology/W11-1209
45. Ji, H.: Mining name translations from comparable corpora by creating bilingual information networks. In: Proceedings of the 2nd Workshop on Building and Using Comparable Corpora:

from Parallel to Non-parallel Corpora, pp. 34–37. Association for Computational Linguistics, Singapore (August 2009), http://www.aclweb.org/anthology/W/W09/W09-3107
46. Kilgarriff, A.: Comparing corpora. Int. J. Corpus Linguist. **6**(1), 1–37 (2001)
47. Kilgarriff, A.: Comparable corpora within and across languages, word frequency lists and the kelly project. In: Proceedings of workshop on Building and Using Comparable Corpora at LREC, Malta (2010)
48. Knight, K., Megyesi, B., Schaefer, C.: The [copiale] cipher. In: Proceedings of the 4th Workshop on Building and Using Comparable Corpora: Comparable Corpora and the Web, pp. 2–9. Portland (June 2011), http://www.aclweb.org/anthology/W11-1202
49. Koehn, P.: Europarl: a parallel corpus for statistical machine translation. In: Proceedings of MT Summit 2005 (2005), http://www.iccs.inf.ed.ac.uk/ pkoehn/publications/europarl-mtsummit05.pdf
50. Koehn, P., Knight, K.: Learning a translation lexicon from monolingual corpora. In: Proceedings of the ACL-02 workshop on Unsupervised lexical acquisition, pp. 9–16 (2002)
51. Lahaussois, A., Guillaume, S.: A viewing and processing tool for the analysis of a comparable corpus of kiranti mythology. In: Proceedings of the 5th Workshop on Building and Using Comparable Corpora, pp. 33–41. ELDA, Istanbul (2012)
52. Langlais, P., Patry, A.: Translating unknown words by analogical learning. In: Proceedings of the 2007 Joint Conference on Empirical Methods in Natural Language Processing and Computational Natural Language Learning (EMNLP-CoNLL), pp. 877–886 (2007)
53. Li, B.: Measuring and improving comparable corpus quality. Ph.D. thesis, Universit de Grenoble, Grenoble (June 2012)
54. Michelbacher, L., Laws, F., Dorow, B., Heid, U.,, Schütze, H.: Building a cross-lingual relatedness thesaurus using a graph similarity measure. In: Proceedings of the Seventh conference on International Language Resources and Evaluation (LREC'10), Valletta (2010)
55. Morin, E., Daille, B., Takeuchi, K., Kageura, K.: Bilingual terminology mining—using brain, not brawn comparable corpora. In: Proceedings of the 45th Annual Meeting of the Association for Computational Linguistics, pp. 664-671. Prague, Czech Republic (2007)
56. Morin, E., Prochasson, E.: Bilingual lexicon extraction from comparable corpora enhanced with parallel corpora. In: BUCC2011 (2011)
57. Munteanu, D.S., Marcu, D.: Improving machine translation performance by exploiting non-parallel corpora. Comput. Linguist. **31**(4), 477–504 (2005)
58. Munteanu, D., Marcu, D.: Extracting parallel sub-sentential fragments from non-parallel corpora. In: Proceedings of International Conference on Computational Linguistics and Association of Computational Linguistics, COLING-ACL 2006. Sydney (2006)
59. Patry, A., Langlais, P.: Identifying parallel documents from a large bilingual collection of texts: Application to parallel article extraction in Wikipedia. In: Proceedings of the 4th Workshop on Building and Using Comparable Corpora: Comparable Corpora and the Web, pp. 87–95. Portland (June 2011), http://www.aclweb.org/anthology/W11-1212
60. Pekar, V., Mitkov, R., Blagoev, D., Mulloni, A.: Finding translations for low-frequency words in comparable corpora. Mach. Transl. **20**(4), 247–266 (2006)
61. Peters, C., Picchi, E.: Using linguistic tools and resources in cross-language retrieval. In: Hull, D., Oard, D. (eds.) Cross-Language Text and Speech Retrieval Papers from the 1997 AAAI Spring Symposium, pp. 179–188. AAAI Press, San Francisco (1997)
62. Picchi, E., Peters, C.: Exploiting lexical resources and linguistic tools in cross-language information retrieval: the EuroSearch approach. In: First International Conference on Language Resources & Evaluation, pp. 865–872. Granada (1998)
63. Prochasson, E., Fung, P.: Rare word translation extraction from aligned comparable documents. In: Proceedings of ACL-HLT, Portland (2011)
64. Rapp, R.: Identifying word translations in non-parallel texts. In: Proceedings of the 33rd ACL, pp. 320–322. Cambridge (1995)
65. Rapp, R.: Automatic identification of word translations from unrelated English and German corpora. In: Proceedings of the 37th ACL, pp. 395–398. Maryland (1999)

66. Rapp, R., Sharoff, S., Babych, B.: Identifying word translations from comparable documents without a seed lexicon. In: Proceedings of the Eighth Language Resources and Evaluation Conference, LREC 2012. Istanbul (2012)
67. Rapp, R., Zock, M.: Automatic dictionary expansion using non-parallel corpora. In: Fink, A., Lausen, B., Ultsch, W.S.A. (eds.) Advances in Data Analysis, Data Handling and Business Intelligence. Proceedings of the 32nd Annual Meeting of the GfKl, 2008. Springer, Heidelberg (2010)
68. Rapp, R., Zock, M.: The noisier the better: identifying multilingual word translations using a single monolingual corpus. In: Proceedings of the 4th International Workshop on Cross Lingual Information Access at COLING. pp. 16–25. Beijing (2010)
69. Resnik, P., Smith, N.: The Web as a parallel corpus. Comput. Linguist. **29**(3), 349–380 (2003), http://www.umiacs.umd.edu/ resnik/strand/
70. Robitaille, X., Sasaki, Y., Tonoike, M., Sato, S., Utsuro, T.: Compiling French-Japanese terminologies from the Web. In: Proceedings of the 11st Conference of the European Chapter of the Association for Computational Linguistics, pp. 225–232. Trento (2006)
71. Rosset, S., Grouin, C., Fort, K., Galibert, O., Kahn, J., Zweigenbaum, P.: Structured named entities in two distinct press corpora: contemporary broadcast news and old newspapers. In: Proceedings of the Sixth Linguistic Annotation Workshop, pp. 40–48. Association for Computational Linguistics, Jeju, Republic of Korea (July 2012), http://www.aclweb.org/anthology/W12-3606
72. Schafer, C., Yarowsky, D.: Inducing translation lexicons via diverse similarity measures and bridge languages. In: Proceedings of CoNLL (2002)
73. Segouat, J., Braffort, A.: Toward categorization of sign language corpora. In: Proceedings of the 2nd Workshop on Building and Using Comparable Corpora: from Parallel to Non-parallel Corpora, pp. 64–67. Association for Computational Linguistics, Singapore (August 2009), http://www.aclweb.org/anthology/W/W09/W09-3111
74. Sharoff, S.: Creating general-purpose corpora using automated search engine queries. In: Baroni, M., Bernardini, S. (eds.) WaCky! Working Papers on the Web as Corpus. Gedit, Bologna (2006), http://wackybook.sslmit.unibo.it
75. Sharoff, S.: In the garden and in the jungle: comparing genres in the BNC and Internet. In: Mehler, A., Sharoff, S., Santini, M. (eds.) Genres on the Web: Computational Models and Empirical Studies, pp. 149–166. Springer, Berlin (2010)
76. Shezaf, D., Rappoport, A.: Bilingual lexicon generation using non-aligned signatures. In: Proceedings of the 48th Annual Meeting of the Association for Computational Linguistics. p. 98107. Uppsala (2010)
77. Skadiņa, I., Vasiljevs, A., Skadiņš, R., Gaizauskas, R., Tufiş, D., Gornostay, T.: Analysis and evaluation of comparable corpora for under resourced areas of machine translation. In: Proc. 3rd Workshop on Building and Using Comparable Corpora. Malta (2010).
78. Tanaka, K., Iwasaki, H.: Extraction of lexical translations from non-aligned corpora. In: Proceedings of the 16th conference on Computational linguistics (COLING96), vol. 2, pp. 580–585 (1996)
79. Tillmann, C.: A beam-search extraction algorithm for comparable data. In: Proceedings of the ACL-IJCNLP 2009 Conference Short Papers, pp. 225–228 (2009)
80. Tsvetkov, Y., Wintner, S.: Automatic acquisition of parallel corpora from websites with dynamic content. In: Proceedings of The Seventh International Conference on, Language Resources and Evaluation (LREC-2010) (2010)
81. Uszkoreit, J., Ponte, J.M., Popat, A.C., Dubiner, M.: Large scale parallel document mining for machine translation. In: COLING '10: Proceedings of the 23rd International Conference on Computational Linguistics, pp. 1101–1109 (2010)
82. Varga, D., Halacsy, P., Kornai, A., Nagy, V., Nemeth, L., Tron, V.: Parallel corpora for medium density languages. In: N. Nicolov, K. Bontcheva, G.A., Mitkov, R. (eds.) Recent Advances in Natural Language Processing IV. Selected papers from RANLP-05, pp. 247–258. Benjamins (2007), http://www.kornai.com/Papers/ranlp05parallel.pdf

83. Wang, R., Callison-Burch, C.: Paraphrase fragment extraction from monolingual comparable corpora. In: Proceedings of the 4th Workshop on Building and Using Comparable Corpora: Comparable Corpora and the Web, pp. 52–60. Association for Computational Linguistics, Portland (June 2011), http://www.aclweb.org/anthology/W11-1208
84. Yu, K., Tsujii, J.: Extracting bilingual dictionary from comparable corpora with dependency heterogeneity. In: Proceedings of HLT-NAACL 2009, pp. 121–124. Boulder (2009)
85. Zhao, B., Vogel, S.: Adaptive parallel sentences mining from Web bilingual news collection. In: Proceeding of the 2002 IEEE International Conference on Data Mining (ICDM 2002) (2002)

Part I
Compiling and Measuring Comparable Corpora

Mining Parallel Documents Using Low Bandwidth and High Precision CLIR from the Heterogeneous Web

Simon Shi and Pascale Fung

Abstract We propose a content-based method of mining bilingual parallel documents from websites that are not necessarily structurally related to each other. There are two existing approaches for automatically mining parallel documents from the web. Structure based methods work only for parallel websites and most of content based methods are either requires large scale computational facilities, network bandwidth or not applicable to heterogeneous web. We propose a novel content based method using cross lingual information retrieval (CLIR) with query feedback and verification and supplemented with structural information, to mine parallel resources from the entire web using search engine APIs. The method goes beyond structural information to find parallel documents from non-parallel websites. We obtained a very high mining precision and extracted parallel sentences improved SMT performance, with higher BLEU score, is comparable to that obtained with high quality manually translated parallel sentences illustrating the excellent quality of the mined parallel materiel

Keywords Parallel document mining · Parallel sentence mining · Parallel corpus · Cross lingual information retrieval · CLIR

S. Shi (✉) · P. Fung
Human Language Technology Center, The Hong Kong University of Science and Technology, Clear Water Bay, Hong Kong
e-mail: eesys@ust.hk

P. Fung
e-mail: pascale@ust.hk

S. Sharoff et al. (eds.), *Building and Using Comparable Corpora*,
DOI: 10.1007/978-3-642-20128-8_2, © Springer-Verlag Berlin Heidelberg 2013

1 Introduction

1.1 Mining Parallel Resource from the Web

Parallel corpora, which contain translated sentences, are essential resources as training data for statistical machine translation and building or extending bilingual lexicons and terminologies. Parallel resources such as bilingual lexicon and sentence translations are typically obtained from translated parallel documents.

In the past, parallel resources from human translations and translations of newswire documents were the main source of parallel corpora for statistical machine translation systems. These kinds of resources are very costly and size limited. However, large amount of documents are publicly available on the World Wide Web under trillions of URLs[1] in hundreds of languages. Thus, the web becomes an important source of parallel documents.

In previous studies, many researches have been done on parallel document mining from the web, using both structure based and content based methods. Previous approaches commonly require downloading all documents in both languages and then performing alignment locally. However, the number of documents on the web is growing so fast that it is not yet feasible for most research institutions to build a local archive containing trillions of documents. There is a need to readdress the problem of how to mine parallel documents from the entire web.

With the development of the search technology, online search engine becomes an efficient tool in finding documents on the web which is heterogeneous in nature. Online search engines such as Google or Bing have built a comprehensive and up-to-date index including almost all public documents on the Internet.

The use of the search engine is simple and fast which only includes one HTTP client server communication and search time for each query which is less than a second. It would be efficient and helpful for research institutions to mine parallel resources from the entire web if we can make use of the search engine APIs to find target document for any source document.

1.2 Objectives

Previous work by [1–3] and [4] produced parallel sentences mainly from pre-downloaded bilingual corpus in both source and target languages. It is very costly in CPU power if the size of the comparable corpus is large, for example, the entire web.

[1] We knew the web was big . . . on the Official Google Blog. http://googleblog.blogspot.com/2008/07/we-knew-web-was-big.html.

We focuses on investigating a new method and implementing a tool for mining parallel resources from the web for training machine translation systems and bilingual lexicon extraction with high precision and low bandwidth.

High Precision

We focus on mining directly translated (parallel) documents rather than comparable ones because the parallel resources are mainly used for statistical machine translation (SMT) systems as training data and bilingual lexicon extraction. The quality of parallel documents must be high enough for these kinds of applications.

Low Bandwidth

The web contains trillions of URLs. Building an archive of these URLs will consume a large amount of bandwidth, storage and CPU power. We want to establish a new method which can be used by most research institutions. So the method should not in the way of creating local archive of all target documents and then conducting alignment. Our novel approach uses online search engines to find, by keywords, the translation of each document.

Using the new method should also save bandwidth if we only find parallel document pairs for a small amount of source documents. In previous works by [5] and [2] they proposed to align parallel documents in a pre-downloaded, closed collection of web documents. The bandwidth and CPU power utilization is very high for downloading documents from the entire web and aligning documents in the corpus.

Improved Recall

The web, however, is heterogeneous in nature. We do not know whether the desired target document exists or not and where the document is. Our method must search the target document in the entire web scope without any restriction on domain name, language, date or etc.

There are a lot of documents containing the same keywords, on the same topics or are the translations of the given document. It is very challenging to search the web for a specific document in another language.

To mine documents from the entire web, the system should be scalable. There must not be any centralized process that limits the system to run only on single CPU core or single machine. The structure based methods, such as URL matching, hyperlinks and HTML structure, introduced in previous works are simple and accurate for mining parallel documents from bilingual parallel websites compared to content based methods. We aim to use URL structure based methods as part of our approach to generate more parallel resources from parallel sites and use them as sources of expansion.

2 Background

2.1 Parallel Resources on the Web

Parallel resources are translated documents and sentences. They are mainly used as training data for a statistics machine translation system or bilingual lexicon extraction. On the World Wide Web, parallel resources reside on a diverse range of websites which can be classified into the following categories:

Parallel Websites

A single website with structurally aligned translated pages is a parallel website. Typically there are websites of institutions, governments and commercial companies. These websites provide information in more than one language. The same pages in different languages usually have similar URLs. For example, as shown in Fig. 1, Financial Times Chinese have the same news in English and Chinese and the two pages have similar URLs.

Comparable Websites

Websites contain parallel contents in different languages without any structural relation in URL, hyperlinks or HTML structure between document pairs. Press agencies have independent content management systems and editors for publishing news in different regions and languages. That is, on comparable websites, parallel documents exist but they are not linked together by website editors.

For example, Reuters China (cn.retures.com) and Reuters.com have independent editors and use identical content management systems. Some news is directly translated from the other site but there is no hyper-link to the original version. As shown in Fig. 2, the two translated pages neither have similar URL nor are they hyper-linked to each other.

Quasi-Comparable Websites

Independent websites that contain some translated parallel contents are quasi-comparable. They may contain stories, documentations and books chapters in many languages on different websites.

For example, in Fig. 3, the book Uncle Tome's Cabin (*湯姆叔叔的小屋*) is available in Chinese (a) and English (b). The Chinese version is hosted on tianya book online and the English version is on the University of Virginia Library. The two websites are not under the same domain name and there is no link between the two pages. We can only find the target document using the content of source document.

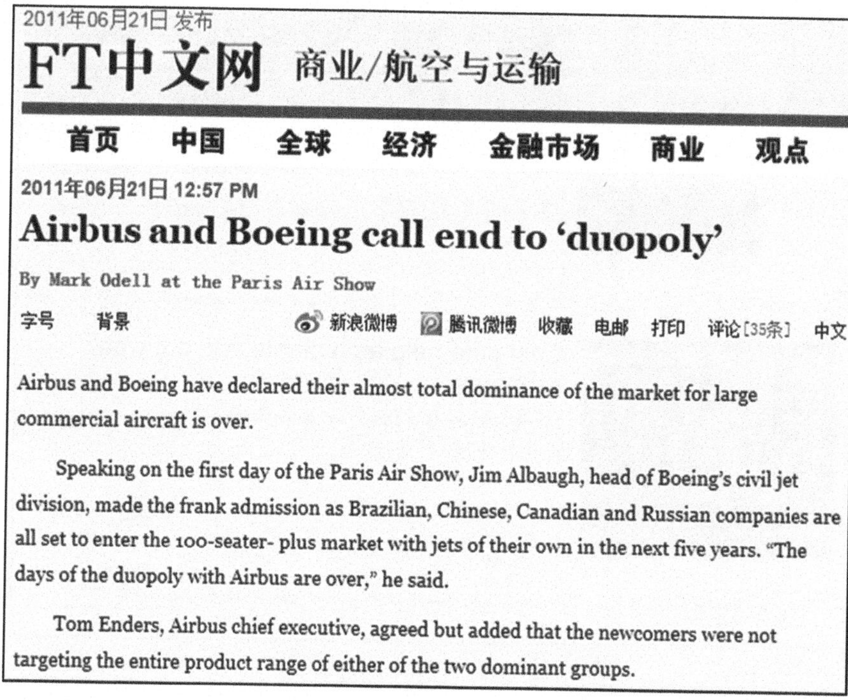

2011年06月21日 发布

FT中文网 商业/航空与运输

首页 中国 全球 经济 金融市场 商业 观点

2011年06月21日 12:57 PM

Airbus and Boeing call end to 'duopoly'

By Mark Odell at the Paris Air Show

字号 背景 新浪微博 腾讯微博 收藏 电邮 打印 评论[35条] 中文

Airbus and Boeing have declared their almost total dominance of the market for large commercial aircraft is over.

Speaking on the first day of the Paris Air Show, Jim Albaugh, head of Boeing's civil jet division, made the frank admission as Brazilian, Chinese, Canadian and Russian companies are all set to enter the 100-seater- plus market with jets of their own in the next five years. "The days of the duopoly with Airbus are over," he said.

Tom Enders, Airbus chief executive, agreed but added that the newcomers were not targeting the entire product range of either of the two dominant groups.

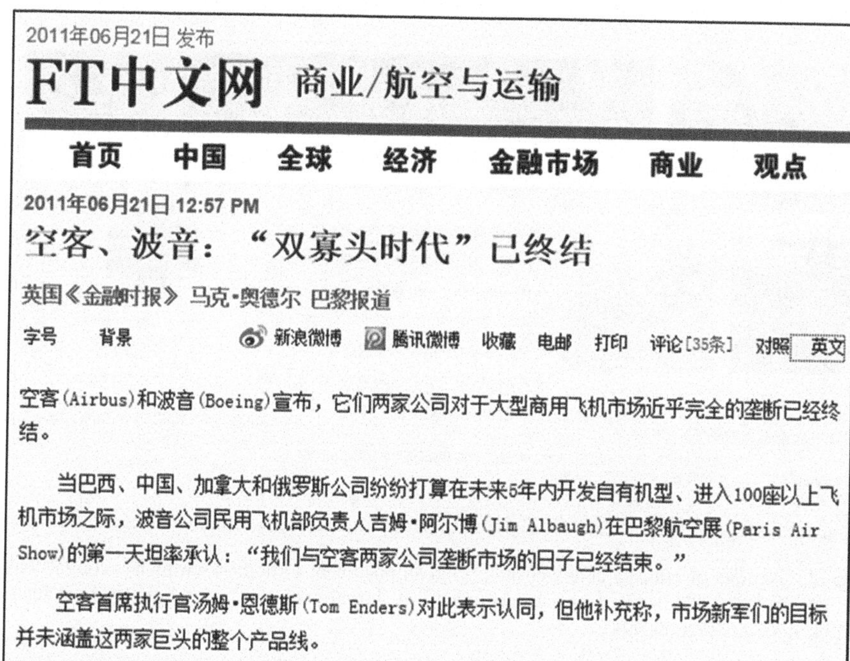

2011年06月21日 发布

FT中文网 商业/航空与运输

首页 中国 全球 经济 金融市场 商业 观点

2011年06月21日 12:57 PM

空客、波音："双寡头时代"已终结

英国《金融时报》马克·奥德尔 巴黎报道

字号 背景 新浪微博 腾讯微博 收藏 电邮 打印 评论[35条] 对照 英文

空客(Airbus)和波音(Boeing)宣布，它们两家公司对于大型商用飞机市场近乎完全的垄断已经终结。

当巴西、中国、加拿大和俄罗斯公司纷纷打算在未来5年内开发自有机型、进入100座以上飞机市场之际，波音公司民用飞机部负责人吉姆·阿尔博(Jim Albaugh)在巴黎航空展(Paris Air Show)的第一天坦率承认："我们与空客两家公司垄断市场的日子已经结束。"

空客首席执行官汤姆·恩德斯(Tom Enders)对此表示认同，但他补充称，市场新军们的目标并未涵盖这两家巨头的整个产品线。

Fig. 1 Example of parallel website. From *Financial Times Chinese*, at http://www.ftchinese.com/story/001039198/en/ and http://www.ftchinese.com/story/001039198/

Fig. 2 Example of comaparable websites (English document at http://www.reuters.com/article/idUSN2332431020090423; Chinese document at http://cn.reuters.com/article/CNTechNews/idCNChina-4323720090424)

第一章　给读者介绍一位好心人

二月的某一天，天气依然比较寒冷。黄昏时分，在P城一间布置典雅兼作餐厅的接待室里，两位绅士相对而坐，喝着酒。他们没有要仆人在旁边侍候。他们紧挨着坐着，好像在商量什么很重要的事情。

为了便于读者阅读，我们暂且称他们“绅士”。其实，如果我们挑剔地观察一下就可看出，其中一位看来不配称为“绅士”。他身材矮小，长相并无独特之处，但神态却是洋洋自得，一看便知他是那种混迹于社会、想方设法向高处爬的势利小人。他的衣服穿着有失风度，一件俗气的杂色背心，一条醒目的黄点蓝底围巾，脖子上是一条色彩艳丽的领带。他的这身打扮与他的派头看来还比较相配。他粗大的手指上套着几枚戒指，一串形状奇特、色彩艳丽的图章缀在那沉沉的表链上。当谈话进行得顺利时，他喜欢把表链弄得叮叮当当地响，俨然一副踌躇满志的神态。他的话语丝毫不符合默里氏语法规则，从他的嘴里经常冒出一些下流、猥陋的单词。尽管作者努力让自己的叙述更加形象，但还是难以正确地转述他的意思。

相反，与他谈话的希尔比先生倒不失绅士风度。室内的摆设和情调都向我们证明这个家庭的生活殷实而且非常安逸。而现在这两个人正在认真地商讨着某件事情。

“我想这件事就这么办吧。”希尔比先生说。

“希尔比先生，这样成交，我实在难以答应。”对方一面回答，一面举起酒杯，对着客厅的灯看着。

Chapter 1

CHAPTER I
In Which the Reader Is Introduced to a Man of Humanity

Late in the afternoon of a chilly day in February, two gentlemen were sitting alone over their wine, in a well-furnished dining parlor, in the town of P -- --, in Kentucky. There were no servants present, and the gentlemen, with chairs closely approaching, seemed to be discussing some subject with great earnestness.

For convenience sake, we have said, hitherto, two *gentlemen*. One of the parties, however, when critically examined, did not seem, strictly speaking, to come under the species. He was a short, thick-set man, with coarse, commonplace features, and that swaggering air of pretension which marks a low man who is trying to elbow his way upward in the world. He was much over-dressed, in a gaudy vest of many colors, a blue neckerchief, bedropped gayly with yellow spots, and arranged with a flaunting tie, quite in keeping with the general air of the man. His hands, large and coarse, were plentifully bedecked with rings; and he wore a heavy gold watch-chain, with a bundle of seals of portentous size, and a great variety of colors, attached to it, -- which, in the ardor of conversation, he was in the habit of flourishing and jingling with evident satisfaction. His conversation was in free and easy defiance of Murray's Grammar, [1] and was garnished at convenient

-42-

intervals with various profane expressions, which not even the desire to be graphic in our account shall induce us to transcribe.

His companion, Mr. Shelby, had the appearance of a gentleman; and the arrangements of the house, and the general air of the housekeeping, indicated easy, and even opulent circumstances. As we before stated, the two were in the midst of an earnest conversation.

"That is the way I should arrange the matter," said Mr. Shelby.

"I can't make trade that way -- I positively can't, Mr. Shelby," said the other, holding up a glass of wine between his eye and the light.

Fig. 3 Example of quasi-comparable websites (Chinese Version: http://www.tianyabook.com/waiguo2005/s/situofuren/tmss/001.htm; English Version: http://etext.lib.virginia.edu/etcbin/toccer-new2?id=StoCabi.sgm\&images=images/modeng\&data=/texts/english/modeng/parsed\&tag=public\&part=1\&division=div1)

2.2 Approaches to Mine Parallel Documents

There are two categories of approaches to mine parallel documents from the web, the structure based methods [6–8] and the content based methods [1–5, 9]. The structure based approaches mainly uses URL, HTML structure, hyper-links to find pre-linked parallel documents on parallel websites whereas the content based approaches can find extra parallel resources from comparable and quasi-comparable websites by aligning the content.

Structure Based Approaches

Parallel websites have different language versions of the same content on the same website. Parallel documents are already aligned on parallel websites with one of the following features:

- Generalizable URL pattern (/en/, /zhs/ . . .) [6]
- Anchor to each other with language identifier (English, 中文. . .) [6]
- HTML DOM Tree structure (same tree structure) [7]

Structure based approaches use the above characteristics to match parallel webpages with very high precision.

In previous works, [6] used (1) parent pages containing links to versions of one document in different languages and (2) sibling pages that contain links to the translation of the current documents. For each webpage on a parallel website, they try to access the same page in another language by directly modifying the URL based on a pre-defined pattern. They also rely on the URL and anchor text to spot language specific version of the documents. If there is a hyper-link with the anchor text which indicates a language, such as English, Françis or *中文版* (Chinese Version), the hyper-linked page will be considered as parallel document of the current one.

A structural alignment using the DOM tree representation was proposed by [7] to align parallel documents by using the HTML structure. They identify the translational equivalent texts and hyperlinks between two parallel DOM trees to find parallel documents.

However, the web is a heterogeneous collection of documents that extend far beyond bilingual and comparable pages with obvious structural features, such as similar URLs or common titles. Structural features only work for bilingual websites or document pairs that are already linked by editors. As only a small portion of websites are in the parallel website category, we cannot use this approach to crawl parallel resources from the entire web. It is also hard to discover new parallel websites automatically.

Content Based Approaches

Content based approaches typically align parallel documents by the contents. Those methods in previous works usually required to create a local archive of all documents in both source and target languages. [2] and [3] proposed to download all source language and target language documents and then perform Cross Language Information Retrieval (CLIR) to extract candidate parallel documents.

[1] and [4] primarily focused on mining parallel documents from a downloaded collection of news articles, using time stamp alignment and content matching. More recently, [10] proposed an adaptive pattern-based bilingual data mining method to mine bilingual web pages for parallel phrases and terms. [5] aligned parallel documents by querying n-gram index built from translation of multilingual documents.

The steps of these approaches can be generalized and shown in Fig. 4. All of them require a large local archive of both source and target documents. This can be very costly when we have to query the entire web.

The local index has to be updated regularly as the World Wide Web is updating continuously. Building and maintaining the local archive of trillions of documents is extremely costly in terms of CPU consumption, network bandwidth usage as well as disk storage utilization. Parallel document alignment algorithm cannot start until all documents in both languages have been downloaded.

Previous content based methods still require downloading all webpages in target languages even if we only want a small portion of source documents under a specific domain to be aligned.

Fig. 4 Typical content based approach used in previous works. *1* Download all documents in both languages, *2* alignment on documents in local archive

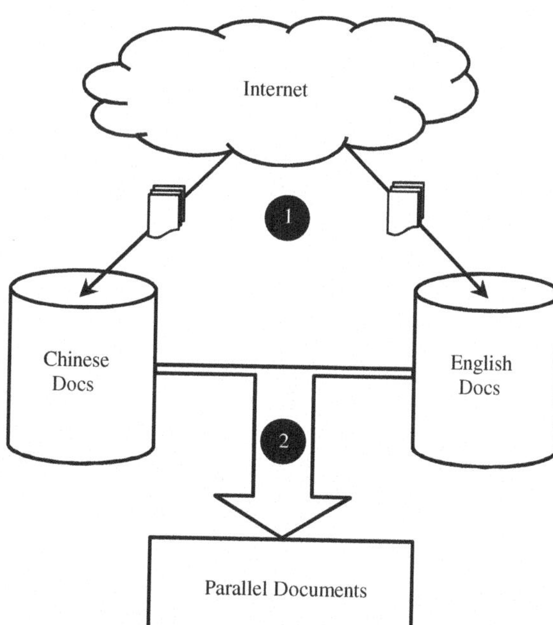

Moreover, [5] uses statistical machine translation (SMT) system to translate all documents into target language to build a query index. Due to the complexity of machine translation algorithms (typically O(n2)), it is waste of resource to download all target language documents, machine translate them, then select the desired candidate parallel documents.

Steps used in finding target documents can be replaced by using search engine APIs, such as Google or Bing, with search query generated from source documents to save CPU and bandwidth consumption in downloading trillions of target documents into a local achieve.

We use the online mode which processes one source document at a time. Most research institutions interested in mining parallel documents do not possess a large number of CPUs or storage on the scale of the world's top search companies. It is also desirable that any site can scale the mining speed and volume according to the computing resources available to them. There is no need for research institutions to crawl all target documents and build a local search engine which saves a lot of computational resources as well as bandwidth.

Changing of ranking algorithm is a new challenge to the approaches that query the web using search engines APIs. We do not want to refine the algorithm frequently in long term mining. The algorithm should communicate with the search engine interactively to adapt a new algorithm. The search query relevance score (SQRS) we proposed is used to evaluate the quality of each search query and monitor the search engine feedback during the search.

To this end, we propose a new low bandwidth and high precision content based method to on the one hand complement structural matching, and on the other hand reduce the complexity of content matching.

[11] proposed a mining approach on selected Chinese news article containing cue phrases, such as "根据外電報道" (According to foreign media). They proposed multiple algorithms to locally rank keywords which are generated from the source documents. Then they search the ranked keyword set in search engine API and find parallel documents. In non-oracle queries, which keywords are extracted from source documents, a maximum 40 % of the documents found were parallel while the remaining pairs were comparable. This is a benchmark in mining precision using search engine APIs.

As the parallel resources mined are often used to improve SMT systems or yield bilingual lexicons, it is desirable that the mining output is of high precision (Fig. 5).

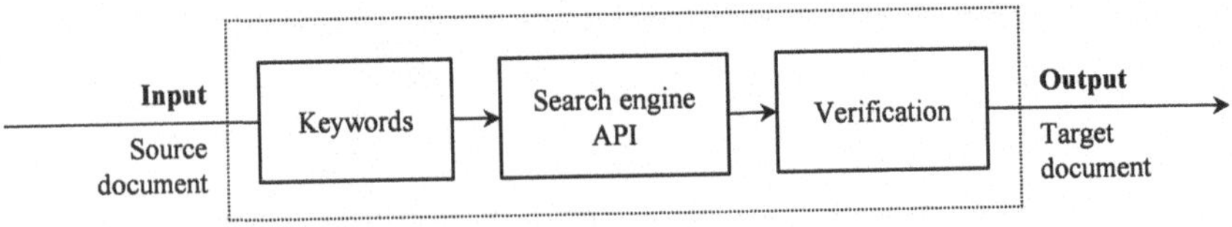

Fig. 5 Basic steps of content based approach using search engine APIs

3 Our Content Based Approach Supplemented with Structural Information

Our proposed content based approach (Fig. 6) primarily aims to discover parallel documents from all kinds of parallel, comparable or quasi-comparable websites on the World Wide Web. We take advantage of online search engines to find candidate documents but against the search result, thereby saving bandwidth and cost and avoiding the need to crawl all target documents and for storage in an archive.

Content based approach queries the document in target language using keywords from documents in the source language. In our approach, queries are generated from source documents and expanded dynamically and interactively by search result quality score (SQRS) as feedback. Our algorithm can automatically adapt to the change of ranking algorithms.

As neither machine translation of the full text nor pre-download of target documents is needed for our approach, mining can start instantly when a batch of source documents from the same domain has been downloaded.

We mainly focus on the precision of output parallel documents. The query expansion feedback score is the key in improving the precision of target documents found. We also use DTW and R2 regression as verification methods to further ensure the quality of parallel document pairs. Although the target documents in search result always have some relationship with the source document in common keywords, but

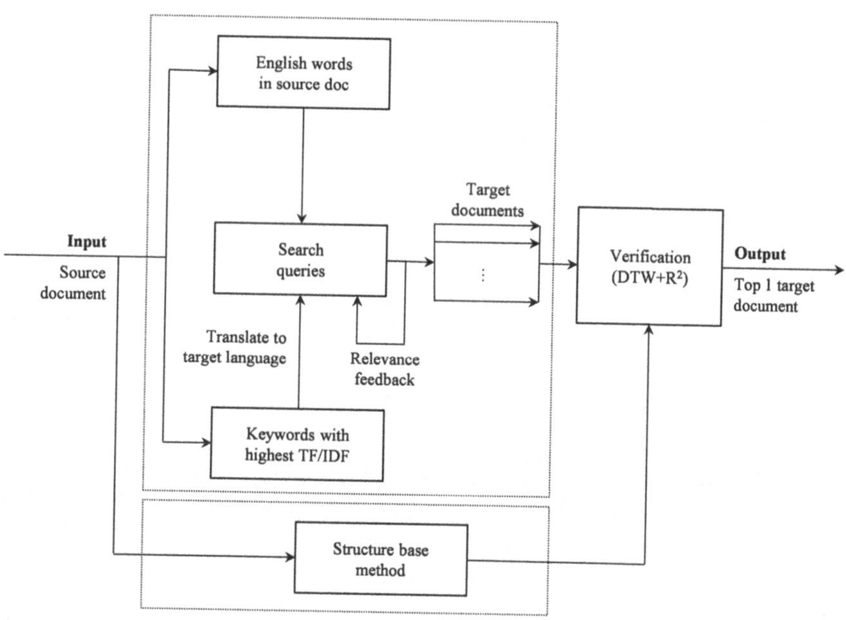

Fig. 6 Our content based approach

they may not in the sense of comparable or parallel. We do not allow non-comparable document pairs in the output. If a source document is found to have no translation in the target language or the target document fails the verification, the system simply returns <not-found>.

3.1 Source Document Representation

Online search engines use keywords to find documents. We cannot enter source documents with thousands of words directly into a search engine. We have to convert text into keywords to perform automated queries. Each single keyword may exist in multiple articles. However, they can identically represent a document if several keywords are put together as a keyword set [10]. To find document in target language, we translate each keyword into target language using bi-lexicon to form the initial query.

A query does not always yield the desired target document by directly using all the translated keywords from source document. The reasons are:

- Keyword translation might not correspond to the actual words in the target document
- Certain keywords in the target documents might have been removed by content editors
- There are errors in keyword translation or selection
- Number of keywords exceeds search engine limit

It is essential to select appropriate keywords and use correct translation to find the desired target document in a search engine. Two conditions that an appropriate keyword set should satisfy are: (1) they should represent the document exclusively [10] and (2) they should have unique or common translation in both languages.

We suggest that words with high Term Frequency and Inverse Document Frequency (TF-IDFs) and English words in Chinese text are usually keywords that fulfill both conditions above.

$$K = K_T \bigcup K_E \tag{1}$$

KT: set of words with high TF-IDF score
KE: set of English words in Chinese documents

To obtain TF-IDFs that are representative of the keyword in the source document, in our experiments, they are trained from all source documents under the same domain name (e.g. www.ftchinese.com). This can help us avoid words that exist in navigation area (banner, footer) and advertisements since they have very low IDF in the website. Sometimes, source documents may not be downloaded from the same website. Domain-wise TF-IDF training is not yet possible. In such cases, it is acceptable to use TF-IDF trained in the general domain.

Keywords in K_E are more important because most of them are words actually used in the target document. There is no translation error since words in K_E are in the target language already. However, in many cases, there are additional words in K_E so that we cannot find any document by directly searching K_E. The effect of additional wrong keywords to search result will be discussed in the next section. Our method combines K_E with K_T to obtain the keyword set and use words in K_E first.

3.2 Search Target Documents with Search Query Relevance Score (SQRS)

Search engine takes multiple criteria, including keyword significance, domain popularity, date, popularity, page rank and etc., to return the most relevant documents that match the query. Each keyword may have more than one possible translation in the bi-lexicon. We do not know the words that are actually used in the target documents when we are generating keywords from source document.

In previous studies, [11] used local ranked keywords and query each document once. However, search engine ranking algorithm and index are changing over time. The same query may not always yield the desired document in the search result. To maximize the possibility of finding the target document, we take search engine feedback into account to interactively amend queries.

We proposed to use SQRSs which evaluate the relevance of the search result and the search query. Then we amend the keywords according to the SQRS feedback. By using SQRS as feedback, we can:

- Choose better keywords set
- Use the correct translation of keywords
- Adopt change of search engine ranking algorithm

Commercial search engines omit some keywords when there is no document in the index containing all keywords. In such cases, the rank of documents usually changes significantly.

The following example shows the ranking change caused by erroneously translated keyword. The search result of two search queries are generated from the Chinese version of *My Space launching new version of website*[2] "|" indicates separator of keywords.

Query 1:
myspace | mike jones | facebook | san francisco | new | website

Query 2:
myspace | mike jones | facebook | san francisco | new | website | fashion

---

[2] Source: http://cn.reuters.com/article/CNTechNews/idCNCHINA-3233720101027 on May 10, 2011.

Table 1 SQRS of Query 1

SQRS	Search engine omitted keyword	Rank in result
7.742	–	1 (target)
5.174	Web(site)	3
4.951	Web(site)	2
4.663	Web(site)	4
4.545	Web(site)	5

Table 2 SQRS of Query 2

SQRS	Search engine omitted keyword(s)	Rank in result
6.155	Fashion	5 (target)
3.951	Web(site) I fashion	1
5.867	Website	3
0.871	mike I new I website I fashion	4
−2.921	mike jones I new I website I fashion	2

In Query 1, the oracle target document was the topmost in the search result. The short summary contains every keyword we entered in the query. The SQRS generally decreases with the rank of webpages (For definition and calculation of SQRS, see Eqs. 2–7) (Tables 1, 2).

In Query 2, we added *fashion* which is the English translation of "新潮" (but the actual English version used hottest). The rank of search result changed and each summary omitted at least one keyword in the query. The SQRS does not align with the search result yet.

This phenomenon implies that the document containing all keywords in Query 2 does not exist on the web. The recently added keyword fashion must be erroneously translated.

In many similar cases, an erroneously translated keyword can pollute the query quality and decrease the rank of the target document. Parallel document mining cannot rely on the document rank of a search engine. The system must have a mechanism to detect the problem when expanding the query. Otherwise, a batch of irrelevant documents will be downloaded which need to be filtered out.

To avoid adding erroneously translated keywords to further reduce the amount of undesirable documents downloaded, we introduced the *search query relevance score (SQRS)* that describes how well the search result is and how we can refine the query if the desired document is not presented.

The SQRS score is determined by comparing the query with highlighted keywords in the search result. Generally, a webpage has higher SQRS if the summary contains more keywords that match the query.

Table 3 Result quality and SQRS

Source documents	Target documents have largest SQRSs	%
112	81	72.3

We ran experiments on 112 randomly selected source documents and observed the SQRSs in the search result. As shown in Table 3, 81 (72.3 %) target documents have the highest SQRS among other URLs in the search results. It implies the SQRS are an effective measure of query formation and keyword translation.

$$Q = (k_1 k_2 \dots k_C) \tag{2}$$

$$k_i = (w_{i1} w_{i2} \dots w_{iJ}) \tag{3}$$

$$\Gamma K_i = \{(w_{ia} \dots w_{ib})\,|0 < a < b \leq J\} \tag{4}$$

$$count(c, t) = \#\text{ of occurrence of } c \text{ in } t \tag{5}$$

$$\text{where } \delta(c, t) = \begin{cases} 0 \; count(c, t) > 0 \\ 1 \; count(c, t) = 0 \end{cases} \tag{6}$$

$$SQRS(Q, T) = \sum_{c=1}^{C} \left[\sum_{s \in \Gamma K_c} [\log(count(s, T) + 1)] - \delta(k_c, T) \right] \tag{7}$$

where Q is the query, k is keyword, w is English word and T is the short text with highlighted keywords in search result.

Although the query may include multiple translations of a keyword in bilingual lexicon dictionary, the SQRS ensures that there is minimum adverse effect from incorrect translations and irrelevant keywords.

SQRS can avoid erroneously translated keywords on one hand. On the other hand, it can also help the system to select keywords because some keywords extracted from source document may not appear in the target document. The negative effect of searching irrelevant keywords is similar to adding erroneously translated keywords (Fig. 7).

There are many permutations of keyword sets for each document. It is unlikely to use the correct keyword set and get the desired document in the first search attempt. To improve the precision of the keyword set, we use SQRS as relevance feedback as shown in Fig. 8.

First, we rank the keywords in K_T by the TF-IDF scores. Next, the query is expanded by SQRS. When keyword w is added to the current query, we compare the maximum $SQRS_c$ among top n results with the previous highest score $SQRS_p$ without w. w will be discarded from the keywords if $SQRS_c < SQRS_p$ or simply caused an empty search result. Otherwise, the query will be expanded by adding w.

The search engine returns the total number of target documents for each query. If this number is less than a threshold M, we verify the parallelity of top-ranked documents in the next step and stop the query expansion. To save network bandwidth,

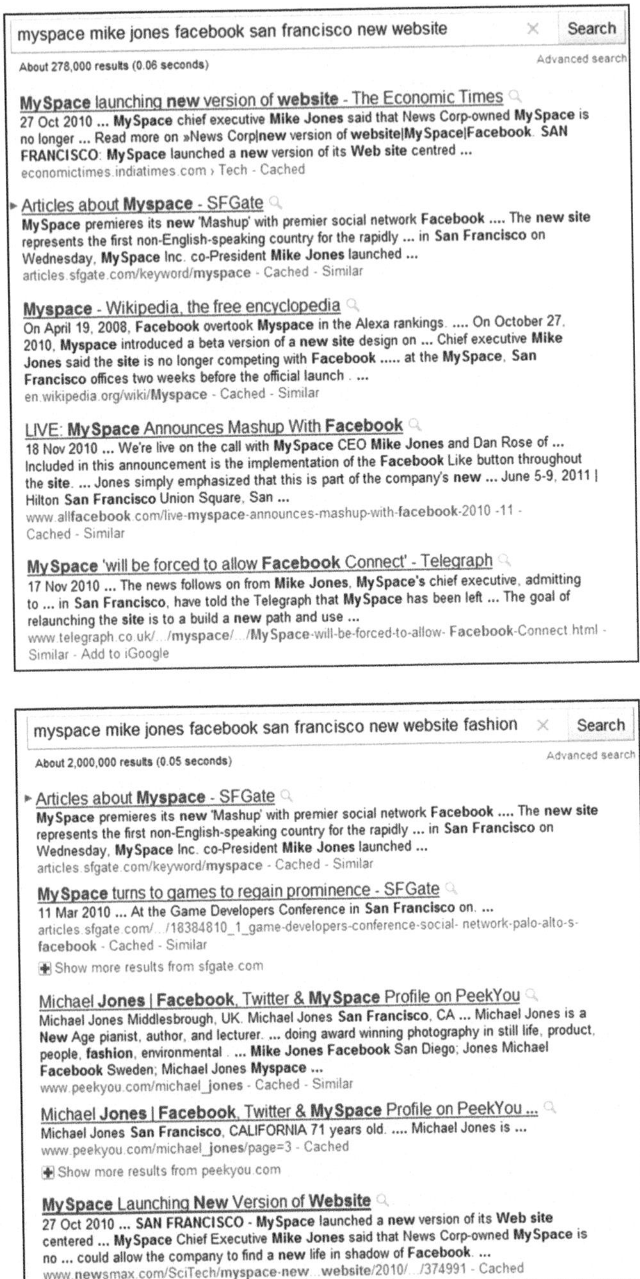

myspace mike jones facebook san francisco new website × Search

Advanced search

About 278,000 results (0.06 seconds)

MySpace launching **new** version of **website** - The Economic Times
27 Oct 2010 ... **MySpace** chief executive **Mike Jones** said that News Corp-owned **MySpace** is no longer ... Read more on »News Corp|**new** version of **website|MySpace|Facebook**. **SAN FRANCISCO**: **MySpace** launched a **new** version of its **Web site** centred ...
economictimes.indiatimes.com › Tech - Cached

Articles about **Myspace** - SFGate
MySpace premieres its **new** 'Mashup' with premier social network **Facebook** .... The **new site** represents the first non-English-speaking country for the rapidly ... in **San Francisco** on Wednesday, **MySpace** Inc. co-President **Mike Jones** launched ...
articles.sfgate.com/keyword/**myspace** - Cached - Similar

Myspace - Wikipedia, the free encyclopedia
On April 19, 2008, **Facebook** overtook **Myspace** in the Alexa rankings. .... On October 27, 2010, **Myspace** introduced a beta version of a **new site** design on ... Chief executive **Mike Jones** said the **site** is no longer competing with **Facebook** ..... at the **MySpace**, **San Francisco** offices two weeks before the official launch. ...
en.wikipedia.org/wiki/**Myspace** - Cached - Similar

LIVE: **MySpace** Announces Mashup With **Facebook**
18 Nov 2010 ... We're live on the call with **MySpace** CEO **Mike Jones** and Dan Rose of ... Included in this announcement is the implementation of the **Facebook** Like button throughout the **site**. ... Jones simply emphasized that this is part of the company's **new** ... June 5-9, 2011 | Hilton **San Francisco** Union Square, San ...
www.all**facebook**.com/live-**myspace**-announces-mashup-with-**facebook**-2010 -11 - Cached - Similar

MySpace 'will be forced to allow **Facebook** Connect' - Telegraph
17 Nov 2010 ... The news follows on from **Mike Jones**, **MySpace's** chief executive, admitting to ... in **San Francisco**, have told the Telegraph that **MySpace** has been left ... The goal of relaunching the **site** is to a build a **new** path and use ...
www.telegraph.co.uk/.../**myspace**/.../**MySpace**-will-be-forced-to-allow- **Facebook**-Connect.html - Similar - Add to iGoogle

myspace mike jones facebook san francisco new website fashion × Search

Advanced search

About 2,000,000 results (0.05 seconds)

Articles about **Myspace** - SFGate
MySpace premieres its **new** 'Mashup' with premier social network **Facebook** .... The **new site** represents the first non-English-speaking country for the rapidly ... in **San Francisco** on Wednesday, **MySpace** Inc. co-President **Mike Jones** launched ...
articles.sfgate.com/keyword/**myspace** - Cached - Similar

MySpace turns to games to regain prominence - SFGate
11 Mar 2010 ... At the Game Developers Conference in **San Francisco** on. ...
articles.sfgate.com/.../18384810_1_game-developers-conference-social- network-palo-alto-s-**facebook** - Cached - Similar
Show more results from sfgate.com

Michael **Jones | Facebook**, Twitter & **MySpace** Profile on PeekYou
Michael Jones Middlesbrough, UK. Michael Jones **San Francisco**, CA ... Michael Jones is a **New** Age pianist, author, and lecturer. ... doing award winning photography in still life, product, people, **fashion**, environmental. ... **Mike Jones Facebook** San Diego; Jones Michael **Facebook** Sweden; Michael Jones **Myspace** ...
www.peekyou.com/michael_**jones** - Cached - Similar

Michael **Jones | Facebook**, Twitter & **MySpace** Profile on PeekYou ...
Michael Jones **San Francisco**, CALIFORNIA 71 years old. .... Michael Jones is ...
www.peekyou.com/michael_**jones**/page=3 - Cached
Show more results from peekyou.com

MySpace Launching **New** Version of **Website**
27 Oct 2010 ... **SAN FRANCISCO** - **MySpace** launched a **new** version of its **Web site** centered ... **MySpace** Chief Executive **Mike Jones** said that News Corp-owned **MySpace** is no ... could allow the company to find a **new** life in shadow of **Facebook**. ...
www.**newsmax**.com/SciTech/**myspace-new**...**website**/2010/.../374991 - Cached

Fig. 7 Search result of Query 1 (*upper*) and 2 (*lower*) on Google.com

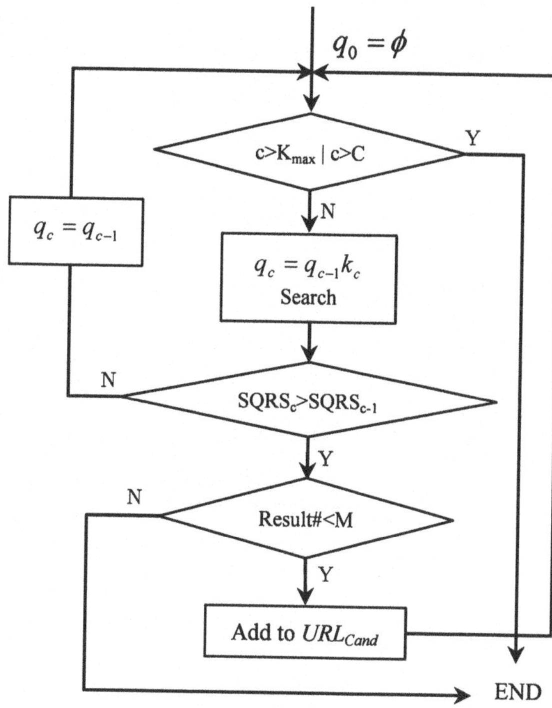

Fig. 8 Flowchart of query expansion algorithm

the system only considers K_{Max} words with the highest TF-IDF scores and directly output <not-found> if no target document is found.

3.3 Document Verification

Documents pairs found in the search engine are candidates of parallel documents. Even though we have SQRS to measure the quality of keywords translation and selection, however, it is based on keywords only. We need to have a step to verify the parallelness of document pairs to guarantee the precision.

All candidate document pairs are subjected to pass the parallelness check before output. Output is <not-found> if all candidate target documents failed the verification process. We propose using both dynamic time warping (DTW) and R2 regression as in our previous work [12] on every pair of the source and targets document to evaluate the parallelness and discard those pairs that failed the verification.

Table 4 Comparison of target document found with and without SQRS

	Source Doc	Doc Pairs	Sentences
Without SQRS	1000	217	2906
With SQRS	1000	243 (+12.0%)	3068 (+5.6%)

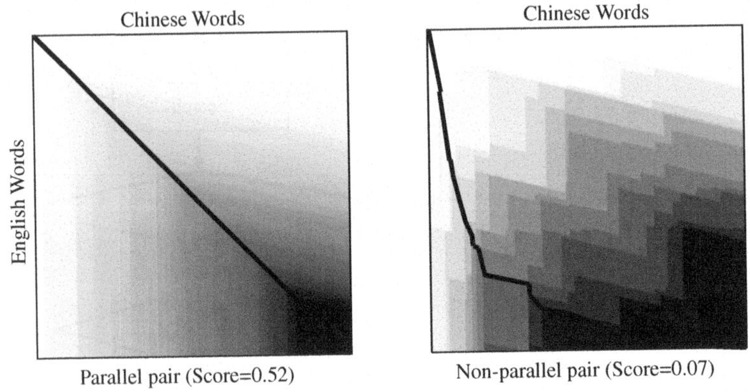

Fig. 9 DTW of parallel and non-parallel document pairs

Dynamic Time Warping (DTW) Score

DTW alignment is much faster than machine translation (MT). We measure the word level DTW score between source document and target document with local constrain of d (Eq. 8). Stop words are removed from the English text before DTW processing.

We use a bilingual lexicon (dictionary) to calculate the DTW path of each pair. If there is an entry in the bi-lexicon for ith word of source document and jth word of the target document (i.e. Chinese and English) respectively, the score of point (i,j) is 1, otherwise 0. The total score is normalized by maximum number of steps (moves) from (0,0) to (m,n) to convert DTW score to a number between 0 and 1 (Table 4).

Parallel document pairs tend to have a path close to the diagonal line with high DTW score. Figure 9 shows the DTW paths and scores of a parallel document pair and a non-parallel pair. The precision of output sentences increases if the DTW score threshold is set higher. The non-parallel documents also have overlapped words but the word sequences are too different. Table 5 is the relationship between DTW scores and the precision of candidate pairs based on our experiment and human evaluation of pairs.

The output pairs are mainly comparable documents of the same topic if the DTW score are upper limited to a threshold.

Table 5 DTW and precision of candidate pairs

DTW	# Pairs	# Parallel	Precision %
>0.45	122	121	99.18
>0.40	224	219	97.77
>0.35	298	288	96.64
>0.30	354	337	95.20
>0.28	389	364	93.57
>0.26	429	389	90.68
>0.25	456	403	88.38
>0.24	488	415	85.04
>0.22	545	417	76.51
>0.20	627	426	67.94

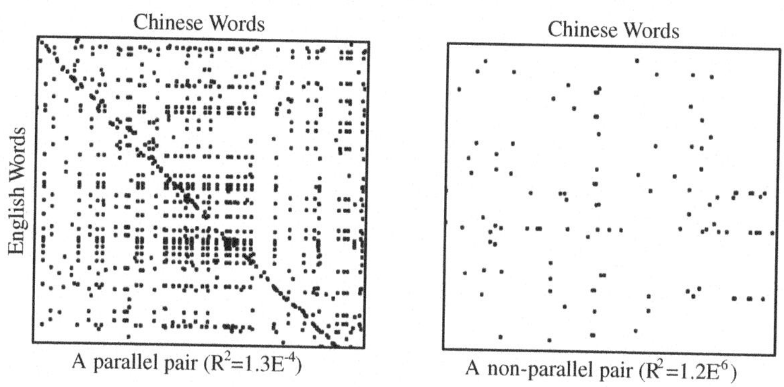

Fig. 10 R^2 of parallel and non-parallel document pairs

$$DTW(i_m, i_n) = c + \min \begin{pmatrix} DTW(i_m, i_n) & \cdots & DTW(i_m - d, i_n) \\ \vdots & \ddots & \vdots \\ DTW(i_m, i_n - d) & \cdots & DTW(i_m - d, i_n - d) \end{pmatrix} \quad (8)$$

R^2 Regression

The parallel documents contain parallel sentences that may have different word orders, especially in the case of English and Chinese. The DTW score may be affected by different word order. We propose to use R^2 regression as an additional feature (Fig. 10).

Score measures the deviation of the matching path of shared words in both documents from the diagonal. Stop words are not included in the bi-lexicon dictionary. The size of source and target documents is usually different. The R^2 score is normalized by the size ratio to make it comparable among document pairs.

Table 6 Mining precision of DTW and R^2

	DTW (>0.22)	R^2 (1.0E-5,1)	DTW+R^2
# Pairs	545	534	481
# Parallel	417	403	399
Precision %	76.51	75.47	82.95

Table 7 Comparison of content based method and URL matching

Source	# of Doc	Content based	Content based + URL
FTChinese	11,009	2,968	9,066
WSJ (CN)	3,327	1,002	3,120

$$R^2 = R^2_{score}/sizeof(T_{src}) * sizeof(T_{tar}) \tag{9}$$

Combining DTW and R^2

DTW score helps filter out non-parallel pairs and R2 is introduced as a supplementary feature to further improve the precision of extracted parallel documents. A comparison of using these measures is shown in Table 6.

Structural Features

The final step of verification uses structural features of the document pair candidates:

- **Language**: mined document should be in the target language
- **Absolute size**: neither document should be too small/large in file length
- **Size difference**: two documents must be of similar size
- **Document type**: both documents must be content page in a website

3.4 Improve Recall by Structure Based Information

Since search engines rank target documents by various criteria, including page rank, some bilingual website documents might be missed by the pure content based approach. We propose to supplement our approach with the URL matching patterns or anchors if the content based method has found several pairs of source and target documents having the same pattern.

We examine the pairs found by the content based method and look for any characteristics that match the structure based patterns, including URL, anchors and etc. Then, we apply the patterns to all source web pages under the same hostname. This will greatly increase the amount of parallel documents in the output as the recall of content based method is usually lower than the structure base methods.

The experiment result shown in Table 7 illustrates the number of target documents found by content based method and URL matching for the same set of source documents on parallel websites. URL matching generally improves the recall by two times.

All pairs found by both content based and structure based methods are also subjected to passing the verification process introduced in Sect. 3.3.

4 Experiment Setup and Result

4.1 Experiment Setup

We evaluate our approach on two sets of experiments. They are conducted on a single machine with eight processing cores. Source and target documents in the experiments are Chinese and English respectively.

Website Parsing

To compare the DTW and R2 of each document pairs, we have extracted the main content of the web page. We find the main content of the document using ExtMainText[3] in python.

However, ExtMainText computes the text density on a HTML page, which defined as:

$$\rho_{text} = \frac{l(Text)}{l(Text) + l(Tag)} \tag{10}$$

A specified threshold of ρ_{tex} is required for different sites or even different documents to get the main content extracted. We also need to adjust the value for different languages.
We have implemented an algorithm to select ρ_{text} automatically.
We calculated ρ_{tex} for each line. Then classified ρ_{tex} from the first line to the last, i.e. lines with similar ρ_{tex} will be grouped into the same class. The class including most number of lines is extracted as the main text of the page.

In our experiment, we use different TF-IDF for each domain name. For example, we train TF-IDF for ftchinese.com using all source documents under this domain name and this TF-IDF which are used for these source documents only.

Words appearing in the header and footer will get very high in both TF and IDF and will be low weighted in the keyword selection.

We use the Chinese-English Translation Lexicon Version 3.0 from LDC[4] The system can directly load the bi-lexicon in the same format for other language pairs.

[3] http://www.elias.cn/En/ExtMainText/

[4] LDC Catalog Number: LDC2002L27.

4.2 Content Based Baseline

As a baseline of content based method, we directly enter English words in the original Chinese document in the search engine (Experiment i). Then, we add keywords ranked by TF-IDF to query the target document but do not perform SQRS to expand the query (Experiment ii).

Finally, SQRS is used to refine each keyword to get better results (Experiment iii). Our approach is not search engine dependent. All experiments use both Google Search API and Bing Search API to search the keyword sets. Results from different search engines are merged together by a unique URL.

We generalize URL patterns (if any) from document pairs when we find some document pairs by the content based method on parallel websites. By structure based information, we extract more parallel webpages from parallel websites that follow those URL patterns.

4.3 Experiment 1: Find Target Documents

Source (Chinese) documents in our experiments are news from the following 5 agencies:

Parallel websites:

1. Financial Times Chinese (ftchinese.com)
2. Wall Street Journal Chinese (cn.wsj.com)

Parallel website contain both Chinese and English documents under the same host and can be aligned with URL matching.

Comparable/quasi-comparable websites:

3. Reuters China (cn.reuters.com)
4. Forbes China (forbeschina.com)
5. Fortune China (fortunechina.com)

Documents on quasi-comparable or comparable websites may have target documents on either the corresponding agencies' global site (e.g. cn.reuters.com and www.reuters.com) or somewhere else. No parallel documents from such websites can be found by URL matching or other structure based methods.

To evaluate the precision of our content based approach, we applied our method to the above sites for target documents. The percentage of parallel documents that we can successfully find is highly dependent on the type of documents and search engine

Table 8 Source documents for pure content based approach

Source	# Chinese Docs
ftchinese.com	11,009
cn.wsj.com	3,327
cn.reuters.com	8,570
forbeschina.com	6,281
fortunechina.com	593
Total	29,780

index. Calculating recall, on the other hand, is only possible for sites we already knew. For comparable or quasi-comparable sites, it is not possible to evaluate recall of the algorithm because:

- Some source documents may not have translation in the target language
- Target language pages may not be indexed by search engines
- Manual evaluation of all documents for recall calculation is not feasible

Thus, we mainly focus on the precision of the output pairs. In the verification process, we discard the document pairs if:

- DTW score < 0.25 (Output document pairs have 88 % precision)
- R^2 score < 1.0E-5
- Article size is too small (Only one or two sentences)
- Size of source and target are too different
- URL is root (/) under hostname (We consider content pages only)
- Text in wrong language

We manually evaluate the effectiveness of our method on randomly selected document pairs from the output of our content based method. Only strict parallel document pairs are considered as correct. The target precision of parallel documents in this experiment is 88 % (Table 8).

The result of Experiment 1 is shown as follows:

We directly searched all English keywords in Chinese documents and found 153 target documents (baseline). Then we search translation of top ranked TF-IDF keywords (ii). Query expansion with SQRS further improved by 23.56% of the output sentences compared to the baseline (Table 9).

We also applied our content based method on larger scale source documents. Among the 29,680 Chinese documents retrieved from the five news agencies, we obtained 7,253 parallel document pairs with 88 % precision by content based approach alone.

In many such cases, parallel document pairs are on different websites could be found neither by URL matching nor by content-based methods that used times stamps for matching.

With structure based information, we increase the output of the parallel documents from parallel websites. Table 10 shows that the URL matching can improve the output quantity a lot, compensating for the missing target documents with low page ranks.

Table 9 Comparison of different methods

	Source Doc	Doc Pairs	Sentences	Improvement
i.	1000	153	2483	Baseline
ii.	1000	217	2906	+17.04%
iii.	1000	243	3068	+23.56%

i. Direct Search of K_E
ii. Top ranked keywords without SQRS
iii. With SQRS

Table 10 Output document pairs of Experiment 1

Source	# of Doc	Content based	Content based + URL
FTChinese	11,009	2,968	9,066
WSJ	3,327	1,002	3,120
Reuters	8,570	1,911	1,911
Forbos	6,281	1,166	1,166
Fortune	593	206	206
Total	29,780	7,253	15,469

For parallel bilingual websites, the content based method can find about 1/3 (33.3 %) of the target documents compared to the content based method URL matching. It shows that, however, our query expansion with relevance feedback approach has higher recall than the 18 % produced by the local ranked keywords in [11].

4.4 Experiment 2: Parallel Sentence Extraction for SMT

In order to obtain a sentence alignment for pairs of document, we first need to extract the proper content of each page and remove the header and footers that are of little interest and are unlikely to be parallel anyway.

For parallel documents, we can use additional features in parallel text to further improve effectiveness of the web parsing algorithm to extract the main content. We first segment the documents in sentences and filter out improper ones, such as English sentences containing Chinese characters, or Chinese sentence containing roman characters only. We then use DTW again to find a continuous path in the documents and extract the longest path. The header and footer will generally not align and will be discarded; only the chunk of true alignable content will be preserved.

Using this method, we manage to find the beginning and the end of source and target content and extract it. Then discarding pairs of document whose number of extracted sentences are too different, sentence alignment is performed on the remaining documents using the Champollion ToolKit [8] which is already trained for Chinese-English document pairs.

Table 11 BLEU score obtained for SMT (Experiment 2)

	BLEU
Baseline	29.54
CLIR	29.88

Finally, we filter all the sentences using a simple word overlap score. Sentences whose lengths are too different or whose word overlap score is too low are discarded, to ensure a high precision at the end.

Among the 15,469 Chinese-English document pairs found in Table 10, we extracted 225,374 parallel sentence pairs with mining precision of over 97 %. We evaluate the quality of those sentences for training machine translation with the Moses SMT engine.

We use Bilingual Evaluation Understudy (BLEU) [13] score to evaluate the quality of parallel sentences. BLEU score is given by comparing the machine translated text with professional human translated text. The closer the result, the higher the quality of the translation is. We compare the BLEU score obtained with a 4,097,357 sentence pairs corpus, manually aligned (baseline) and the BLEU score obtained with the same corpus, replacing 225,374 sentence pairs by the ones we extracted. The results are presented in Table 11. They are evaluated using the NIST MT06 evaluation set.

These results show that our set of sentences, together with a larger parallel corpus, yield results similar to the one obtained with manually aligned sentences only.

4.5 Experiment 3: Bilingual Lexicon Extraction

The extracted sentences from 15,469 documents have been processed for rare word translation extraction [13]. Jaccard similarity is used to measure words association in documents of comparable corpus. Log-likelihood and cosine similarity are used in the extraction process to normalize and compare the context vector respectively.

As a result of the experiment using document pairs mined by our system, 80 % of F-Measure is yielded for Chinese-English rare word lexicon extraction. This result is comparable to the F-Measure of another language pair of French English using aligned document pairs from Wikipedia.

4.6 Experiment 4: Parallel Document Mining for Hindi-English

We have also expanded our tool to mine parallel documents of Hindi-English pairs. The source documents are in Hindi and the target documents are in English. We changed the lexicon of Chinese-English to Hindi-English. All the other setup is the same as that used in Experiment 1.

Table 12 Result of Experiment 4

DTW	Source documents	Correct	Precision (%)	Precision for zh-en (%)
>0.25	22	19	86.4	88.4
>0.24	24	20	83.3	85.0
>0.22	33	24	72.7	76.5
>0.20	59	34	57.6	67.9

The preliminary result showed the tool works on Hindi-English parallel document extraction. For Hindi-English pairs, we have yielded the precision of 86.4 % with DTW > 0.25 on 187 human evaluated documents pairs. See Table 12 for details.

The precision of Hindi-English pairs are slightly lower than the Chinese-English pairs. It is affected by the size and quality of bilingual lexicon.

The recall of the experiment is about 20 %. However, the recall does not reflect the performance of the tool because it depends on search engine index coverage and types of source document.

4.7 System Performance and Scalability

We carried out our mining experiments on the workstation with 8 state-of-art CPU cores. The average time taken for each source document is 30 s which is only bottle-necked by the limitation of search engine APIs.

As the TF/IDF scores are pre-trained only from the source documents and our content based approach mines the target document for each source document individually, the system can be easily scaled to run in parallel on multiple servers.

File storage can be solved by using distributed file system or storage servers to enable large scale mining from the entire web.

The database manages the state of each URL and outputted document pairs. Mainstream database management systems such as MySQL, SQL Server and Oracle can also be scaled using existing solutions.

4.8 Future Works and Discussion

We have shown that the system works on finding English target documents by using source document in Chinese or Hindi to mine Chinese-English or Hindi-English parallel documents. The mining precisions are similar for these two language pairs. We only extract parallel sentences from aligned Chinese-English parallel documents and evaluate the result by using the machine translation system and BLEU scores. We will also implement a sentence extractor for the Hindi-English sentences and evaluate the quality of those parallel sentences.

In future works, we may extent the system to other languages, for both source and target documents. For some language pairs, the quantity of parallel resources available is very limited. The mining of parallel documents in those language pairs will have significant contribution to the size of parallel corpus and improve the quality of machine translation of the two languages.

It is noticeable that the recall of our content based approach is not as high as the structure based approach. For comparable and quasi-comparable websites, it may miss some parallel documents that can only be found by the content based method. We may improve the recall and further increase the precision by:

Increase bilingual lexicon/dictionary size. The dictionary we used is directly downloaded from the LDC which is small compared to the words in documents we have processed. We can use a large dictionary to increase the accuracy of translation and DTW verification.

Use other keyword extraction methods. We use keywords in target language (KE) and keywords with highest TF-IDF (KT) to form queries and search for target documents. Other keyword sets such as name entities and temporal information from the source documents also can be keywords in this step. They may need to have additional training for the source document domain in order to be used by the system.

Add other search engines. Google and Bing are used in the system for finding target documents. However, these two largest commercial search engines rank webpages mainly rely on criteria other than keywords relevance. This may cause difficulty in finding target document which is not new. We may integrate other search engines to search target documents and rank by keywords relevance only.

There is a usage limitation for search engine APIs of Google and Bing which is the bottleneck of our mining system. The speed of the system in parallel documents mining can also be improved if there is another search engine which has looser usage limitation for its API.

5 Conclusion

We have proposed a content based CLIR approach which searches any part of the web to find parallel documents without the limitation of URL-matched bilingual websites.

We use search query relevance score (SQRS) to ensure translation correctness and measure relevance between keywords and search result to further ensuring the keywords we use represent the source document. Using a supplementary verification process, the web documents are then filtered by dynamic time warping and regression scores.

Experimental results show an 88 % mining precision on the parallel documents extracted from parallel, comparable and quasi-comparable web sites.

Another experiment on extracting bilingual sentences from the mined documents shows that the sentence extraction adds another layer of verification which further improves precision from 88 to 97 %.

SMT experiments on using our mined parallel sentences, together with a larger baseline training set, to train an SMT system show comparable performances from using our data to that of using manually aligned bilingual sentences. BLEU score is improved from 29.54 to 29.88.

Moreover, the precision of parallel resources mined can be adjusted by users to balance between quality and quantity of document pairs. It is also possible to mine comparable documents by limiting the DTW and regression score to a certain range.

The parallel documents mined by our approach are also used as a part of the corpus to extract translations of rare words. 80 % F-Measure are obtained for the extraction of rare Chinese-English lexicon.

Our proposed method does not require full-text machine translation, nor does it require downloading all documents in the target language into an archive for document matching, thereby saving lots of network bandwidth and computational resources. In addition, Mining process can start once a batch of source documents is downloaded and the TF-IDFs are trained using the source documents on the same domain name.

The new content based approach works on the heterogeneous web. We use Google and Bing APIs to automatically search parallel documents by keywords with SQRS for search result feedback.

The experiment result shows the supplementary structured base method of URL matching mined parallel documents from parallel websites to be twice as many as the pure content based method. Our system works on multiple language pairs in parallel documents mining and it is extendable to more language pairs. We ran experiment to mine Hindi-English parallel documents and the mining precision is similar to that we get in Chinese-English pairs.

Our system is scalable to run on multiple servers simultaneously and is linear in time to the number of input source documents. It can also be run continuously to discover and mine for newly added web documents that were not there previously. It is also extendable to mine for parallel documents in multiple target languages at the same time.

Acknowledgments This project is partially funded by a subcontract from BBN, under the DARPA GALE project.

References

1. Munteanu, D., Marcu, D.: Improving machine translation performance by exploiting non-parallel corpora. Comput. Linguist. **31**:477–504 (2005)
2. Nie, J.-Y., Simard, M., Isabelle, P., Durand, R.: Cross-language information retrieval based on parallel texts and automatic mining of parallel texts from the web. In: Proceedings of the 22nd Annual International ACM SIGIR Conference on Research and Development in, Information Retrieval, pp. 74–81 (1999)
3. Grefenstette, G.: Cross-Language Information Retrieval. Kluwer Academic, New York (1998)
4. Munteanu, D., Marcu, D.: Extracting parallel sub-sentential fragments from nonparallel corpora. In: Proceedings of the 21st International Conference on Computational Linguistics and

the 44th Annual Meeting of the Association for Computational Linguistics, Morristown, NJ, USA, pp. 81–88 (2006)
5. Uszkoreit, J., Ponte, J., Popat, A., Dubiner, M.: Large scale parallel document mining for machine translation. In: Proceedings of the 23rd International Conference on Computational Linguistics (COLING 2010), Beijing, China, pp. 1101–1109 (2010)
6. Resnik, P., Smith, N.: The web as a parallel corpus. Comput. Linguist. **29**:349–380 (2003)
7. Shi, L., Niu, C., Zhou, M., Gao, J.: A dom tree alignment model for mining parallel data from the web. In: Proceedings of the 21st International Conference on Computational Linguistics and the 44th Annual Meeting of the Association for Computational Linguistics, Morristown, NJ, USA, pp. 489–496 (2006)
8. Ma, X.: Champollion: a robust parallel text sentence aligner. In: Proceedings of the Fifth International Conference On Language Resources and Evaluation (LREC 2006), ELRA. Genoa, Italy (2006)
9. Chen, J., Nie, J.-Y.: Parallel web text mining for cross-language information retrieval. In: Recherche d'Informations Assistée par Ordinateur (RIAO), pp. 62–77 (2000)
10. Jiang, X., Hu, Y., Li, H.: A ranking approach to keyphrase extraction. In: Proceedings of the 32nd International ACM SIGIR Conference on Research and Development in Information Retrieval, SIGIR'09, New York, NY, USA (2009)
11. Hong, G., Li, C.-H., Zhou, M., Rim, H.-C.: An empirical study on web mining of parallel data. In: Proceedings of the 23rd International Conference on Computational Linguistics (COLING2010), Beijing, China, pp. 474–482 (2010)
12. Cheung, C., Fung, P.: Unsupervised learning of a spontaneous and colloquial speech lexicon in Chinese. Int. J. Speech Technol. **7**, 173–178 (2004)
13. Papineni, K., Roukos, S., Ward, T., Zhu, W.-J.: BLEU: a method for automatic evaluation of machine translation. In: Proceedings of the 40th Annual meeting of the Association for Computational Linguistics (ACL 2002), pp. 311–318 (2002)
14. Prochasson, E., Fung, P.: Rare word translation extraction from aligned comparable documents. In: Proceedings of the 49th Annual Meeting of the Association for Computational Linguistics, Portland, USA (2011)
15. Carpuat, M., Fung, P., Ngai, G.: Aligning word senses using bilingual corpora. ACM Trans. Asian Lang. Inform. Process. **5**(2):89–120 (2006)
16. Abdul-Rauf, S., Schwenk, H.: On the use of comparable corpora to improve SMT performance. In: Proceedings of the 12th Conference of the European Chapter of the Association for, Computational Linguistics (EACL'06), pp. 16–23 (2006)
17. Akamine, S., Kato, Y., Kawahara, D., Shinzato, K., Inui, K., Kurohashi, S., Kidawara, Y.: Development of a large-scale web crawler and search engine infrastructure. In: Proceedings of the 3rd international Universal Communication, Symposium (IUCS'09), pp. 126–131 (2009)
18. Fung, P., Prochasson, E., Shi, S.: Trillions of comparable documents. In: Proceeding of the 3rd Workshop on Building and Using Comparable Corpora (BUCC'10), Language Resource and Evaluation Conference (LREC2010), Malta, pp. 26–34 (2010)
19. Gleim, R., Mehler, A., Dehmer, M.: Web corpus mining by instance of wikipedia. In: Proceedings of the 2nd International Workshop on Web as Corpus (WAC'06), Morristown, NJ, USA, pp. 67–74 (2006)

Automatic Comparable Web Corpora Collection and Bilingual Terminology Extraction for Specialized Dictionary Making

Antton Gurrutxaga, Igor Leturia, Xabier Saralegi and Iñaki San Vicente

Abstract In this article we describe two tools we have built, one for compiling comparable corpora out of the Internet and the other for bilingual terminology extraction out of comparable corpora, and an evaluation we have subjected them to: bilingual terminology has been extracted out of automatically collected domain-comparable web corpora, in Basque and English, and the resulting terminology lists have been validated automatically using a specialized dictionary, in order to evaluate the quality of the extracted terminology lists. Thus, this evaluation measures the usefulness of putting these two automatic tools to work together in a real-world task, that is, specialized dictionary making.

Keywords Comparable corpora · Corpora building · Terminology extraction

1 Introduction

1.1 Automatic Corpora Building and Dictionary Extraction

Any language aiming to survive in a world that is becoming more intercommunicated and global day by day, and to be used normally in education, media, etc., must

A. Gurrutxaga (✉) · I. Leturia · X. Saralegi · I. San Vicente
Elhuyar Foundation, Zelai Haundi kalea 3, Osinalde Industrialdea, 20170 Usurbil, Spain
e-mail: a.gurrutxaga@elhuyar.com

I. Leturia
e-mail: i.leturia@elhuyar.com

X. Saralegi
e-mail: x.saralegi@elhuyar.com

I. San Vicente
e-mail: i.sanvicente@elhuyar.com

S. Sharoff et al. (eds.), *Building and Using Comparable Corpora*,
DOI: 10.1007/978-3-642-20128-8_3,

necessarily have at its disposal language resources such as dictionaries or corpora, preferably in digital form. The ever-growing presence of ICTs in everyday life adds to these requisites the existence of language technologies and NLP tools for that language, which in turn also need electronic dictionaries and corpora in order to be developed. Therefore, the need for lexical resources and corpora of any language aspiring to be modern is undeniable.

Besides, modern lexicography and terminology is hardly ever done based solely on experts' knowledge or intuition; empirical evidence is needed or previous use at least is studied, and these are provided by corpora. And there are many tools that ease the process of building thesauri or lexical or terminological dictionaries by making use of NLP and statistical methods to automatically extract candidates out of corpora.

So it is clear that corpora of any kind (monolingual, parallel, comparable...) are a very valuable resource for many aspects of the development of a language. And generally, the bigger the corpora, the better the results obtained from them.

But less resourced languages are not exactly rich in corpora, let alone big corpora: on the one hand, building a corpus in the classical way, i.e. out of printed texts, is normally a very costly process; on the other, the number of language experts or researchers dealing with these languages is much smaller than that of major languages.

However, the Internet provides a huge number of texts in a digital and easy to manipulate standard format. For any less resourced language there are bound to be many more texts on the web than in any corpus. That is why turning to the Internet to build corpora is a very logical choice for less resourced languages. And by combining these corpora with the aforementioned automatic terminology extraction methods, thesauri and dictionaries can be built at a cost very attractive for a less-resourced language like Basque, or, why not, for any other language as well.

1.2 *Automatic Bilingual Terminology Extraction Out of Comparable Corpora for Basque*

In the particular case of bilingual terminology extraction, although automatic tools for bilingual lexical acquisition have traditionally used parallel corpora as raw material, due to the scarceness of these kinds of corpora—especially in certain domains or language pairs–, recent years have seen a significant increase in the use of comparable corpora for this task. The reason is that they are usually easier to obtain in large sizes and therefore, as recent research has shown, they have the potential to overcome the limitations of parallel corpora not only in bilingual terminology extraction [34, 55] but also in fields such as machine translation [52], or cross-language information retrieval [66].

Basque, as is the case for other less-resourced languages, lacks specialized dictionaries in many areas, and neither is it rich in parallel corpora; so even if tools for Basque terminology extraction out of parallel corpora exist [5], they do not entirely solve the problem. So we have developed a tool for extracting bilingual terminology out of domain-comparable corpora, based on context similarity and cognate detection.

Regarding the obtaining of domain-comparable corpora, this has traditionally been done in a supervised or directed way: out of news agencies, established research corpora (e.g. TREC or CLEF collections), by crawling certain web sites, etc. All these approaches present some problems for the case of Basque. First, they all need a human choice of the sources, which renders the corpora biased at least and often not very diverse. Besides, for small languages like Basque, in many domains, it would not be easy to identify good sources that would contain a significant amount of documents on the topic. Furthermore, competition corpora do not usually include such languages. Finally, focused crawling for specialized corpora often requires domain filtering, usually based on machine learning, which needs special training for each topic, so building a generic tool for any domain is not possible. That is why we have developed a tool that makes use of search engines for building comparable corpora from the web out of many different sources.

This article describes in more detail the aforementioned tools (the comparable corpora building tool and the bilingual terminology extraction tool), and evaluates their usefulness in a real-world task: bilingual terminology has been extracted out of automatically collected domain-comparable web corpora, in Basque and English. The resulting terminology lists have been evaluated automatically using a specialized dictionary, in order to evaluate the validity of this approach for automatic specialized dictionary making.

2 Automatic Comparable Corpora Collection from the Internet

2.1 Related Work

Surprisingly, there is not much literature about the process of collecting comparable corpora. Most of the literature concerning comparable corpora deal with the exploitation of such resources, and simply mention that the comparable corpus has been obtained, as we have already mentioned, from news agencies [13, 52] or by crawling certain sites.

Talvensaari et al. [67] do describe a system for obtaining comparable corpora based on focused crawling—the idea of focused crawling for monolingual specialized corpora was first introduced by Chakrabarti et al. [22].

Some other works deal with converting comparable corpora from "light" to "hard" [15, 18, 64, 67]. The "light" and "hard" comparability levels for corpora were first introduced by Bekavac et al. [15]. A light comparable corpus would be composed of corpora from two (or more) languages composed according to the same principles (i.e. corpora parameters) which are defined by features such as domain, size, time-span, genre, gender and/or age of the authors, etc. The hard type comparability is dependent on already collected and established light comparable corpora. This type is drawn from the light corpora by applying certain language technology tools/techniques and/or document meta-descriptors to find out which documents in lightly comparable corpora really deal with the same or similar topic. A subset of lightly comparable corpora which has been selected by those tools/techniques, whether document-level aligned or not, can be regarded as hard comparable corpora. Our interest, for the moment, relies on obtaining the light corpora, which the aforementioned studies again treat very superficially.

However, any process that sets out to build domain-comparable corpora from the Internet has unavoidably to deal with certain problems. We will be dealing with these problems and the different ways of tackling them in the following subsections.

Obtaining Texts on the Domain. The BootCaT tool [9], although it has also been used to build big general corpora [61], has become the de facto standard for collecting smaller specialized corpora. It introduced a new methodology for collecting monolingual domain-specific corpora from the Internet: a list of words is given as input, APIs of search engines are queried for random combinations of these seed words and the pages returned are downloaded. Although BootCaT is a monolingual corpora collector, we can expect that, by applying it to word lists on the same subject but in different languages, one could obtain light multilingual comparable corpora.

If we use crawling methods instead of search engines, it is necessary to implement some sort of domain filtering. This is usually done by training ad hoc machine learning filters [22]. The features used for training are content words or terms from the domain [62].

Length Filtering. Fletcher [32] proved that filtering web documents by their size improved the quality of the web corpora. Those that do not reach a minimum (Fletcher put the threshold in 5 KB) tend to have little textual content once page headers, menus, etc. are removed or are usually error messages from web servers. On the other hand, those that are too large (according to Fletcher, larger than 200 KB) are not good for linguistic corpora, since they are often not representative of real language and tend to be lists, catalogues, spam and such things. This method has been used in the enormous corpora collected by the WaCky initiative [11, 12, 30].

Language Filtering. When building a corpus, we are usually looking for texts in a language. When using search engines, we filter by language telling the search engine to return only results in a specific language. But if we are crawling, or if search engines do not offer filtering by the language we want, it is up to us to do the language filtering. The most well known method (and probably the most used) for language detection is the one used by TextCat, based on n-gram frequencies [21]. Another technique is to use a list of the most frequent function words and allow only documents with a minimum proportion of them, because according to Baayen [7] real language (the one of interest for a corpus) fulfills this requirement. This last method is the one used in the very large corpora of the WaCky initiative.

Spam and Porn Filtering. The web is full of spam, porn and other kinds of noise. When we build a corpus out of web documents it is essential to get rid of these, but it is not always easy. The size filter proposed by Fletcher [32] decreases this kind of noise but does not eliminate it completely. In the case of collecting specialized corpora where some kind of machine-learning filter is included for domain detection, most noise will be removed. If we use search engines we will most probably get less spam and porn, since they already do this filtering. But it is always desirable, and in the case of crawling methods necessary, that we implement the detection of spam and porn.

The WaCky initiative, for example, gets rid of spam with the function words list they use for language detection, since spam pages are often made of nonsense word lists or links. For porn, they have a black list of words usual in porn pages and remove pages that have more than a certain number or proportion of them.

Boilerplate Removal. Web pages are full of "*boilerplate*", which is the linguistically uninteresting material that web server software automatically creates and which is repeated throughout every page in a website: headers, navigation menus, copyright notices, ads, etc. It is advisable to remove this boilerplate for various reasons: it makes ugly KWiCs, it distorts word frequencies and it makes the work of other filters (duplicate filtering, domain filtering...) more difficult. When crawling a few websites, it is possible to build filters adapted to the structure of those sites with a performance of 100 %, but this is not valid for a web-wide corpus collection.

The reference in most projects [9, 11, 12, 30, 61] is the BTE (Body Text Extraction) algorithm developed by Finn et al. [31]. It uses the HTML tag density to detect real content, since this contains proportionally less tags than boilerplate.

The other main reference in this area is the Cleaneval competition organized in 2007 in the 3rd Web as Corpus workshop [10]. 10 systems took part in it, some based on heuristics and some based on machine learning, but they all took into account more features apart from tag density (punctuation signs, function words...).

Near-Duplicate Detection. The detection of exact duplicates is a trivial task easily accomplished by hashing techniques. But much content is repeated across different websites (news from agencies in media sites, CC licensed articles in many blogs...) which are not exact duplicates, and these cannot be detected by hashing methods.

The most used method for this job is Broder's algorithm [20]. It takes all the "*shingles*" or n-grams of a document, and if two documents share many of these, it means they are very similar. This simple albeit efficient technique owes its popularity to the level of optimization achieved, which makes it usable even at web scale. Keeping all the shingles of many web documents and comparing all of them for each document pair is computationally very expensive. But Broder takes a fingerprint for each shingle, orders them and groups them again into "*supershingles*" which are again fingerprinted. Just a few numbers must be kept for each document, and the coincidence in only one of them is enough to ensure a very high similarity.

Containment Detection. It is very common that a web page containing an article with its own URL is included in its entirety in the main page of the newspaper or blog. Broder also implemented an algorithm for detection of already contained documents [19]. It is not as optimized as the near-duplicate detection, but it is possible to use it for small- and medium-sized corpora building.

2.2 Our Approach

Obtaining Texts on the Domain. We have commented above that it is not easy to obtain comparable corpora involving the Basque language from news agencies or research collections or by crawling certain websites. Our approach uses search engines applied to domain-specific words to gather the two sub-corpora of each language. This is similar to what BootCaT does, but we apply some significant improvements to its methodology (Fig. 1).

The first improvement we have had to make is due to the bad domain precision obtained by BootCaT (less than 16 %) when trying to build a specialized corpus in Basque [46]. This poor performance is not BootCaT's fault, it is caused by the poor treatment search engines give to the Basque language: no lemmatization or stemming is applied when the pages are indexed, which is crucial in an inflectional and morphologically rich language, and they do not offer the possibility of restricting their results to pages in Basque alone.

We try to solve the former by means of morphological query expansion, which consists of querying for different word forms of the lemma, obtained by morphological generation, within an OR operator. In order to maximize recall, the most frequent word forms are used, and recall is improved by up to 60 % in some cases. For the latter, we use the language-filtering words method, consisting of adding the four most frequent Basque words to the queries within an AND operator, which improves language precision from 15 % to over 90 % [45]. These techniques are in common use in IR or web-as-corpus tools for Basque [42, 44].

Simply by applying these methods plus implementing our own language, size, near-duplicate, etc. filters to a process otherwise almost identical to BootCaT, we raise the domain precision to almost 66 %, which is the baseline BootCaT obtains for other languages.

The other improvement we have made to BootCaT's methodology is due to the fact that, for any language, the domain precision obtained by it is far from perfect. It has scarcely been measured, but a small evaluation performed on the original BootCaT paper hints that one third of the texts could be unrelated to the topic.

So what we do is add a new stage at the beginning of the process and another one at the end. We do not use a list of seed words as a starting point, but a bunch of seed documents representative of the domain of interest that covers as many sub-areas of the domain as possible, which we will call the sample mini-corpus. Depending on the domain, 10–20 small documents may be enough. The seed words are automatically extracted from it taking the content words with the highest Log-Likelihood Ratio (LLR) between the frequency of the word in the mini-corpus and in a general corpus. This list can optionally be revised manually.

Then, after a process similar to BootCaT is carried out (searching for combinations of the words, downloading the pages, various filtering stages...), we apply a final domain-filtering stage. We represent both the downloaded documents and each of the documents of the sample corpus with a vector of the most significant keywords, i.e. nouns, proper nouns, adjectives and verbs. The keywords are again selected and

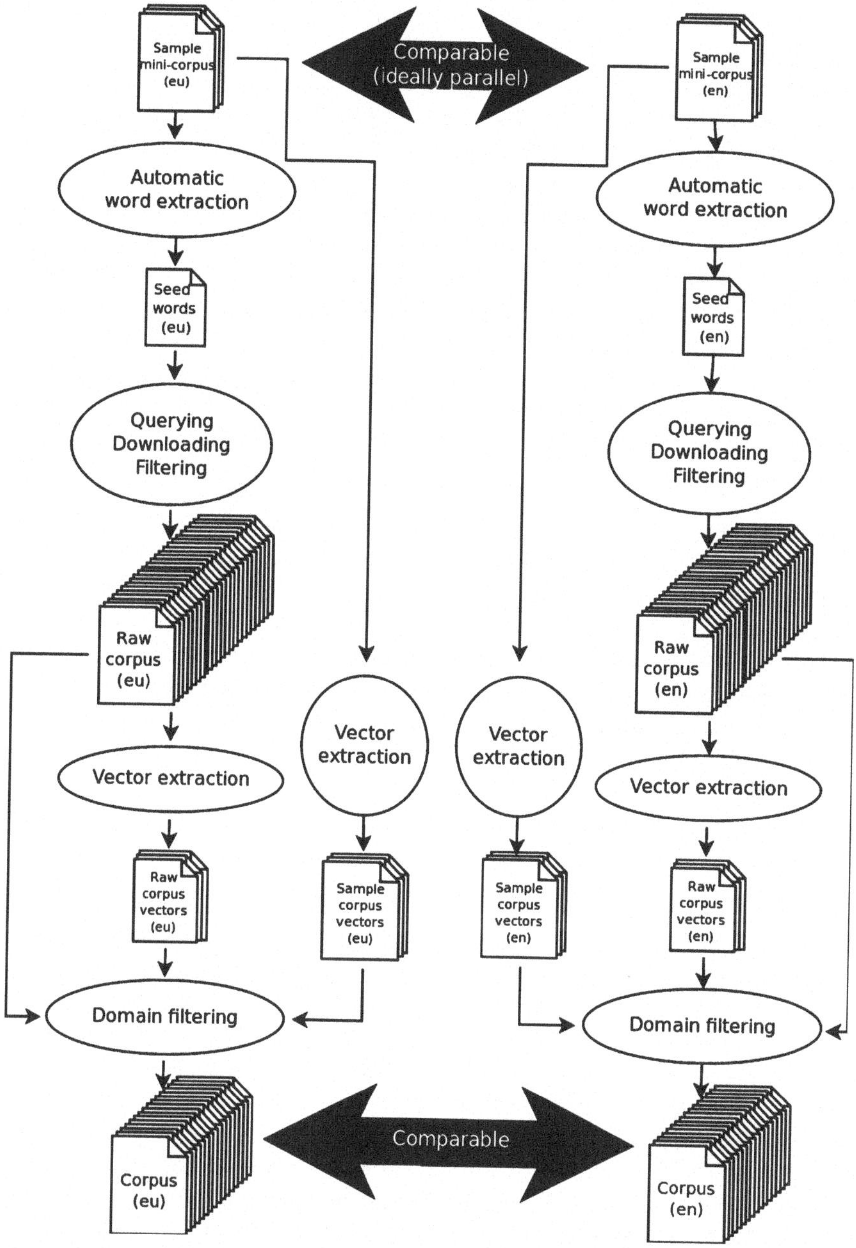

Fig. 1 Diagram showing the process to compile comparable corpora from the Internet

weighted by their LLR. For measuring the similarity we use the cosine, one of the most widely used ways to measure the similarity between documents represented in the vector space model. A document is accepted in the corpus if the maximum of its cosine measures with each of the documents in the sample mini-corpus reaches an empirically defined threshold, and rejected otherwise.

Length Filtering. We do apply a length filter but it is not based on the size of the downloaded file. We reject documents the length of which after conversion to plain text is under 1,000 characters or over 100,000 characters.

Language Filtering. The language filtering of the English part of the comparable corpora is done by using the search engines' filter by language option. For the Basque part we use LangId, a language identifier based on character trigram and word frequencies specialized in Basque, applied at paragraph level so that we can extract content from bilingual documents.

Spam and Porn Filtering. We do not apply any specific filter for spam and porn. There is hardly any in Basque, and for English, the final domain filtering does the job.

Boilerplate Removal. For boilerplate removal, we use Kimatu [43], a language-independent system developed by our team that scored second (74.3 %) in the Cleaneval competition [10].

Near-Duplicate Detection. We have included a near-duplicate detection module based on Broder's shingling and fingerprinting algorithm [20].

Containment Detection. We have also included a containment detection method also based on Broder's works [19].

2.3 Performance

As we have explained, the process of collecting a domain-comparable corpus from the Internet consists of collecting two domain-specialized corpora, one for each of the languages. So in order for the final corpus to be comparable, both corpora need to have a high domain precision.

The system decides if a document will be included in the corpus if it has a minimum similarity with one of the documents in the sample mini-corpus. Higher thresholds ensure 80–90 % precision rates, if recall is not an issue. Otherwise, the threshold can be set to achieve the desired precision/recall balance.

Regarding the domain precision of the corpora, there is an important aspect to point out. Obtaining high topic precision does not imply that the corpus obtained will be highly representative of the universe. In fact, since we are filtering by applying similarity measures using the documents of the sample mini-corpus, if this is not wide enough, that is, if not all the sub-areas of the topic are represented there, we might be missing areas without ever being aware of it. So the quality, heterogeneity and size of the sample mini-corpus are key issues in the method proposed. However, it is not easy to say what a minimum or optimum size of the sample mini-corpus is to ensure good representativeness; it greatly depends on whether the topic is very specialised, or quite general, etc.

However, even if both corpora were 100 % domain-specific, this would not be enough to guarantee a good comparability. Out of two corpora strictly on computer science, one could be mostly made out of texts on hardware and databases and the other on programming and networks; they could not be considered very comparable, and they would most definitely not be very practical for any of the tasks comparable corpora are used for. This is why the sample mini-corpora of the two languages have to be as comparable or as similar as possible (ideally, a parallel corpus would be best), so that the corpora obtained would be more comparable.

We cannot give a datum for the degree of comparability of the corpora our method obtains, simply because there is no standard measure for measuring the comparability of multilingual corpora, and there is little literature on the topic. There are some works on measuring corpus similarity based on word-frequency lists [40, 41] and others that use also POS and semantic tag frequencies [56], but they deal with monolingual corpora (they could be applied to multilingual comparable 'corpora using dictionaries). Some others propose using genre or register similarity [51] or the number of semantic similarities at the document level [58]. However, none of them could be regarded as a standard measure.

3 Automatic Bilingual Terminology Extraction Out of Comparable Corpora

3.1 Related Work

The first algorithms for automatic extraction of translation pairs from comparable corpora were developed around the middle of the 1990s by Rapp [54] and Fung [33]. These algorithms are mainly based on the idea that translation equivalents tend to co-occur within similar contexts, the same hypothesis that is used for the identification of synonyms. In order to do this we need to model the contexts of the candidate terms, translate the contexts and calculate the degree of similarity [34, 55]. An alternative is to detect translation equivalents by means of string similarity or cognate detection [2].

Monolingual Candidate Term Extraction. As in other areas of automatic language processing, two main approaches have been proposed for terminology extraction from texts: linguistic and statistical approaches.

On the one hand, linguistic techniques rely on the assumption that terms present specific morphosyntactic structures or patterns [17]. The basic strategy of these techniques is to detect and extract the word sequences whose structure matches some given pattern. Since these patterns are in most cases language-dependent, linguistic techniques demand specific language knowledge processing.

On the other hand, statistical approaches take into account the fact that terms have different statistical features from normal words for the purpose of identifying them (for example, the high association grade of multiword constituents). More precisely, in order to estimate the termhood of the candidates, we can use statistical models which analyse observed counts of linguistic information related to the candidates.

Most of the statistical approaches focus on the extraction of multiword terms, mainly by means of calculating association measures [25, 28, 65]. Most of the association measures proposed in the literature are intended to rank bigrams and rely on different concepts. For example, Mutual Information (MI), introduced in this field by Church and Hanks [25], was taken from Information Theory. Other measures such as the LLR introduced by Dunning [29], t-score and Chi-square are based on hypothesis testing. In order to rank multiword units composed of two or more words, Dias et al. [28] introduced Mutual Expectation (ME), a measure based on Normal Expectation, which is a generalization of Dice coefficient for N-grams. Blaheta and Johnson [16] use measures based on parameters of certain Log-linear models to rank verbs composed of two or more words.

Regarding one-word terms, there are several methods to obtain the termhood of a one-word candidate. For example, tf-idf, which is used regularly, compares the frequency of a given word in the document with its frequency in other documents of the corpus. Matsuo and Ishizuka [48] use statistical information about word co-occurrence. LLR can also be used to identify relevant one-word terms of a corpus. It measures the degree of association of the one-word candidates with respect to the corpus by using an open domain corpus as a reference. Those with a high association degree are taken as terms [56].

Some authors adopt hybrid approaches, combining linguistic and statistical techniques. Some of them apply syntactic filters after statistical processing, in order to extract the statistically significant word combinations that match some given morphosyntactic pattern [65]. In other cases, statistical measures are calculated for a list of term candidates previously selected through linguistic techniques [26, 39].

Modelling of Contexts. There are several techniques for determining which words make up the context of a word: distance-based window and syntax based-window [35] are two examples. Different models have been proposed to represent the context of words. The most widely used one combines the Vector Space Model and Association Measures (AM) for establishing the weight of the context words with regard to a word: LLR, Mutual Information, Dice coefficient, Jaccard measure, frequency, tf-idf, etc. Another way of representing the contexts [60] is by using probabilistic or language models, as used in the IR field. In those cases the representation of the contexts is implicit in the retrieval framework.

Translation of Context Words. To be able to compute the similarity, the contexts have to be put in the same space by translating one of them. This is a similar problem to the one posed when having to translate the query in a Cross-Language Information Retrieval (CLIR) system. MT, parallel corpora and dictionary-based approaches are proposed to deal with this task [38]. In the context of comparable corpora exploitation usually only seed dictionaries are available. However, the dictionary approach poses two main problems: it has to solve the ambiguous translations and it has a coverage problem (out-of-vocabulary words or OOV).

Regarding ambiguity, many algorithms have been proposed for dealing with the translation disambiguation resulting from query translation guided by bilingual dictionaries. The simplest method is to select the first translation given by the dictionary as the best since the translations are often sorted by frequency of use. However, this approach fails to take into account the domain of the query, so the disambiguation can be very rigid. Other more flexible and better performing approaches [53] take all the translations and group them as a unique word when the TF and DF values of the document words are calculated by the ranking method. The syn operator offered by the Indri and Inquery query languages allows this type of grouping [53]. Other more complex approaches [8, 23, 36, 47], which also use statistical information of monolingual word concurrences, are those based on the degree of cohesion or association between the translation candidates. They try to obtain the combination of translation candidates that maximize the coherence between them. A corpus in the target language is used to compute association scores.

Context Similarity Calculation. After representing the words by their contexts and translating the contexts in one of the languages to the other, the proposed algorithms compute for each word the similarity between its context vector and all the context vectors in the other language by means of measures such as Cosine, Jaccard or Dice. According to the hypothesis, the correct equivalent should be ranked in the first positions.

Alternatively, the context similarity calculation tasks could again be seen as an IR problem. Therefore, all the paradigms proposed in the IR literature could be useful in this context. Nevertheless, the number of works that exploit the recent advances obtained in the IR community for this kind of task is limited, in particular, works involving probabilistic models [6, 57]. [60] could be an example of the use of probabilistic models for terminology extraction; it represents the contexts by using language models.

Cognates. Another technique proposed in the literature, also useful for the treatment of out-of-vocabulary words (OOV), is the identification of translations by means of cognate detection [2]. This method could be appropriate in a science domain where the presence of cognates is high. In fact, using a Basque-English technical dictionary we were able to calculate automatically that around 30 % of the translation pairs were cognates. Dice coefficient or Longest Common Subsequence Ratio (LCSR) [49] measures are proposed for computing string similarity.

3.2 Our Approach

In order to define which of the different methods and techniques mentioned in the literature best suited the task and languages we usually work with, we carried out various experiments with monolingual term candidate extraction and also with bilingual equivalents calculation (described in [58] and [59]). Implementing the methods

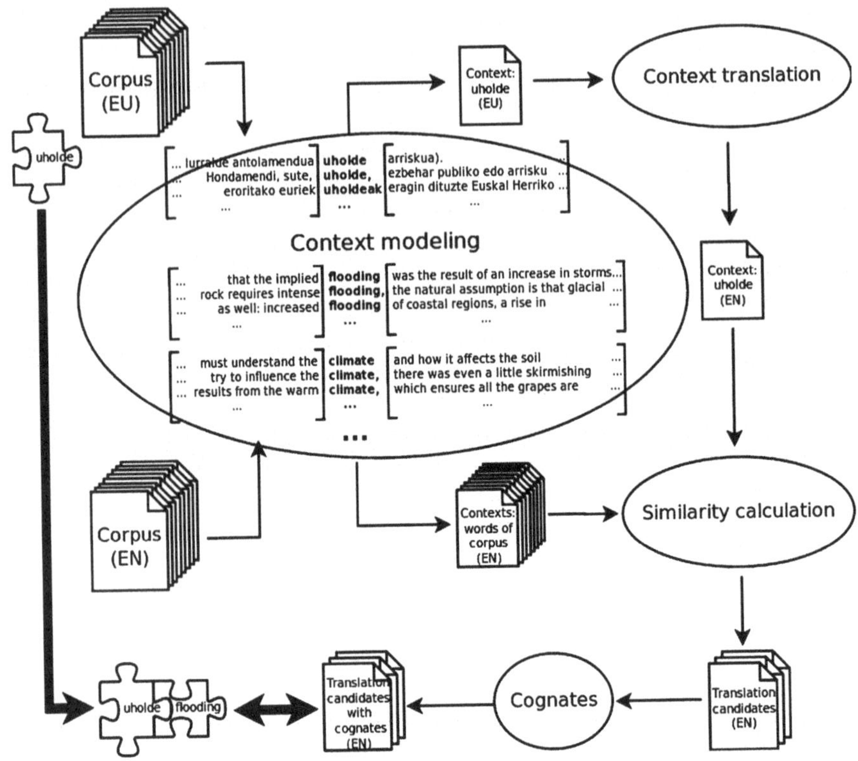

Fig. 2 Diagram showing the process of searching for the translation of a word by context similarity and cognate detection

that performed best, we built a tool for bilingual equivalents identification called AzerHitz (Fig. 2). The methods we used are described in this subsection.

Monolingual Candidate Term Extraction. The monolingual term candidate extraction, both in Basque and English, is done by using a hybrid approach that first looks for some linguistic patterns and afterwards applies some statistical processing to them. Our system only considers noun phrases as terminological units.

For Basque, we use the term extraction tool Erauzterm (which is described in [4]). It uses Eustagger (formerly Euslem [1]) for lemmatization and POS tagging, and a grammar identifies the terms corresponding to the most usual NP structures of Basque terms (described in [3]). The candidate terms are retrieved in their canonical form, so basic normalization is carried out. Most of the typographical variations like capitalization, hyphen insertion, etc. are treated too. We also deal with nested terms, which increases recall considerably. Some kinds of morphosyntactic variations are treated, but not semantic variations.

For English, the corpus is parsed with the chunker of the TreeTagger [68] tool and then we look for the most used morphosyntactic patterns of terms.

Then, different statistical methods are applied to one-word and multi-word term candidates. For one-word terms, we use the LLR as termhood measure for calculating the domain relevance of the term with respect to an open domain corpus; for multi-word terms, we use unithood as a clue to termhood; to measure unithood, we use LLR as the association measure. The ones with the highest measures are taken as the final term candidates.

Modelling of Contexts. Only content words are included in the contexts, that is, nouns, adjectives and verbs. In our experiments, adverbs were found to produce noise.

POS tags were used for this task. TreeTagger is the tagger we chose to tag the English corpus and Eustagger in the case of the Basque corpus.

We established a distance-based window to delimit the contexts of the words. The window size was determined empirically: 10 words for Basque (plus and minus 5 around a given word) and 14 for English (plus and minus 7). The smaller window for Basque is due to the agglutinative nature of the Basque language, where articles, prepositions, etc. are appended to content words. Furthermore, our experiments showed that using punctuation marks to delimit the window improved the results, so we also included this technique in our system.

The contexts are modelled using the Okapi probabilistic model offered by the Lemur Toolkit. We indexed the context words of a word in this toolkit as if it were a document, that is, the words that make up the context of a word throughout the collection are included in the document that is indexed, referred to hereinafter as the "context document" of a word.

Translation of Context Words. To compute the translation of a Basque word, we translated its contexts in order to make them comparable with English contexts. A bilingual dictionary was used for this purpose. The first translation approach was taken in the case of ambiguity.

The treatment of OOVs is very important in this task, since the recall of the dictionary determines the representativity of the context document. In our experiments with a general dictionary, the average translation recall by context was 55 %. To increase the recall of our translated contexts, we try to find equivalents not included in the dictionary by means of cognate detection. For all the OOV words, we look for cognates among all the content words in the target language. The identification of these cognates is carried out by calculating the LCSR between the Basque and English content words. Before applying the LCSR, we processed some typographic rules to normalize equal phonology n-grams (e.g., c → k, factor = faktore) or regular transformation ones (e.g., -tion → -zio, reaction = erreakzio) in both equivalent candidates. The candidates that exceeded the empirically determined threshold of 0.8 were taken as translations.

Context Similarity Calculation. IR systems rank documents according to the topic similarity with respect to a query. Our approach is based on the "topic similarity = context similarity" equivalence. We perform a retrieval by sending the translated context document of the source word as query to the Lemur toolkit. The Okapi probabilistic relevance model is used. The highest-ranked documents returned (which are actually contexts of words) will be the most similar contexts, and therefore, the corresponding words will be the most probable translations.

To prevent noisy candidates, we pruned those that had a different grammatical category from that of the word to be searched.

Cognates. In addition to context similarity, string similarity between source words and equivalent candidates is also used to rank candidates. LCSR is calculated between each source word and its first 100 translation candidates in the ranking obtained after context similarity calculation.

The candidates that exceeded an empirically established threshold (0.8) are ranked first, while the position in the ranking of the remaining candidates remains unchanged. A drawback to this method is that cognate translations are promoted over translations based on context similarity.

4 Evaluation of the Combination of Both Tools for Specialized Dictionary Making

4.1 Experimental Setup

The two tools described above, the comparable corpora collecting tool and the bilingual terminology extraction tool, were each subjected to individual evaluation beforehand. However, we have already said that it is not easy to evaluate the comparability of multilingual corpora. Besides, we think that the concept of comparability is closely related to the task for which the corpora are intended.

Therefore, we put both tools working together: we collected comparable corpora with our automatic tool and automatically extracted bilingual terminology out of them. Then we evaluated the quality of the obtained terminology lists to measure their domain precision, and verify the validity of these two tools working together for the task of specialized dictionary making.

As the proposed method comprises two processes, the performance of the whole process and the quality of the final bilingual terminology lists are affected by both, and it is not easy to tell to what extent each of them has influenced the result. So in order to measure the performance of the comparable corpora collection tool more effectively, we repeated the process with manually compiled comparable corpora similar to the automatically collected ones (same domain, similar size) and we also extracted bilingual terminology out of them. This way, we ended up with a reference against which to measure the performance of the automatic corpora collection tool and also the appropriateness of the whole proposed terminology extraction process.

The domains chosen were two: Computer Science and Physics. We prepared sample mini-corpora for each of them. The mini-corpus of Computer Science in Basque consisted of the 41 articles about computer science in the *Basic Dictionary of Science and Technology* (a dictionary containing 23,000 concepts from all the subdomains of Science and Technology; all concepts are provided with a definition in Basque and equivalents in Spanish, English and French, and 600 of them with an encyclopaedic article in Basque) [14]. In addition to that, 33 news items from Zientzia.net, a popular science website in Basque [69], were included in the mini-corpus. The sample mini-corpus of Physics in Basque consisted of the 76 articles on physics from the *Basic Dictionary of Science and Technology* (hereinafter BDST). For the mini-corpora in English, the articles in Wikipedia that defined the same concepts as in the articles of the BDST were taken, and we googled for news that dealt with the same subjects as the news items from Zientzia.net.

We did not have a target size for the automatically collected corpora: for bilingual terminology tasks based on context similarity, the larger the corpora, the better the results are; and the Internet in Basque is not as big as that in other languages and the number of pages we would want to collect for a particular domain might not exist. So we simply launched the collecting processes for the Basque part and stopped them when the growing speed of the corpora fell to almost zero, thus obtaining corpora that were practically as large as possible. Then we obtained English corpora that were roughly 40 % larger in words than the Basque corpora (Basque appends articles and prepositions to content words, so for the same texts in Basque and English, the English ones are about that percentage larger in terms of words). The sizes of the sample mini-corpora and the obtained corpora are shown in Table 1.

The corpora collected manually are roughly the same size and were obtained from different sources: books, media, websites, etc. We are aware that the manual corpus building implemented cannot be compared with a standard reference-corpus building process, which is designed to guarantee or achieve to a great extent a high level of representativeness and balance. Nevertheless, we made a careful selection of the Basque books on Physics and Computer Science freely available, and tried as much as possible to include texts from the different sub-domains. In some sub-domains, there are few Basque publications (for example, Optics in Physics, or Artificial Intelligence in Computer Science). This fact must be taken into account in the selection of English texts for the corresponding manual corpora. So we endeavoured to collect English texts that, at least a priori, would ensure that the manual corpora were as comparable as possible.

Out of these comparable corpora, we first extracted the monolingual terms out of the sub-corpora of each language and evaluated their domain precision against a dictionary reference. Then we tried to find the English translations for the most relevant Basque terms using the context similarity and cognate detection method.

Table 1 Sizes of the sample mini-corpora and the obtained corpora

		Computer Science		Physics	
		Basque	English	Basque	English
Sample mini-corpora	Documents	74	74	76	76
	Words	66,461	193,406	73,760	306,263
Automatically collected corpora	Documents	1,920	1,568	780	442
	Words	3,696,601	4,990,914	1,155,995	1,710,219

The extracted translation pairs were automatically evaluated using the dictionary reference. Those terms not found in the dictionary were evaluated manually.

The dictionary reference is made up of two sources: (a) the BDST (27,084 English terms and 25,143 Basque terms) and (b) a terminological database which includes terms from terminological dictionaries published by Elhuyar, specialized terms published in Elhuyar's general dictionaries, and terms extracted from translation memories. This second resource includes terms from a large variety of domains, not only from Science and Technology. In order to simplify the presentation of the results, and bearing in mind that the experiments are dealing with Physics and Computer Science corpora, we gathered together all the domains not belonging to Science and Technology as "Other domains". The total amount is 62,043 for English and 101,479 for Basque terms.

4.2 Results

Monolingual term extraction We performed an evaluation of the domain precision of the corpora by evaluating monolingual term extraction. Taking into account that most terms are single words and bigrams, in the interest of simplification, evaluation was focused only on single word and bigram terms. For that purpose, all extracted term candidates with an LLR score above 10 and with a frequency over 25 in Com-

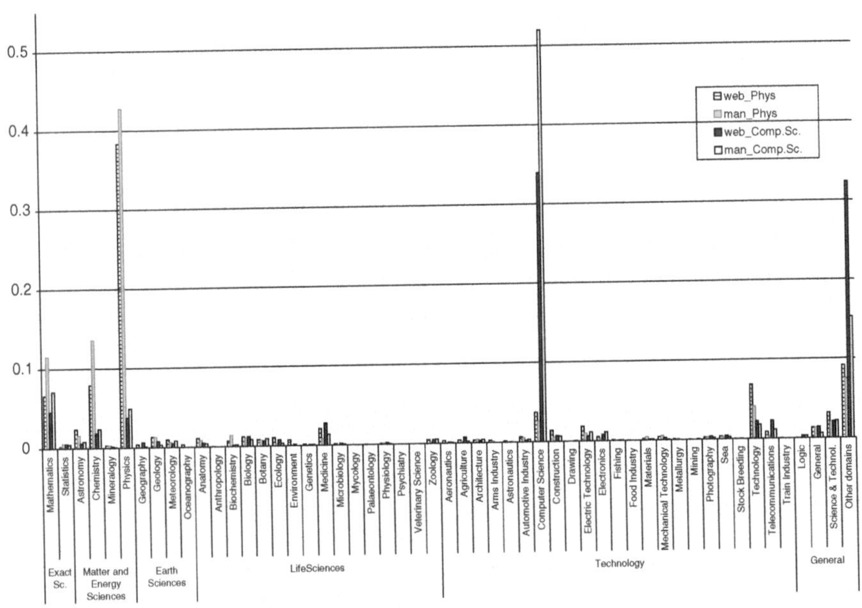

Fig. 3 Domain distribution of extracted Basque terms

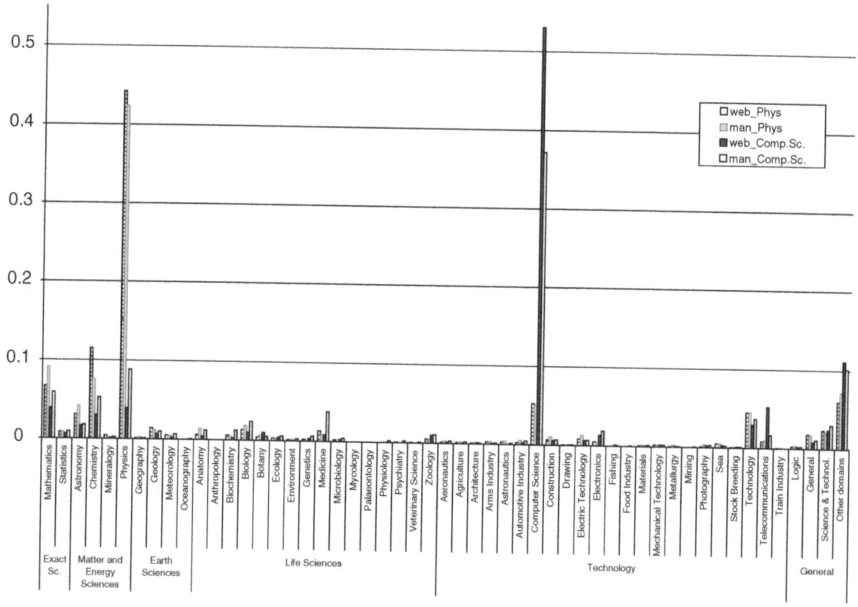

Fig. 4 Domain distribution of extracted English terms

puter Science and 10 in Physics were evaluated against a dictionary reference, in order to measure the overall term precision and the domain precision. The difference in the minimum frequency required for the different domains is due to the smaller size of the corpora of the Physics domain. The domain distribution of the extracted terms validated against the dictionary reference can be seen in Figs. 3 and 4.

We can observe that all corpora logically show peaks in their respective domains or in domains closely related to them. This is true for the automatically collected web corpora too (in fact, in the English extraction the web corpora behave better than the manual ones). The proportion of dictionary-validated terms belonging to the desired domain is generally over 40 %. We can compare those figures with the proportions of Physics and Computer Science terms in the dictionary (6.08 and 3.64 % respectively), and conclude that even though there is no objective reference to draw a comparison with, the corpora have a clear domain-profile and can be considered as specialized. Nevertheless, a more detailed analysis of the domain-distribution results previously presented, alongside the domain precision results for n-best candidate lists selected by LLR and ordered according to decreasing frequency shown in Figures 5 and 6, reveal that the corpora have some different characteristics. In general, a domain precision close to 0.4 is achieved in every extraction when the 1,000 best candidates are selected. The precision is a bit lower in Basque extractions perhaps due to the worse quality of the Basque corpora. This difference decreases as the amount of selected candidates increases.

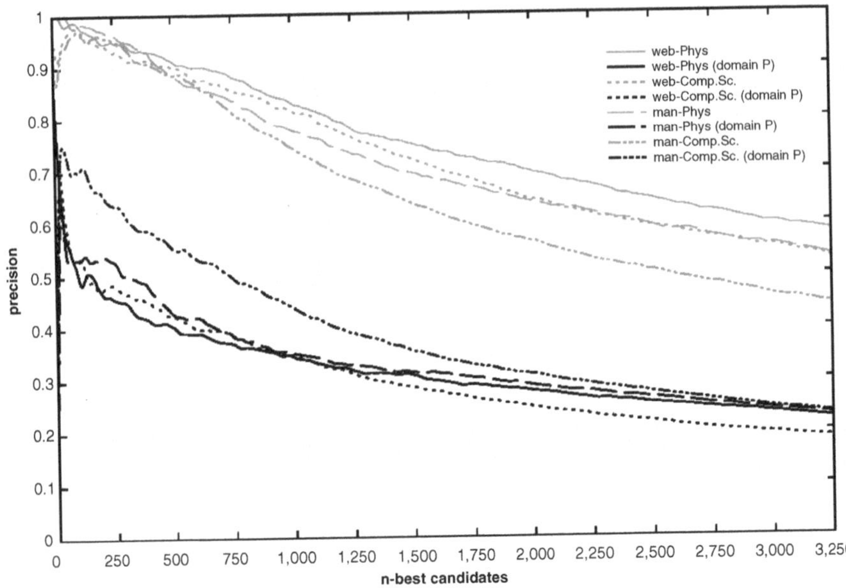

Fig. 5 Term precision and domain precision of terms extracted from Basque corpora

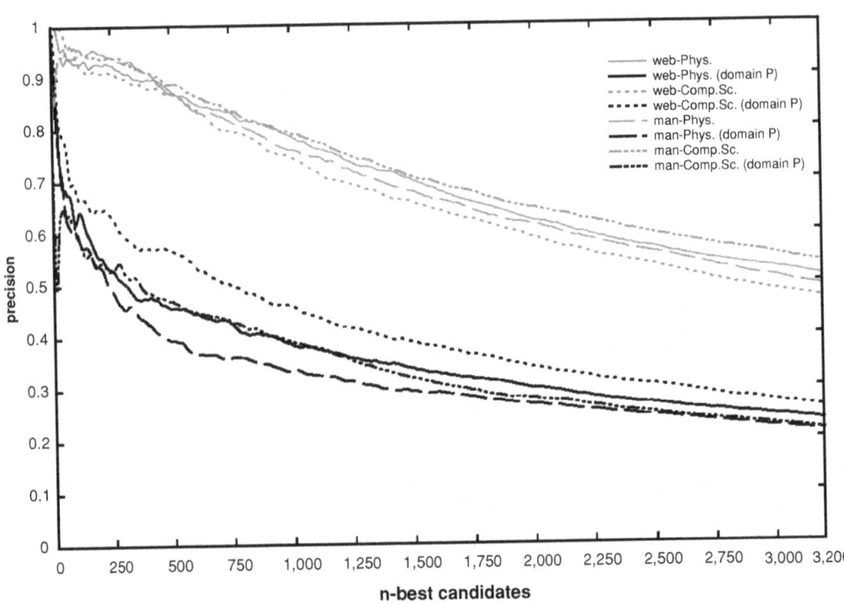

Fig. 6 Term precision and domain precision of terms extracted from English corpora

Another observation we can make by looking at the domain distribution of the Basque terms (Figs. 3 and 4) is that the computer science web corpus peaks sharply in “Other domains”. A possible explanation for this is that computer science is an applied science that is used in many areas such as linguistics, education, etc., and so texts and terms from those domains appear, and in the manual corpus too [50]. But the proportion of these in the Basque web corpus is higher most likely because we tried to make the corpus as large as possible and, therefore, we might have lowered the domain precision requirements too much (the Basque web is not as rich in specialized content as we would wish).

Furthermore, the domain precision of the terms of the web corpora is lower for Basque than for English, both in Computer Science and in Physics (Figs. 5 and 6). This might again be due to the scarcity of specialized content in the web in Basque, compared to English.

If we compare the domain precision of the Basque terms from the web corpora with those of the manual corpora, we can observe that the Physics corpora are more similar to each other than the Computer Science ones. The reason for this could be, as we have already pointed out, that we might have forced the size of the Basque Computer Science web corpus too much and sacrifice domain precision.

Looking at English terms, surprisingly the domain precisions of the web corpora are better than those of the manual corpora. This phenomenon is also reflected in the bilingual extraction, which we will be discussing in the next subsection.

Finally, it must be noted that due to the completely automatic evaluation using the dictionary, it is possible that the real precisions are higher as the automatic process will most likely have found terms that are not present in the dictionary reference, as we observed in previous experiments where a manual evaluation was performed [37].

Multilingual term extraction We also evaluated the precision of the multilingual term extraction. As for the source term candidates, we took the 600 single-word Basque terms and the 400 multi-word Basque terms with the highest LLR, all of them having a frequency of over 25 in computer science and 10 in physics, due to the smaller size of the physics corpora. For the English translation candidates we took all the extracted terms with a minimum frequency of 10. We applied the context similarity and cognate detection method for translation term extraction to these source candidates.

The translation pairs obtained were automatically evaluated using the dictionary reference, and the pairs that were not present in this dictionary were manually evaluated by a professional lexicographer. Tables 2 and 3 show the precision of the bilingual term extraction for different top n candidates, using cognate detection and without using them, and also separating the results for domain-specific terms and those of other domains.

We consider that the results obtained in the bilingual terminology extraction task are acceptable. It must be noted that performance is better when dealing with terms in the desired domain (improvement between 11 and 28 %). The cognate detection method also offers a significant improvement. In the Physics domain, our method obtains a precision of 0.31 in the top-1 (0.47 using cognate detection) and 0.63 in the top-30 (0.68 with cognate detection). In the Computer Science domain, it achieves

0.25 precision in the top-1 (0.40 with cognate detection) and 0.60 in the top-30 (0.65 with the help of cognate detection).

Nevertheless, we cannot really compare these results with other results from the literature, because the experimental setups vary greatly from one to another: the language pair, the size and domain specificity of the corpora, the size and methodology to create the reference term list to translate, etc. Experiments conducted with large news corpora (several million words) report results of up to 90 % precision for the top 10–20 candidates [34, 55]. Research carried out with domain specific corpora have been mostly conducted with small medical corpora (several hundred thousands words). Reported precisions range from 50 % [51] to 74 % [24] for the top 10–20 candidates. Moreover, few works deal with multi-word terms [27, 51, 63], and the accuracy is below the results achieved with single word terms. With respect to the language pair, the experiments conducted with the Basque-English pair [58] obtained a maximum precision of 79 % for high frequency words (frequency > 50) and 43 % for lower frequency words, when dealing only with single word terms, and with a 2-million-word popular science corpus.

However, if we compare them with the results corresponding to manually collected corpora, the results of the web corpora are surprisingly better. It is true that they are not reference corpora, that is, fully balanced, randomly chosen and representative of the domain, but for the domains we are working with and the sizes needed for this task, there are no corpora in Basque of this kind, and in English we are not aware of any either. Reference corpora would most likely beat our automatically collected web corpora. Nevertheless, we believe that we have built good quality corpora with a reasonable amount of effort and cost. Even with an opportunistic approach, the job has not been easy and, particularly in the Basque part, might not be a possible alternative for many organizations (the Elhuyar Foundation, apart from doing research into language technologies, is a media group for popular science in Basque, and also does a considerable amount of scientific and technical book translations into Basque).

In order to find the reasons for the lower performance of the manual corpora, we manually analysed the domain precision of the first 1,000 monolingual candidates extracted from both web and manual corpora, and we found that English manual corpora are richer in terms corresponding to other science and technology domains. For example, in the computer science manual corpus we detected a high presence of terms corresponding to Medicine and Chemistry. These terms came from articles

Table 2 Precision of bilingual term extraction for the Computer Science corpora

Computer Science		Top 1		Top 5		Top 10		Top 20		Top 30	
		–	Cog.	–	Cog.	–	Cog.	–	Cog.	–	Cog.
Web	Domain terms	0.25	0.40	0.42	0.51	0.47	0.55	0.56	0.61	0.60	0.65
	Non-domain terms	0.11	0.19	0.16	0.24	0.19	0.26	0.24	0.30	0.26	0.32
	All terms	0.18	0.30	0.30	0.38	0.34	0.41	0.40	0.46	0.44	0.49
Manual	Domain terms	0.08	0.21	0.21	0.31	0.26	0.35	0.33	0.41	0.37	0.44
	Non-domain terms	0.03	0.08	0.06	0.11	0.08	0.12	0.10	0.14	0.11	0.15
	All terms	0.07	0.02	0.19	0.30	0.23	0.33	0.30	0.39	0.34	0.41

Table 3 Precision of bilingual term extraction for the Physics corpora

Physics		Top 1		Top 5		Top 10		Top 20		Top 30	
		–	Cog.	–	Cog.	–	Cog.	–	Cog.	–	Cog.
Web	Domain terms	0.31	0.47	0.51	0.60	0.57	0.64	0.61	0.66	0.63	0.68
	Non-domain terms	0.17	0.29	0.27	0.37	0.32	0.41	0.37	0.45	0.40	0.48
	All terms	0.23	0.36	0.38	0.47	0.43	0.51	0.47	0.54	0.50	0.56
Manual	Domain terms	0.18	0.32	0.29	0.40	0.34	0.43	0.40	0.48	0.43	0.50
	Non-domain terms	0.09	0.20	0.14	0.23	0.18	0.26	0.22	0.29	0.24	0.31
	All terms	0.16	0.31	0.25	0.37	0.31	0.41	0.37	0.45	0.40	0.48

whose topics are applied computer science such as bioinformatics. The manually collected English corpora happened to contain a non-negligible amount of such texts. This shows that manually built corpora are also prone to errors and not necessarily better than automatically collected ones.

In view of the results, we have no other choice but to conclude that, for bilingual terminology extraction, the method of using search engines for automatic comparable corpora collection from the web, and the randomness and width of scope this involves, is at least comparable to an opportunistic manual collection made with a reasonable amount of effort.

Another aspect to note is that, in all cases, the domain-specific terms obtained much better results than those that belonged to other domains. This confirms that all the corpora obtained can be considered as belonging to the domain to a high degree.

We can also observe that the results for Physics are better than those for Computer Science, although the latter corpora are much larger. This phenomenon was also observed in previous evaluations of monolingual term extraction [37], and might be attributable to the more applied nature of Computer Science, which leads to the appearance of more non-domain and general terms and therefore polysemy.

The manual evaluation of the extracted candidate pairs that were not in the dictionary reference was also useful to show that the method is also valid for obtaining new terminology that is not in dictionaries: in the Physics web corpus, out of the 220 term candidates that were not in the reference, 18 new terms were found (23 with cognate detection) in the top-1, and 39 in the top-30; and in the Computer Science web corpus, out of 413, 31 new terms were found in the top-1 (46 with the help of cognate detection) and 76 in the top-30. In both cases, about 10 % of the terms not in the dictionary are new valid terms in the top-1 and about 18 % in the top-30.

5 Conclusions

The combination of our automatic methods of domain-comparable corpora collection from the Internet and bilingual terminology extraction out of them achieve good precision results, especially for terms belonging to the domain of the corpus. The results are at least as good as those of a corpus collected manually at a reasonable cost, but involving a much smaller effort.

However, bilingual terminology extraction out of comparable corpora based on context similarity needs a minimum size of corpora for it to work properly. Although the results for the domains chosen in this paper are satisfactory, for a language like Basque, it might not be possible to obtain enough texts for some domains, depending on their level of specificity, the production, etc.

References

1. Aduriz, I., Aldezabal, I., Alegria, I., Artola, X., Ezeiza, N., Urizar, R.: Euslem: A lemmatiser/tagger for basque. In: Proceedings of 7th EURALEX International Conference, vol. 1, pp. 17–26. EURALEX, Göteborg, Sweden (2002)
2. Al-Onaizan, Y., Knight, K.: Machine transliteration of names in arabic text. In: Proceedings of the ACL-02 workshop on Computational approaches to Semitic languages, pp. 1–13. ACL, Philadelphia, USA (2002)
3. Alegria, I., Gurrutxaga, A., Lizaso, P., Saralegi, X., Ugartetxea, S., Urizar, R.: Linguistic and statistical approaches to basque term extraction. In: Proceedings of GLAT 2004. Barcelona, Spain (2004)
4. Alegria, I., Gurrutxaga, A., Lizaso, P., Saralegi, X., Ugartetxea, S., Urizar, R.: An xml-based term extraction tool for basque. In: Proceedings of the 4th International Conference on Language Resources and Evaluations (LREC). ELRA, Lisbon, Portugal (2004)
5. Alegria, I., Gurrutxaga, A., Saralegi, X., Ugartetxea, S.: Elexbi, a basic tool for bilingual term extraction from Spanish-Basque parallel corpora. In: Proceedings of Euralex 2006, pp. 159–165. Euralex, Torino, Italy (2006)
6. Amati, G., Van Rijsbergen, C.: Probabilistic models of information retrieval based on measuring divergence from randomness. Trans. Inform. Syst. **20**(4), 357–389 (2002)
7. Baayen, R.: Word Frequency Distributions. Kluwer, Dordrecht (2001)
8. Ballesteros, L., Croft, W.: Resolving ambiguity for cross-language retrieval. In: Proceedings of SIGIR Conference, pp. 64–71. ACM, Melbourne (1998)
9. Baroni, M., Bernardini, S.: Bootcat: Bootstrapping corpora and terms from the web. In: Proceedings of LREC 2004, pp. 1313–1316. ELRA, Lisbon, Portugal (2004)
10. Baroni, M., Chantree, F., Kilgarriff, A., Sharoff, S.: Cleaneval: a competition for cleaning web pages. In: Proceedings of LREC 2008. ELRA, Marrakech, Morocco (2008)
11. Baroni, M., Kilgarriff, A.: Large linguistically-processed web corpora for multiple languages. In: Proceedings of EACL 2006, pp. 87–90. EACL, Trento, Italy (2006)
12. Baroni, M., Ueyama, M.: Building general- and special purpose corpora by web crawling. In: Proceedings of the 13th NIJL International Symposium. Tokyo, Japan (2006)
13. Barzilay, R., Lee, L.: Learning to paraphrase: an unsupervised approach using multiple-sequence alignment. In: Proceedings of HLT/NAACL, pp. 16–23. NAACL, Edmonton, USA (2003)
14. Basic dictionary of science and technology, http://zthiztegia.elhuyar.org
15. Bekavac, B., Osenova, P., Simov, K., Tadić, M.: Making monolingual corpora comparable: a case study of Bulgarian & Croatian. In: Proceedings of LREC 2004, pp. 1187–1190. ELRA, Lisbon, Portugal (2004)
16. Blaheta, D., Johnson, M.: Unsupervised learning of multi-word verbs. In: Proceedings of the 39th Annual Meeting of the ACL, pp. 54–60. ACL, Toulouse, France (2001)
17. Bourigault, D.: Lexter, a natural language processing tool for terminology extraction. In: Proceedings of 7th EURALEX International Conference. Göteborg, Sweden (1996)
18. Braschler, M., Schäuble, P.: Multilingual information retrieval based on document alignment techniques. In: Proceedings of the 2nd European Conference on Research and Advanced Technology for Digital Libraries, pp. 183–197. Springer, Heraklion, Greece (1998)

19. Broder, A.: On the resemblance and containment of documents. In: Proceedings of Compression and Complexity of Sequences 1997, pp. 21–29. IEEE, Salerno, Italy (1997)
20. Broder, A.: Identifying and filtering near-duplicate documents. In: Proceedings of Combinatorial Pattern Matching: 11th Annual Symposium, pp. 1–10. Montreal, Canada (2000)
21. Cavnar, W., Trenkle, J.: N-gram-based text categorization. In: Proceedings of Third Annual Symposium on Document Analysis and Information Retrieval, pp. 161–175. Las Vegas, USA (1994)
22. Chakrabarti, S., Van der Berg, M., Dom, B.: Focused crawling: a new approach to topic-specific web resource discovery. In: Proceedings of the 8th International WWW Conference, pp. 545–562. W3C, Toronto, Canada (1999)
23. Chen, H., Bian, G., Lin, W.: Resolving translation ambiguity and target polysemy in cross-language information retrieval. In: Proceedings of the 37th annual meeting of the Association for Computational Linguistics, pp. 215–222. ACL, College Park, USA (1999)
24. Chiao, Y., Zweigenbaum, P.: Looking for candidate translational equivalents in specialized, comparable corpora. In: Proceedings of the 19th International Conference on Computational Linguistics (COLING 2002), pp. 1208–1212. ACL, Taipei, Taiwan (2002)
25. Church, K., Hanks, P.: Word association norms, mutual information and lexicography. In: Proceedings of the 27th Annual Meeting of the ACL, pp. 76–83. ACL, Vancouver, Canada (1989)
26. Daille, B.: Combined approach for terminology extraction: lexical statistics and linguistic filtering. Tech. Rep. UCREL Technical Papers 5, UCREL (1995)
27. Daille, B., Morin, E.: French-english terminology extraction from comparable corpora. Natural Language Processing—IJCNLP, p. 707G718 (2005)
28. Dias, G., Guilloré, S., Lopes, J.: Mutual expectation: a measure for multiword lexical unit extraction. In: Proceedings of VExTAL—Venezia per il Trattamento Automatico delle Lingue. Venezia, Italy (1999)
29. Dunning, T.: Accurate methods for the statistics of surprise and coincidence. Comput. Linguist. **19**(1), 61–74 (1994)
30. Ferraresi, A., Zanchetta, E., Baroni, M., Bernardini, S.: Introducing and evaluating ukwac, a very large web-derived corpus of English. In: Proceedings of WAC4 Workshop. ACL SIGWAC, Marrakech, Morocco (2008)
31. Finn, A., Kushmerick, N., Smyth, B.: Fact or fiction: content classification for digital libraries. In: Proceedings of Personalisation and Recommender Systems in Digital Libraries Workshop. Dublin, Ireland (2001)
32. Fletcher, W.: Corpus Linguistics in North America 2002. In: Making the Web More Useful as a Source for Linguistic Corpora. Rodopi, Amsterdam (2004)
33. Fung, P.: Compiling bilingual lexicon entries from a non-parallel English-Chinese corpus. In: Proceedings of the Third Workshop on Very Large Corpora, pp. 173–183. Boston, USA (1995)
34. Fung, P., Yee, L.: An ir approach for translating new words from nonparallel comparable texts. In: Proceedings of COLING-ACL, pp. 414–420. ACL, Montreal, Canada (1998)
35. Gamallo, P.: Learning bilingual lexicons from comparable English and Spanish corpora. In: Proceedings of Machine Translation Summit XI, pp. 191–198. Copenhagen, Denmark (2007)
36. Gao, J., Nie, J.: A study of statistical models for query translation: finding a good unit of translation. In: Proceedings of SIGIR Conference, pp. 194–201. ACM, Seattle, USA (2006)
37. Gurrutxaga, A., Leturia, I., Saralegi, X., San Vicente, I.: Evaluation of an automatic process for specialized web corpora collection and term extraction for basque. In: Proceedings of eLexicography 2009. Presses Universitaires de Louvain, Louvain-la-Neuve, Belgium (2009)
38. Hull, D., Grefenstette, G.: Querying across languages: a dictionary-based approach to multilingual information retrieval. In: Proceedings of the 19th annual international ACM SIGIR conference on Research and development in information retrieval, pp. 49–57. ACM (1996)
39. Justeson, J.: Technical terminology: Some linguistic properties and an algorithm for identification in text. Tech. Rep. IBM Research Report RC 18906 (82591), IBM (1993)
40. Kilgarriff, A.: Using word frequency lists to measure corpus homogeneity and similarity between corpora. In: Proceedings of workshop on very large corpora, pp. 231–245. ACL SIGDAT, Beijing and Hong Kong, China (1997)

41. Kilgarriff, A., Rose, T.: Measures for corpus similarity and homogeneity. In: Proceedings of EMNLP-3, pp. 46–52. ACL SIGDAT, Granada, Spain (1998)
42. Leturia, I., Gurrutxaga, A., Alegria, I., Ezeiza, A.: Corpeus, a 'web as corpus' tool designed for the agglutinative nature of basque. In: Proceedings of the 3rd Web as Corpus Workshop, pp. 69–81. Presses Universitaires de Louvain, Louvain-la-Neuve, Belgium (2007)
43. Leturia, I., Gurrutxaga, A., Alegria, I., Ezeiza, A.: Kimatu, a tool for cleaning non-content text parts from html docs. In: Proceedings of the 3rd Web as Corpus Workshop, pp. 163–167. Presses Universitaires de Louvain, Louvain-la-Neuve, Belgium (2007)
44. Leturia, I., Gurrutxaga, A., Areta, N., Alegria, I., Ezeiza, A.: Eusbila, a search service designed for the agglutinative nature of basque. In: Proceedings of Improving non-English web searching (iNEWS'07) workshop, pp. 47–54. SIGIR, Amsterdam, The Netherlands (2007)
45. Leturia, I., Gurrutxaga, A., Areta, N., Pociello, E.: Analysis and performance of morphological query expansion and language-filtering words on basque web searching. In: Proceedings of LREC 2008. ELRA, Marrakech, Morocco (2008)
46. Leturia, I., San Vicente, I., Saralegi, X., Lopez de Lacalle, M.: Basque specialized corpora from the web: language-specific performance tweaks and improving topic precision. In: Proceedings of the 4th Web as Corpus Workshop, pp. 40–46. ACL SIGWAC, Marrakech, Morocco (2008)
47. Liu, Y., Jin, R., Chai, J.: A maximum coherence model for dictionary-based cross-language information retrieval. In: Proceedings of SIGIR Conference, pp. 536–543. ACM, Salvador, Brazil (2005)
48. Matsuo, Y., Ishizuka, M.: Keyword extraction from a document using word co-occurrence statistical information. Trans. Jpn. Soc. Artif. Intell. **17**(3), 217–223 (2000)
49. Melamed, I.D.: Bitext maps and alignment via pattern recognition. Comput. Linguist. **25**(1), 107–130 (1999), http://portal.acm.org/citation.cfm?id=973215.973218
50. Milos, E., Zhang, Y., He, B., Dong, L.: Automatic term extraction and document similarity in special text corpora. In: Proceedings of the Sixth Conference of the Pacific Association for Computational Linguistics, pp. 275–284. Halifax, Canada (2003)
51. Morin, E., Daille, B., Takeuchi, K., Kageura, K.: Bilingual terminology mining—using brain, not brawn comparable corpora. In: Proceedings of the 45th Annual Meeting of the Association of Computational Linguistics, pp. 664–671. ACL, Prague, Czech Republic (2007)
52. Munteanu, D., Marcu, D.: Improving machine translation performance by exploiting non-parallel corpora. Comput. Linguist. **31**(4), 477–504 (2005)
53. Pirkola, A.: The effects of query structure and dictionary setups in dictionary-based cross-language information retrieval. In: Proceedings of SIGIR Conference, pp. 55–63. ACM, Melbourne, Australia (1998)
54. Rapp, R.: Identifying word translations in non-parallel texts. In: Proceedings of the 33rd Annual Meeting of the Association for Computational Linguistics, pp. 320–322. ACL, Cambridge, USA (1995)
55. Rapp, R.: Automatic identification of word translations from unrelated English and German corpora. In: Proceedings of the 37th annual meeting of the Association for Computational Linguistics, pp. 519–526. ACL, College Park, USA (1999)
56. Rayson, P., Garside, R.: Comparing corpora using frequency profiling. In: Proceedings of the Workshop on Comparing Corpora, pp. 1–6. ACL, Hong Kong, China (2000)
57. Robertson, S., Walker, S., Beaulieu, M.: Okapi at trec-7: automatic ad hoc, filtering, vlc and interactive track. In: Proceedings of 7th Text REtrieval Conference (TREC-7), pp. 199–210. Gaithersburg, USA (1998)
58. Saralegi, X., San Vicente, I., Gurrutxaga, A.: Automatic extraction of bilingual terms from comparable corpora in a popular science domain. In: Proceedings of Building and using Comparable Corpora workshop. ACL, Marrakech, Morocco (2008)
59. Saralegi, X., San Vicente, I., Lopez de Lacalle, M.: Mining term translations from domain restricted comparable corpora. Procesamiento del Lenguaje Natural **41**, 273–280 (2008)
60. Shao, L., Ng, H.: Mining new word translations from comparable corpora. In: Proceedings of the 20th International Conference on Computational Linguistics (COLING 2004), pp. 618–624. ACL, Geneva, Switzerland (2004)

61. Sharoff, S.: WaCky! Working papers on the Web as Corpus, chap. Creating general-purpose corpora using automated search engine queries, pp. 63–98. Gedit, Bologna, Italy (2006)
62. Sharoff, S.: Classifying web corpora into domain and genre using automatic feature identification. In: Proceedings of the 3rd Web as Corpus Workshop, pp. 83–94. Presses Universitaires de Louvain, Louvain-la-Neuve, Belgium (2007)
63. Sharoff, S., Babych, B., Hartley, A.: 'irrefragable answers' using comparable corpora to retrieve translation equivalents. Lang. Resour. Eval. **43**(1), 15–25 (2007), http://www.springerlink.com/content/8k6631431pl35381/
64. Sheridan, P., Ballerini, J.: Experiments in multilingual information retrieval using the spider system. In: Proceedings of the 19th Annual International ACM SIGIR Conference, pp. 58–65. ACM, Zurich, Switzerland (1996)
65. Smadja, F.: Retrieving collocations from text: Xtract. Comput. Linguist. **19**(1), 143–177 (1993)
66. Talvensaari, T., Laurikkala, J., Järvelin, K., Juhola, M., Keskustalo, H.: Creating and exploiting a comparable corpus in cross-language information retrieval. ACM Trans. Inform. Syst. **25**(1), 4 (2007)
67. Talvensaari, T., Pirkola, A., Järvelin, K., Juhola, M., Laurikkala, J.: Focused web crawling in acquisition of comparable corpora. Inform. Retr. **11**, 427–445 (2008)
68. Treetagger, http://www.ims.uni-stuttgart.de/projekte/corplex/TreeTagger/
69. Zientzia.net, http://www.zientzia.net

Statistical Comparability: Methodological Caveats

Reinhard Köhler

Abstract The notion of comparable corpora implies the notion of comparability. The present paper aims at explicating this notion with respect to statistical methods because statistical comparison requires the use of statistical tests, which again require certain properties of the data under analysis. Linguistic data, however, do not automatically meet these requirements. In corpus linguistics and other linguistic fields, statistical methods are often applied without any previous check of their applicability. The paper will give some warnings and show some examples of corresponding test procedures. A number of other frequently used terms and concepts, such as representativeness, homogeneity, and balanced corpora, play a central role in corpus-linguistic argumentations and will be analysed in the paper, too, as they concern compilation and use of comparable corpora.

Keywords Comparability · Statistical tests · Representativeness · Homogeneity · Methodology

1 Introduction

A large and growing part of corpus-linguistic activities is based on statistical methods. However, most researchers (and textbooks) in this field lack the methodological knowledge of the particular properties of linguistic data as opposed to data from fields such as the natural sciences, sociological, economical, and other disciplines. Many commonly applied methods as presented in statistical textbooks are designed for the use of data which meet certain requirements. Therefore, these properties are taken for granted in most studies in these fields. Linguistic data, however, do not meet those requirements. One example of such a requirement is the following: The

R. Köhler (✉)
FB II, Linguistische Datenverarbeitung, Universität Trier, Postfach, 54286 Trier, Germany
e-mail: koehler@uni-trier.de

S. Sharoff et al. (eds.), *Building and Using Comparable Corpora*,
DOI: 10.1007/978-3-642-20128-8_4, © Springer-Verlag Berlin Heidelberg 2013

Chi-square test for significance relies on the fact that deviations from the mean are distributed according to the Normal (Gauss) distribution. As a rule, researchers in the above-mentioned fields may also presuppose that the original data, too, follow this distribution at least approximately (abiding by the "law of large numbers"). Figure 1 shows the shape of the symmetrical normal distribution. As opposed to this common form of statistical data, Fig. 2 displays typical linguistic data. It is a plot of the distribution of the number of different sentence structures in German as determined from the valency of frequent verbs. Most data instances can be found extremely close to the x-axis and cannot be visually distinguished from it whereas very few data points can be found at the beginning of the curve, but these have extremely large values. Again, they cannot be seen because they are too close to the y-axis. In quantitative linguistics, bi-logarithmic transformations (cf. Fig. 3) are used for optical inspection of such data. The data of this example abide by the famous Zipf-Mandelbrot law, which is best known from the rank-frequency distributions of words in texts (cf. e.g. [1–3]).

In quantitative linguistics, not a single case of linguistic data was reported where the condition that the data be normal-distributed was complied with. Researchers who want to test the Gauss distribution on their data can use the Shapiro–Wilk test [4] if the sample size is in the interval <3, 5000> (originally, only sample sizes in the interval <7, 30> were possible. The test procedure is integrated in statistical software packages such as SPSS and R). If the samples are larger than 5000, a regular Chi-square goodness-of-fit test with the normal distribution would be in order if the Chi-square value does not increase with increasing N (yet another problem). Besides the Shapiro–Wilk test, there are almost 50 other tests for normality. As a

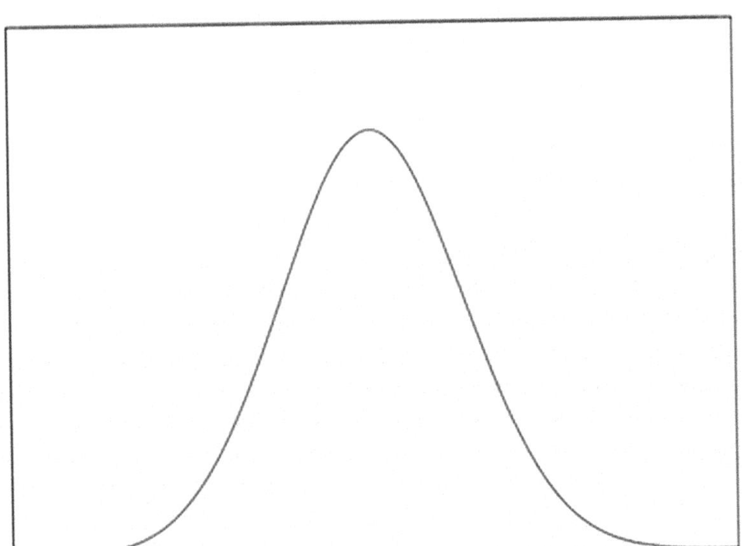

Fig. 1 The bell-shaped, symmetrical normal (or Gauss) distribution

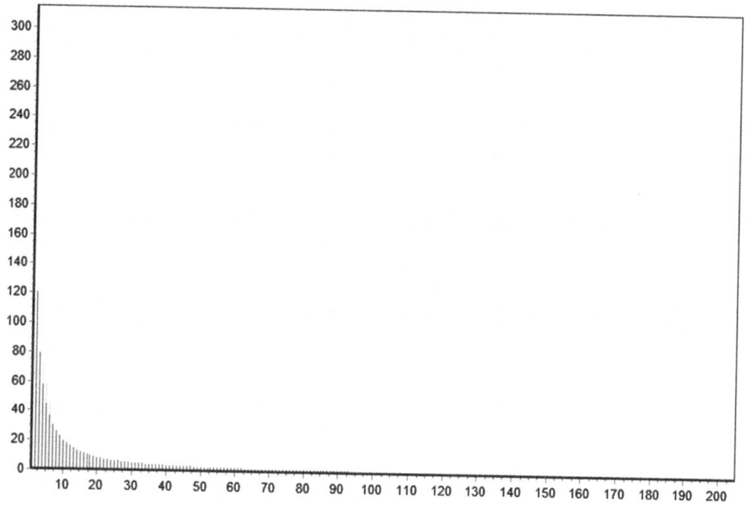

Fig. 2 Typical shape of a distribution of linguistic data

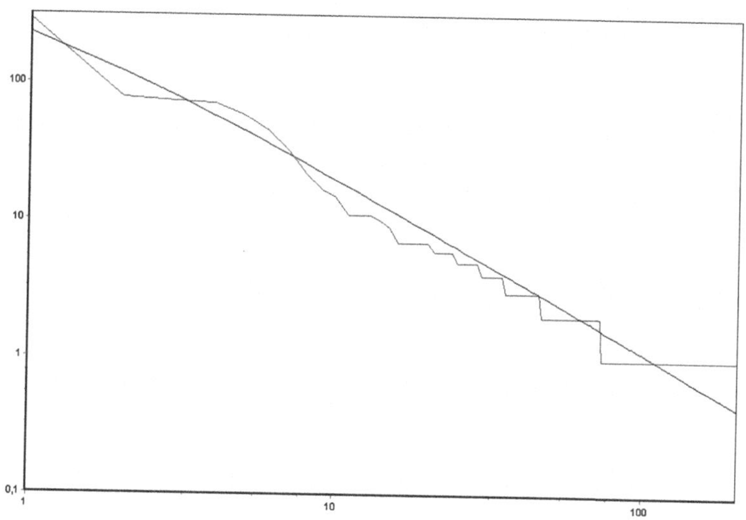

Fig. 3 Bi-logarithmic transformation of the data in Fig. 2

rule, however, one can rely on the general experience that linguistic data have an asymmetric, skew distribution. When asking the question as to whether two corpora may be considered as 'comparable' and if this question is to be answered on the basis of a statistical method, one has to keep in mind this fact about properties of linguistic data.

Another trap that must be avoided is the tendency to implicitly identify the relation between sample and population with the relation between text and corpus or even the relation between corpus and language. Statistical methods enable us to draw reliable conclusions in spite of fragmentary information. Samples, if randomly drawn, large enough and meeting some other requirements allow statistical induction, i.e. reliable statements about the population. An individual text, however, is a semiotic system, not a random sample. Moreover, the relation between texts or corpora and language is of an entirely different nature. Texts and corpora are instances of (observable) linguistic behaviour whereas language is an (unobservable) abstract concept, whence a statistical induction from texts or corpora to language "as a whole" is impossible for methodological and epistemological grounds. This fact can simply be expressed by Orlov's [5] dictum: there are no populations in language.

2 Testing Comparability

The notions of 'comparison' and consequently that of 'comparability' are predicates with at least three arguments: comparable (A, B, C), where A represents the object to be compared, B the object A is compared with, and C is the respect with which A and B are compared to each other. Further arguments might represent the purpose of the comparison or other boundary conditions which could affect the criteria of comparison. This logical analysis is important insofar as the third argument, which plays a crucial role, is not always taken into account at all. A given text or corpus may be comparable to another one with respect to the frequency distribution of letters but not with respect to its genre or its length; it may be comparable with respect to its age but not with respect to its topics. We will concentrate here, of course, on quantitative criteria.

When two existing corpora are to be tested for comparability with respect to some property this task does not cause any principled problem. However, the situation may be different if 'comparable corpora' are concerned. Such corpora are set up and used because the hypothetical original corpora, e.g. parallel corpora, do not exist or are not available. Comparability is in this case rather itself a precondition for using such a corpus as a substitute of another, not available or even not existing one. If we now are not content with taking comparability for granted—how can we assess the comparability of a corpus with one which is not available? It crucially depends, of course, on what we want to do with the two corpora. As a rule, an intuitive selection of texts according to appropriate criteria of text sorts, authors, topics, intended audience, linguistic/stylistic properties (e.g. vocabulary, sentence length or complexity) etc. will be performed—with all the shortcomings of intuitive or opportunistic collections. If a systematic test for comparability is intended, a number of predicates come into play which are logically connected to comparability and must be discussed before.

2.1 *Representativeness*

If our aim is to show that a given corpus shares some property with a non-existent corpus we really claim that the given corpus shares this property with a hypothetical population and that both corpora, the given one and the non-existent one, can be dealt with as if they were samples from the common population. In such cases, statistical induction can be performed if the samples are representative. **Representativeness** is one of the features which are often claimed for large corpora by their creators or users. More often than not, representativeness is used as a predicate with just one argument ("my corpus is representative") whereas the concept turns out to have at least three arguments: the corpus, the object it is (or is not) representative of, the respect in which it is claimed to be representative. One should take account also of the significance level, i.e. the probability of false decisions which is tolerated. This claim confronts the researcher with various problems:

(a) A corpus cannot be large enough to represent the infinite variety of linguistic behaviour even if the scope of a corpus is limited. One of the reasons for this problem is the fact that enlarging a corpus causes an increase in diversity of the data.
(b) It is not possible to assess representativeness of a corpus because we lack the theoretical previous knowledge about the hypothetical population that would be needed.
(c) Every corpus is a stratified object consisting of elements with a multitude of different text parameters. We do not only lack the empirical knowledge of the 'correct balance' between corpus elements with given parameters—it is even impossible to acquire such knowledge since the number of potentially relevant parameters is infinite. For these and a number of other reasons no corpus can be representative in a scientifically meaningful sense, in particular not with respect to statistical methods.

What can be done under these circumstances to find a way to make statements about comparability (or usability for a given purpose)? As we cannot rely on representativeness the only method that is left is comparison of the relevant parameters of the corpus under analysis with at least some texts that would constitute the wanted but unavailable corpus. Imagine we are concerned with a corpus of Hawaiian business letters and want to use it with a parallel text corpus in Mansi but do not have a corresponding Mansi corpus. We would then compile a 'comparable' Mansi corpus consisting of Mansi texts that *could* be translations of the Hawaiian letters with respect to the parameters we are interested in. Imagine the parameter we are specifically interested in is sentence length, e.g. to make sure that our translation system cares for output texts which resemble as much as possible real translations made by native speakers of Mansi. We would decide to include only those Mansi texts in our comparable corpus which behave with respect to sentence length in the same manner as real translations do. Therefore, we need as many real translations as possible but at least a few. With these data sets, we can perform statistical tests and check every text whether it conforms to the sentence length behaviour of the original translations. This would be done according to a fixed scheme:

1. The hypothesis is set up that a text under consideration has the same sentence length distribution as a translation would have.
2. Determine the sentence length distribution of the original translations. From quantitative linguistics it is well-known that there is a great variance in sentence lengths in individual texts so that a single text never has enough instances of all the possible sentence lengths (if measured in terms of the number of words) to obtain a 'smooth' distribution. There are two possibilities to overcome this problem: (a) pooling the frequency classes: Set up frequency classes of 5, 6, 7, or 8 adjacent values, i.e. for $x = 1$ sum up all the numbers of sentence of length 1 to 5, for $x = 2$ all sentences of length 6 to 10 etc. (b) determining sentence length in units of a higher level of linguistic analysis, viz. in the number of clauses.
3. Find a theoretically justified theoretical probability distribution for sentence length. In quantitative linguistics, the 1-displaced negative binomial distribution and the 1-displaced hyper-Pascal distribution (for Chinese also the hyper-Poisson distribution) were deduced and empirically tested as adequate models for the case that sentence length is measured in the number of words. The same distributions proved later as good models also for sentence length distributions if measured in the number of clauses [6]. Fit the best fitting alternative to your data and determine the values of the parameters of the given distribution.
4. Fit the same distribution to the frequency distributions of the sentence lengths in the text under consideration (which is to be determined in the same way as in step 2). If the fitting procedure fails, i.e. if the goodness-of-fit test (e.g. Chi-square) indicates that the data and the theoretically assumed distributions differ significantly, the given text must not be included into the corpus. If, however, the test result is positive it is worthwhile to run another test:
5. The assumption that two texts 'behave similarly' with respect to sentence length can be formulated as the (H_0) hypothesis that the two means of the length distributions do not differ from one another ($m_1 = m_2$). This can be tested against the H_1 hypothesis that the values are different. An appropriate test can be performed, if the means are normal-distributed, with the t statistic, which can be found in every statistics textbook. This test is not exact as we do not know the parameters of the population but is suffices if variance and the degrees of freedom are estimated from the data. The same test can be performed for the variances of the two distributions (H_0: $\sigma_1 = \sigma_2$).
6. The researcher has to decide which level of 'uncertainty', i.e. which significance level s/he tolerates. A significance level of $\alpha = 0.05$ (a value that is often used in philology and corpus linguistics) means that 5 % wrong decisions are accepted. It is, by the way, always possible to use a function instead of a distribution as a model of the data behaviour if a corresponding theoretically justified hypothesis is available. Functions have the advantage that there are no degrees of freedom. The Chi-square test is replaced in such a case by the coefficient of determination R^2.

In many cases, much simpler but not less powerful methods can be applied viz. parameter-free tests. An example is the Mann–Whitney U test or Wilcoxon rank sum test [7]. For this test, only the ranks of the variables (in our case the length classes) according to their frequencies are used. This test is easy to perform but there are some conditions for its applicability. To test whether the values of a relevant property in n texts are distributed according to the same probability distribution (we must assume that they are not normal-distributed) the Kruskall–Wallis [8] test can be used. A much more comfortable way is to directly test a given hypothesis on the data using the software program "Altmann-Fitter" [9] or to determine the best-fitting theoretical probability distributions using the automatic fit mode of the program if a hypothesis is not available.

In this way, a reliable comparable corpus can be established but the procedure described so far must be performed for each variable which is expected to be relevant for the given purpose of the corpus. There are other statistical procedures to achieve analogous results; researchers without sufficient expertise in linguistic statistics are recommended to co-operate with an expert in the field.

2.2 Homogeneity

Representativeness is not the only stumbling block on the way. Most of the commonly applied statistical methods and tests demand that the data be **homogeneous** [10]. There is, of course, no absolute homogeneity in language or text; however, the data should be homogeneous with respect to the property under study; otherwise a naïve application of statistical standard instruments would yield misleading or simply wrong results. Since we know that linguistic data tend, as a rule, towards an inhomogeneous picture, we have to make sure that the data are at least fairly homogeneous with respect to the variables relevant for the purpose of the corpus. This can be achieved in a rather simple way: The text under examination should be subdivided into several (depending on text length but at least 10) subdivisions. Then, these text partitions can be tested against each other in the same way as described above for entire texts. Certainty about homogeneity can be achieved if the procedure is repeated with pooled text blocks, i.e. the 10 or more partitions can be put together pairwise. After repeating the tests, the remaining 5 parts could be pooled into 2 blocks. Finally, the results should be compared with the entire text. The same procedure can be conducted the other way round: First, determine the properties of the entire text, then subdivide it into two parts etc. If the parameters of the distributions do not differ significantly (or if the parameter free tests indicate identical behaviour of the data) homogeneity can be assumed.

2.3 Homoscedasticity and Skewness

A third issue that has to be taken into account is **homoscedasticity** of the data. This property concerns equal variance of the data in all the classes. If we e.g., are interested in texts with a lexical ambiguity which is comparable to the ambiguity of the words in some given texts, homoscedasticity would be given if the distribution of lexical ambiguity (i.e. the number of words with ambiguity 0, 1, 2, ...) has the same variance in the classes 0, 1, 2, ... This condition for the application of statistical standard instruments is absolutely unknown in corpus linguistics but is, notwithstanding this unfortunate fact, a prerequisite for correct decisions.

Another problem in quantitative linguistics and hence in corpus linguistics that may have an effect on the compilation of comparable corpora is the above-mentioned extreme **skewness** of the frequency distribution of words, parts-of-speech, syntactic constructions, semantic roles, speech act types—of virtually every linguistic unit. A typical example was given in Figs. 2 and 3, where a characteristic aspect of this kind of distributions can be observed. The curves begin with a more or less smooth course while they form ever-growing steps toward the end. This absolutely regular behaviour makes visible an effect of the properties of the frequency spectrum. There is a very large portion of the data (the exact size depends on the kind of the unit under study) which consists of elements which occur only once. These data form the first step from the right: they share the frequency value 1 on the y-axis (they are also called *hapax legomena*) and have therefore a horizontal shape; the next step from the right is formed by the elements with two occurrences (*dis legomena*) and so on. One can see that the lengths of the steps decrease from right to left. Taking into account that the diagram (Fig. 3) is a logarithmic transformation one can imagine that the real proportion of the hapax legomena might amount to, say one half of the data, the *dis legomena* to about one third of the rest and so on (this is the appropriate proportion when words are considered, first observed and tentatively explained by Zipf [11], later treated with mathematical rigour and a more satisfying theoretical deduction by Mandelbrot [12]).

An even much larger skewness is well-known from the mentioned frequency distribution of words in texts. Still steeper yet almost completely unknown among linguists are other relations, which we can illustrate with an example from syntax: The frequency distribution of syntactic construction types (in other words the frequency of the application of syntactic rules) in a corpus forms the steepest curve ever observed in linguistics. In the Susanne Corpus [13], the following observations have been made [14]: Of the 4621 different types of constituents with 90821 tokens, 2710 types occur only once. 615 of the rest occur twice; 288 types occur three times, 176 four times (cf. Table 1) etc. In other words, around 60% of all the constituent types (or rules) correspond to not more than a single occurrence; 13% are used just two times. Less than 20% of all constituent types can be found more often than four times in a corpus. Table 1 summarizes the top four ranks of this distribution; Fig. 4 shows the complete distribution of analogous data from the German Negra corpus in a bi-logarithmic transformation.

Table 1 The first four classes of the frequency spectrum (Susanne corpus)

Frequency of constituent type	No. of occurrences	Percentages
1	2710	58.6
2	615	13.3
3	288	6.2
4	176	3.8

A consequence of this fact is that rare elements are inevitably underrepresented in texts and corpora of any shape and size; enlarging the data base does not improve the situation because the frequent elements (think of, e.g. functions words vs. autosemantics) increase their frequency faster than the other ones, which even aggravates the problem. Another consequence of the extremely skew distributions together with the fact that every linguistic property is interrelated with countless others is that significance tests and text comparison are affected: Regardless of which property is studied—the shape of the frequency distribution of the units which carry the given property has to be taken into account. This thought should be kept in mind when it comes to discussions about the effort that has to be taken for the compilation of a corpus with respect to testing properties such as homogeneity. Thus, practical consequences of the skewness of linguistic distributions for comparable corpora affect corpus compilation and use, e.g. the planning of text coverage (enormous amounts of

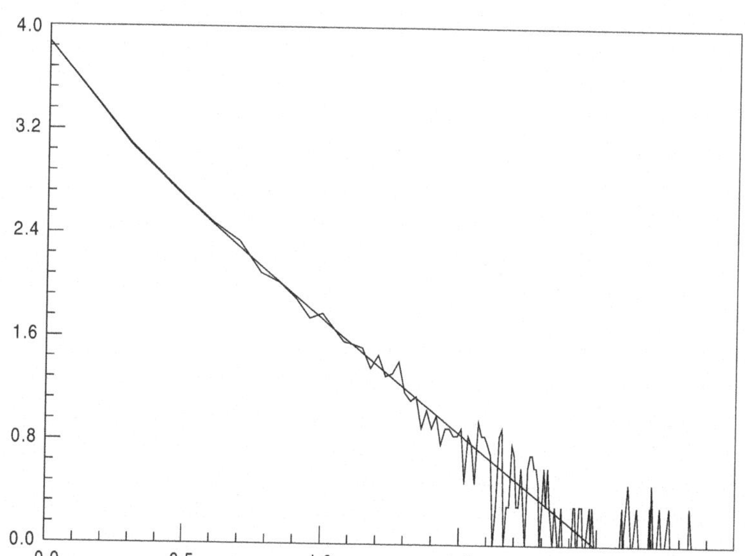

Fig. 4 Fit of the Waring distribution to the data from the Negra Corpus. Both axes are logarithmic (X: frequency class, Y: number of occurrences)

additional linguistic material are needed to slightly increase the probability of occurrence of one of the rare phenomena) and the evaluation of statistical data describing properties which are connected to other properties with a skew distribution.)

2.4 Corpus Balancing

After all, another property should be mentioned which is frequently attributed by providers of corpora to their material: It is often claimed that a corpus be **balanced**. We will discuss it here because this feature may play a role for the compilation of comparable corpora, in particular in combination with representativeness. In corpus linguistic works and in corpus descriptions the attribute *balanced* is often treated as a kind of quality feature. If specified at all, balance is identified with the given proportion between text kinds or genres of the texts the individual corpus may consist of (in percentages or absolute numbers). There is, of course, no objectively correct or best proportion of text kinds as many corpus linguists seem to believe. A single example may suffice for our purposes: Hunston [15, p.156] writes: "[..] a corpus that is a balanced representation of the language as a whole [..]". Such a statement makes some implicit assumptions:

(1) There is a "language as a whole".
(2) It consists of certain text kinds with given proportions between them.
(3) These proportions are constant.
(4) The proportions can be determined.
(5) A text corpus can be considered a sample from the population "language as a whole".
(6) A text corpus that reflects the proportions between the text kinds of the "language as a whole" is better than a corpus that does not reflect them.

Assumption (1) is clearly a confusion of a scientific model with reality. "Language" is of course a simplification and an abstraction of what linguists observe when studying the linguistic behaviour of people. Language is not observable as such. Therefore, also the other assumptions cannot make any sense either; in addition, each of them is problematic in itself. One of the additional problems is connected with assumption (2), which claims that "language" is a (very, very large) collection of texts of different kinds because only in this case a proportion of text kinds or genres could exist, which a corpus could be a balanced representation of. Another example is assumption (4): How could anyone know the "true" proportion of text kinds even if the other assumptions were realistic? For all these reasons, balance has some of its problems in common with other attributes such as representativeness. Comparable corpora, as opposed to some other corpus types, are certainly not disadvantageous if they are claimed to be balanced but they are obviously not better than other ones, i.e. "balance" is a useless and even often meaningless attribute.

2.5 An Example

Finally, a practical example shall be given of how a distribution test for a variable can be performed. For the sake of simplicity, word length is chosen for this purpose. First of all, we will have to decide which of the possible operationalisations of the concept of word length is appropriate for the given study: word length can be measured in terms of the number of morphs, syllables, sounds/phonemes, or letters/graphemes a word is formed of, in terms of duration in milliseconds, of physical length in cm or inch (e.g. for purposes of content analysis when the weight of a topic in a newsletter is to be measured). As we do not have another purpose here than giving an example we will apply a simple measuring technique: we will count the number of vowel letters in a word and consider it as a sufficiently exact estimation of the number of syllables. Next, the concept of word must be defined, and we choose again a simple way: We define a word as a string of letters between white spaces, punctuation marks, and text boundaries (begin or end of a text).

The source of the data we use is an opportunistic comparable corpus collected by an editor of this volume [16]. First, we determine the word length distributions of the first ten texts in the corpus and obtain the mean values 1.6809, 1.8137, 1.8545, 1.8303, 1.7888, 1.8586, 1.8813, 1.8826, 2.2958, and 1.9229. Now, we test the distribution of these mean values for normality using the Shapiro–Wilks test, which forms the quotient between the theoretically expected variance of a corresponding normal distribution and the empirically found variance of the data. Due to space limitations, we do not show the formula and the details of this test. We obtain $W = 0.7578$, $p = 0.004442$; as $p \ll 0.05$ the hypothesis of a normal distribution must be rejected. Thus, the application of the t-test to determine whether the difference between the means of the two data sets is significant would not yield reliable results. Hence, we will find out whether the empirical word length distributions are compatible with the same theoretical frequency distribution. We compare for this demonstration only the first text with a combination of the first ten texts. The empirical distributions are as shown in Table 2. We know that word length distributions abide by a specific group of theoretical probability distributions, among them the Poisson family. Therefore, we test if the two data sets from the corpus are compatible with the Mixed Poisson distribution. Parameter estimation, iterative optimisation and the Chi-square goodness-of-fit test are conducted with the help of the software tool "Altmann-Fitter 3.0"). Tables 3 and 4 show the results, and Figs. 5 and 6 visualise them.

Both test statistics, the probability of the obtained Chi-square value and C indicate good agreement of the data with the hypothesis. In this way, if performed as described in the 6 steps above, homogeneity of the corpus with respect to the kind of word length distribution can be attested. Including a documentation of every performed test and its results in the meta-data section of the corpus is recommended.

It goes without saying that this short contribution cannot provide an overview over the available statistical methods that may be relevant for comparable corpora, nor can it give detailed instructions for correct applications of these methods and their results. It can, hopefully, draw attention to an important part of the work with

Table 2 Word lengths frequencies in two data sets

Text 1		Text 1-10	
x_{1i}	$f(x_{1i})$	x_{2i}	$f(x_{2i})$
1	749	1	7622
2	307	2	4056
3	147	3	2485
4	59	4	1121
5	21	5	423
6	0	6	81
7	1	7	13
8	1	8	4
N	1351	**N**	16622
$\mathbf{m}_1$	1.5988	$\mathbf{m}_2$	1.8287
s_{12}	1.0574	s_{22}	1.3803

Table 3 Results of fitting the mixed Poisson distribution to data set 1

$\mathbf{x}_i$	$\mathbf{f}_i$	$\mathbf{NP}_i$		
1	749	752.49		
2	307	306.18		
3	147	146.67		
4	59	57.31		
5	21	17.21		
6	0	4.14		
7	1	0.83		
8	1	0.17		
Parameters:				
a	1.2046			
b	0.1258			
alpha	0.5090			
$\mathbf{X^2}$	$\mathbf{P(X^2)}$	**DF**	**C**	$\mathbf{R^2}$
2.8265	0.2434	2	0.0022	0.9999
N	$\mathbf{m}_1$	$\mathbf{m}_2$	$\mathbf{m}_3$	$\mathbf{m}_4$
1285	1.6809	0.9737	1.5699	5.6491
Ord I			**Ord S**	
0.5793			1.6124	
Skewness		**Excess**	**Entropy**	**Repeat rate**
1.634		2.725	0.5408	0.6717

Table 4 Results of fitting the mixed Poisson distribution to data set 2

x_i	f_i	NP_i		
1	7622	7613.44		
2	4056	4062.48		
3	2485	2504.11		
4	1121	1118.73		
5	423	376.69		
6	81	101.5		
7	13	22.79		
8	4	5.25		
Parameters:				
a		1.3472		
b		0.0751		
alpha		0.6679		
X^2	**$P(X^2)$**	**DF**	**C**	**R^2**
14.5071	0.0058	4	0.0009	0.9999
N	**m_1**	**m_2**	**m_3**	**m_4**
15805	1.9233	1.2698	1.7433	6.5865
Ord I			**Ord S**	
0.6603			0.6603	
Skewness		**Excess**	**Entropy**	**Repeat rate**
1.2183		2.8449	0.6305	0.7669

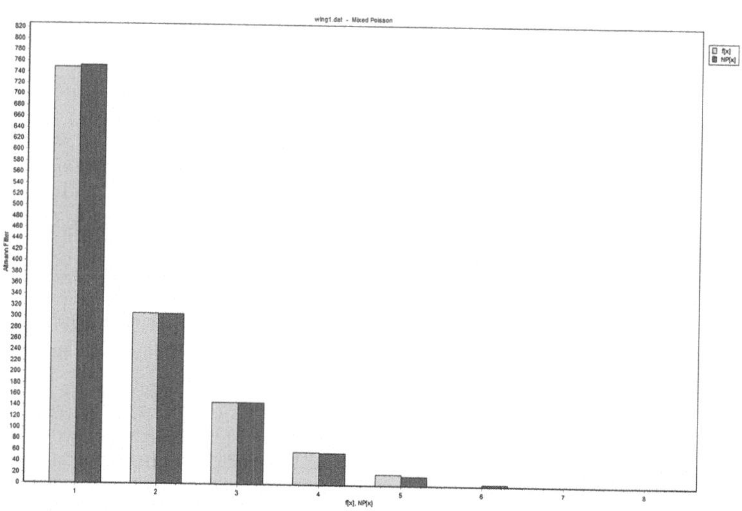

Fig. 5 Plot of the distribution in Table 3

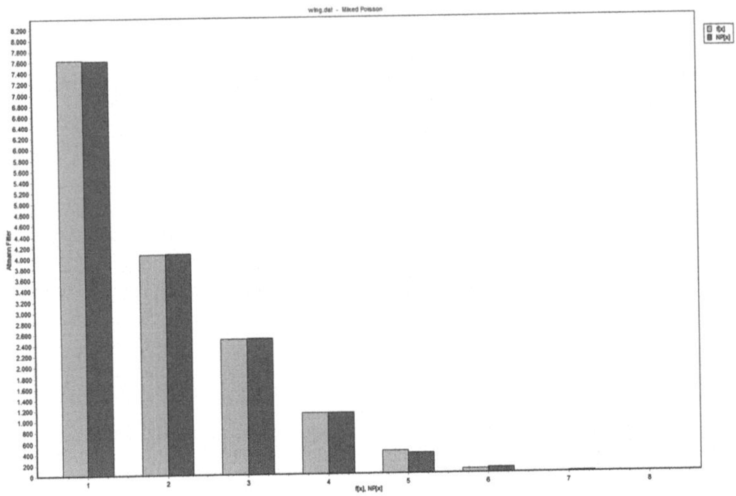

Fig. 6 Plot of the distribution in Table 4

such corpora and give some warnings as not to blindly employ standard procedures which may not be applicable to linguistic data at all.

In any case: methods and results of quantitative tests, criteria etc. should be taken down in the meta-data of the corpus and of the individual texts.

References

1. Altmann, G., Hřebíček, L. (eds.): Quantitative Text Analysis. WVT, Trier (1993)
2. Guiter, H.; Arapov, M.V. (eds.): Studies on Zipf's Law. Brockmeyer, Bochum (1982)
3. Popescu, I.-I., Altmann, G., Gabriel, K., Köhler, R.: Zipf's law—another view. Qual. Quant. **44**(4), 713–731 (2010)
4. Shapiro, S.S., Wilk, M.B.: An analysis of variance test for normality (complete samples). Biometrika **52**(3/4), 591–611 (1965)
5. Orlov, J.K.: Linguostatistik: Aufstellung von Sprachnormen oder Analyse des Redeprozesses? (Die Antinomie, Sprache-Rede" in der statistischen Linguistik). In: Orlov, J.K.; Boroda, M.G.; Nadarejšvili, I.S. (eds.) Sprache, Text, Kunst. Quantitative Analysen, pp. 1–55. Brockmeyer, Bochum (1982)
6. Best, K.-H.: Satzlänge. In: Köhler, R., Altmann, G., Piotrowski, R.G. (eds.) Quantitative Linguistik. Ein internationales Handbuch. Quantitative Linguistics. An International Handbook, pp. 298–304. de Gruyter, Berlin (2005)
7. Mann, H.B., Whitney, D.R.: On a test of whether one of two random variables is stochastically larger than the other. Ann. Math. Stat. **18**(1), 50–60 (1947)
8. Kruskal, W.H., Wallis, W.A.: Use of ranks in one-criterion variance analysis. J. Am. Stati. Assoc. **47**, 583–621 (1952)
9. Altmann Fitter 3.0: Iterative fitting of theoretical probability distributions (Software). Lüdenscheid: RAM-Verlag (2011)
10. Altmann, Gabriel: Das Problem der Datenhomogenität. Glottometrika **13**, 287–298 (1992)

11. Zipf, G.K.: Human Behavior and the Principle of Least Effort, 2nd edn . Addison-Wesley, Cambridge. 1972. Hafner reprint, New York (1949)
12. Mandelbrot, B.: A Note on a Class of Skew Distribution Functions. Analysis and Critique of a Paper by H. Simon. Inform. Control **2**, 90–99 (1959)
13. Sampson, G. (ed.): English for the Computer. Clarendon, Oxford (1995)
14. Köhler, Reinhard: Syntactic structures: properties and interrelations. J. Quantit. Linguist. **6**(1), 46–57 (1999)
15. Hunston, S.: Collection strategies and design decisions. In: Anke Lüdeling, M.K. (eds.) Corpus Linguistics. An International Handbook, pp. 154–168. de Gruyter, Berlin (2008)
16. Sharoff, S.: Creating general-purpose corpora using automated search engine queries. In Baroni, M., Bernardini, S. (eds.) WaCky! Working papers on the Web as Corpus, Gedit, Bologna (2006). http://wackybook.sslmit.unibo.it

Methods for Collection and Evaluation of Comparable Documents

Monica Lestari Paramita, David Guthrie, Evangelos Kanoulas, Rob Gaizauskas, Paul Clough and Mark Sanderson

Abstract Considerable attention is being paid to methods for gathering and evaluating comparable corpora, not only to improve Statistical Machine Translation (SMT) but for other applications as well, e.g. the extraction of paraphrases. The potential value of such corpora requires efficient and effective methods for gathering and evaluating them. Most of these methods have been tested in retrieving document pairs for well resourced languages, however there is a lack of work in areas of less popular (under resourced) languages, or domains. This chapter describes the work in developing methods for automatically gathering comparable corpora from the Web, specifically for under resourced languages. Different online sources are investigated and an evaluation method is developed to assess the quality of the retrieved documents.

Keywords Comparable corpora · Retrieval methods · Evaluation

M. L. Paramita (✉) · D. Guthrie · E. Kanoulas · R. Gaizauskas · P. Clough · M. Sanderson
University of Sheffield, Regent Court, 211 Portobello Street, Sheffield S1 4DP, UK
e-mail: m.paramita@sheffield.ac.uk

D. Guthrie
e-mail: d.guthrie@sheffield.ac.uk

E. Kanoulas
e-mail: e.kanoulas@sheffield.ac.uk

R. Gaizauskas
e-mail: r.gaizauskas@sheffield.ac.uk

P. Clough
e-mail: p.d.clough@sheffield.ac.uk

M. Sanderson
e-mail: m.sanderson@sheffield.ac.uk

S. Sharoff et al. (eds.), *Building and Using Comparable Corpora*,
DOI: 10.1007/978-3-642-20128-8_5,

1 Introduction

The Web contains a vast number of texts authored in a multitude of languages. Crucially, some of these texts are available in multiple languages with varying degrees of correspondences, ranging from parallel versions to describing similar concepts or themes. Texts with a high degree of correspondence can be used to improve Statistical Machine Translation (SMT) systems and work has been undertaken in the past decade to develop methods to automatically retrieve such texts in order to build parallel and comparable corpora. However, most of these methods have only been tested in the context of retrieving document pairs for well resourced languages. The performance and applicability of these methods can differ significantly when they are applied to under resourced languages.

ACCURAT,[1] which stands for Analysis and evaluation of Comparable Corpora for Under Resourced Areas for machine Translation, is an EU project which aims to investigate the use of comparable documents when parallel corpora are not readily available. This project focuses on under resourced European languages, which include Croatian, Estonian, Greek, Latvian, Lithuanian, Romanian and Slovenian. The aim of ACCURAT is to analyse the use of comparable documents for under resourced languages in order to improve the performance of machine translation. The work presented in this chapter is one of the tasks in the ACCURAT project which specifically focuses on developing methods to locate and download comparable documents from the Web.

In the ACCURAT project three levels of comparability have been used. The first level (parallel corpora) represents parallel documents, which are direct translations, although minor language-specific variations are allowed. The second level (strongly comparable) contains pairs of documents about the same topic or derived from the same source. The third level (weakly comparable) represents documents about different topics, but from similar domains or genres.

This chapter starts with a review of related research: previous retrieval approaches are detailed and assessed for their suitability for under resourced languages in Sec. 2. In Sect. 3, a series of novel approaches to retrieve comparable documents and preliminary results from each Web source are discussed. In Sect. 4, we describe our evaluation methods of the comparable corpora, which are focused on measuring the effectiveness of the retrieval methods.

2 Literature Review

There has been a range of previous work related to the tasks of gathering or identifying comparable corpora. The methods involved in the retrieval process can be categorised into two major processes: Web crawling and alignment. The use of Web crawling techniques is described in Sect. 2.1. The crawling process results in a large

[1] http://www.accurat-project.eu/.

collection of unaligned multilingual texts. Several methods are then performed to identify comparable documents. These methods are described in Sect. 2.2.

2.1 Web Crawling

Many tools and approaches have been developed to build comparable corpora using techniques based on retrieval. For example, BootCat [4] retrieves documents from a list of seed words. Outputs can then be used to bootstrap the process by inserting more seed words to improve the recall of the document retrieval stage. This approach assumes that the retrieved results are relevant and satisfy the requirements of the query. Other approaches, on the other hand, perform an evaluation to check the relevance of results. If relevant, the result is used to enhance some underlying language model, or included in the collection to generate a query; otherwise, the results are not considered. This approach is referred to as focused crawling [6] and is shown to retrieve relevant documents in narrow domains more effectively than using general purpose crawlers.

Talvensaari et al. [23] implemented focused crawling using keywords as the input seeds. In an approach that differed from BootCat, they did not specifically look for relevant documents; rather they used the retrieval result to look for websites that consistently produced top results over the majority of these queries. These websites were seen as good resources for that particular domain and were crawled to retrieve all documents within it. Language was detected using a simple n-gram based algorithm [5].

Ghani et al. [10] implemented a different approach, which they call CorpusBuilder. Instead of using query seeds, they used a set of documents previously judged as relevant and non-relevant to a given query (or set of queries). To focus retrieval on documents of under resourced languages, in this case Slovenian, they used Slovenian documents for the relevant documents and those from other languages as the non-relevant documents. They investigated the performance of several query generation methods; and found that an approach based on odds-ratio resulted in the highest performance, compared to term frequency or random sampling baselines. The odds-ratio of each word is calculated by using the probability of the word occurring in a relevant and non-relevant document. A further difference of this method compared to others was that the query used both inclusion and negation of terms. Highest performance was obtained by using three positive and three negative keywords, each chosen based on the highest odds-ratio score of the sets of relevant and non relevant documents. After each retrieval operation, the first document was passed to a language filter. If this document was identified to be in Slovenian, the set of documents was updated, and query generation was performed again. In case the new document did not change the query, the next ranking document in the result was taken as a result, and this process was performed iteratively. This method managed to retrieve general corpora from minority languages effectively.

2.2 Identifying Comparable Text

Given a large collection of unaligned multilingual texts, a range of approaches have been used to align parallel documents or sentences automatically. For example, [17] and [24] used the HTML structure and URL paths of documents in order to find parallel texts on the Web. These approaches are language-independent, however they are not applicable to retrieve comparable documents since such documents do not always share the same URL paths or contain similar HTML structure. Other approaches to align comparable documents require a range of linguistic resources, from bilingual dictionaries, parallel corpora to machine translation systems.

Dictionaries can be used in a straightforward manner to translate the words (and phrases) in a document and these terms can be used as the query to an IR system in the target language. However, ambiguity can be an issue if a word (or phrase) has multiple interpretations, and therefore, translations. Problems also occur when a word does not exist in the dictionary, i.e. out-of-vocabulary term. To solve the latter problem, cognate matching can be used to identify the translation of a word (e.g. "colour" in English and "couleur" in French). However, this method only applies for languages with the same etymological roots and using the same writing system. If multilingual parallel corpora exist alignments can be computed and the resulting aligned texts can be used to build a statistical machine translation system. Unfortunately, this approach is computationally expensive and still has to deal with the problem of out-of-domain vocabulary. In addition, it can also be difficult to gather enough resources for machine translation because of the limited amount of parallel corpora available and accessible on the Web, particularly for under resourced languages. Most parallel corpora only cover a specific domain, such as law, which may cause problems with translation as the system may perform poorly when used to translate documents from a different domain [23].

Munteanu and Marcu [15] attempt to identify pairs of parallel sentences from large collections of comparable news documents in several languages. They first align corpora of Arabic and English news documents by building a query from each Arabic document based on translating every word in the Arabic document to English using a bilingual dictionary. The highest ranked 100 English documents were retrieved for each Arabic document. Documents published outside of a specified time window from the Arabic collection were filtered out. Each sentence in the Arabic document and those in each of the English documents were paired, and certain features were evaluated in each sentence pair to identify which sentences were parallel.

A different approach to align comparable news is by using the metadata information. Argaw and Asker [3] aligned news articles in Amharic and English published on the same date and occurring in the same place. No lexical resources were used to translate the words, instead [3] performed transliteration on the titles and calculated the edit distance between words in the titles. Pairs of documents which scored above a certain threshold were considered as comparable. Other approaches which have been used to align documents are based on overlapping named entities [12] and clustering documents [22].

Fung and Cheung [9] aligned non-parallel corpora by using parallel sentences from bilingual corpora to retrieve new documents. These documents were likely to have different topics and therefore not found by standard keyword searching on the topic, named entities or dates. However, as they contained similar sentences they tend to share similar terminologies which could be used as a source of parallel data.

3 Retrieval of Comparable Documents

In this section, we describe new techniques that have been used to gather comparable documents from the Web. First, we identified different Web sources considered as promising sources of comparable documents, such as news, Wikipedia and Twitter. The characteristics of each source were explored in detail to enable retrieval methods to be created effectively. We then developed different retrieval techniques to gather comparable documents from each source. Techniques which we use to collect articles from news sites are described in Sect. 3.1. In Sect. 3.2, we focus on techniques to retrieve comparable documents from Wikipedia. Techniques developed to extract data from Twitter are described in Sect. 3.3.

3.1 News

News articles are continuously being published on the Web from news agencies across the globe in a variety of languages. These news stories can be highly similar, or even parallel across languages, because they are produced by the same agency written from the same newswire feed (e.g. Associated Press, Press Association or Reuters) or simply because they are reporting the same topic or story. The availability of large amounts of similar texts across languages makes the news domain an extremely promising area in which to perform comparable document mining. In this section we give an overview of our methods to automatically retrieve news data for the construction of comparable corpora. We show how collections with aligned document pairs can be produced and that these methods are useful for under resourced languages. We also show how it is possible to dramatically decrease the number of documents that must be compared using information retrieval and by using additional processing how it is possible to increase the accuracy of the method.

Although there exists an abundance of news articles on the Web, and many of these may be comparable in some way, identifying particular news articles that are the most similar can, in practice, be problematic. For example, often news stories are running stories rather than one-off events: they describe on-going events that can proceed over the course of days, weeks, or even years with many updates that have little difference in their focus or content. Take, for instance, the March 2010 news article coverage of the Icelandic volcano eruptions. News articles concentrated on a range of subjects: when eruptions occurred, the drifting cloud of ash, the environmental impacts, and the assorted disruptions to air travel. Many of the articles during this period are very similar and contain pieces of overlapping information, so they could be considered weakly comparable. However, a smaller subset of these documents may actually be nearly identical in their focus and have a much stronger level of comparability. It is this special subset of similar documents that we aim to identify and match as they contain large sections of information that match across languages and are thus the most useful for improving SMT systems.

We approach this problem of identifying comparable news documents as a type of Cross-Language Information Retrieval (CLIR) task where the goal is to find, for each news article in a certain language, news articles that report the same information in other languages. Some articles may not have matching information in one or more languages and for these articles we would either like to identify articles that are close matches, if they exist, or else judge that the article does not have a match. This setup requires that we perform the CLIR task for a set of seed documents in one language (e.g. English) and find all matches in the target language for each of these documents. This can be computationally intensive, but it critically does not require us to perform the much more expensive and impractical computation of comparing every document in one language with every other document in another language, as long as we choose our method of performing CLIR carefully.

Cross-language information retrieval often involves translation of the query into the target language before standard information retrieval in the target language. This requires translation of every word (or phrase) in the query either using a dictionary or using a machine translation system. The use of a dictionary only allows for words (or phrases) to be translated independently and often gives results that are unsatisfactory, since it is difficult to pick the correct translation for a word due to ambiguity in both languages and limited context given the short length of the query. Machine translation can be more accurate and simpler, but is computationally expensive and requires parallel corpora for training the system. Both of these methods are not very suitable for languages where few electronic resources are available, which is one of the focus areas for our method of gathering comparable corpora. We instead adopt an approach that uses only limited language resources by making use of the unique properties of news articles.

We make use of a useful feature of news articles: the high frequency of proper nouns, such as people, places and organizations. News articles often contain a large number of such entities and identifying them and classifying them as to their type is a well-established problem called Named Entity Recognition (NER) and can be performed with high accuracy [2, 11]. We make use of these entities to query for documents that may be comparable. For this purpose it is not necessary to perform the full Named Entity Recognition task, as the class of these entities is not needed. It is only important to extract these entities from every news article and not label them with their semantic category (e.g. person, place, organization, etc.).

We also exploit the fact that news articles typically have dates associated with them: possibly the date the story was written, the date of some event, or the date it was published to the Web. We propose to make use of a search method that uses the date articles are published to the Web. Publication date is available for every

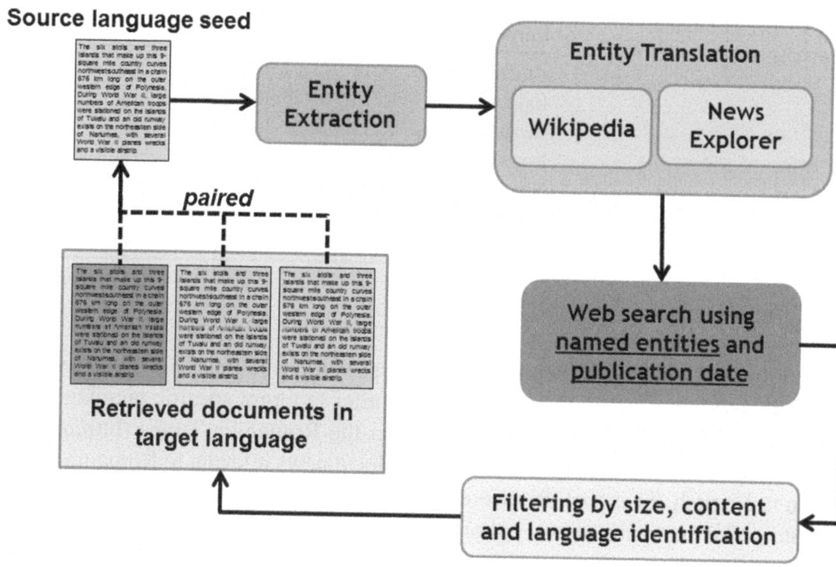

Fig. 1 Retrieval process of news documents

document and does not have the resolution issues associated with dates found in the article, which can refer to past or future events. Making use of these dates allows us to significantly reduce the search space for documents. Instead of performing a CLIR search over all similar news documents on the Web, we search only the news documents published at the same time or within a few days of each other. It seems very likely that limiting the search increases the precision of the search because similar articles will have been written at similar times. It has been shown in previous work that even with these restrictions it is still possible to obtain a large collection of results [8, 16]. The process we developed to retrieve news documents is represented in Fig. 1 and details involved in each process are described in more details in the following sections.

3.1.1 Collecting Source Language Seeds

To build this large collection of comparable text we first gathered collections of news articles in English that we believed likely to have comparable documents in some other project languages. We make use of News Explorer to identify these popular news articles. News Explorer (http://emm.newsexplorer.eu/NewsExplorer/), developed at the European Commission's Joint Research Centre (JRC), is a tool for multilingual news gathering, analysis, and exploration. News Explorer monitors news from approximately 1,500 news portals worldwide. We gathered popular news articles by collecting the top 15 news stories in newsexplorer.com for every day from January 2008 to November 2010. These stories have the most news coverage for

Table 1 Size of English documents seeds

Corpora source	Size in words (excluding markup)
Athens News	4,984,796
Baltic Times	3,307,234
Croatian Times	355,809
NewsExplorer	7,589,978
Nine O'Clock (Romanian)	13,150,159
Romanian Times	196,426
Total	29,584,402

that day (across languages) according to NewsExplorer. Other English collections were gathered from more focused sources in order to target specific languages, for example, a collection of all news stories from the Romanian Times (http://www.romaniantimes.com) newspaper which covers local news relevant to Romania, but is written in English. These articles were then post processed to extract only the main news story. We disregard any very short articles whose cleaned text is less than 1,600 characters long. For each article we store the original HTML page, the date it was published, and the cleaned main text from the web page. In total, 29 million words of cleaned English text was gathered. The exact sizes of the corpora and their sources are shown in Table 1.

3.1.2 Named Entity Extraction

Our method works by taking every news document in one language (the source language) and extracting all proper nouns and the date the article was published. We then use these entities to perform a search over all documents in the target language published within two days of the source document. To increase the recall of results we do not use all entities that occur in the document, but instead choose only the top *n* most frequent entities. This procedure works well when searching languages with similar writing systems, as these named entities are often written exactly the same across languages with no translation. Some entities are written slightly differently across languages even with the same writing system, for example, Băsescu, the president of Romania, is often written in English without the breve as Basescu. In languages with very different writing systems, often either translation or transliteration occurs. Google has a transliteration tool available online (http://www.google.com/transliterate/) with an API available to programmers, but in tests this tool did not perform particularly well on proper names. Instead we can achieve much better translation of named entities by making use of resources specifically generated for entities that have had the benefit of human involvement.

3.1.3 Named Entity Translation

We then implemented methods to make use of this collection of English documents by searching the Web for all documents in other languages that are likely to be comparable with each document in the English collection. The basic method we developed was to identify all proper nouns (entities) in each English document and use the most frequent of these to search for comparable documents written in another language that contained the same entities and were published within 5 days of each other. Specifically, to find news articles comparable to an English article, D, in another language, say L2, we first identify the five most frequent entities in D and then attempt to translate them into L2 using one of four resources: Wikipedia, Google translate, Bing translate, and NewsExplorer. We make use of the translations recourses in the order listed and only use then next resource if no translation was found (or the resource was unresponsive). Wikipedia is used by checking if the English entity has a wikipedia entry with this title in English and if so we check to see if there is a link to a Wikipedia article in L2 and use its title. We use these translated entities to perform a focused search for News articles in the L2 language.

3.1.4 Web Search

We make use of Google News to search once for every document in our source language collection using the most common entities from each source document and limiting the search to within two days of the original source document. We propose to use the top five most frequent entities, but this parameter can be tuned depending on the number of entities that occur in the document and the level of recall required. The use of Google News search allows us to dramatically reduce the search space for every document and with this reduced set of documents we can perform a more exhaustive comparison to determine if they are likely to be comparable. We thus take the documents returned by this search procedure to be a set of candidate comparable document pairs and run additional tests to further limit this set.

An alternative to using Google News search is to use a fixed set of news sites either by crawling and indexing all these sites (as mentioned previously), or simply using this list of sites to restrict a Web search engine, like Google or Bing to these domains. This would require having a fixed set of news sites and then sending a separate search request to every site, whereas Google News searches all sites at once. The use of Google News does not require this list of fixed sites and the sites indexed by Google are likely to be much broader and more comprehensive than a fixed list would provide and are also continuously updated. This allows for a large amount of current news documents to be searched easily without requiring a fixed list of sites.

To perform the language specific focused web search, we currently make use of several available news search engine services that allow us to search for news documents by date as well as more general search engines that allow us to search by date. Just as in entity translation we rely on multiple services but use them only as backups for cases when the primary service is unresponsive or fails to find any results. The services we use in order are: Google Search, Bing Search, and Google News Search. We query each service for documents that contain all of the the top five translated entities and were published within 5 days of the English document and constrain the search to only the desired language (e.g. Croatian) and if no results are found in any service we perform the searches again using only the top three entities and then the top two entities.

3.1.5 Filtering

This procedure will undoubtedly return some documents that are not relevant, so we perform some additional checks before storing a document. We first extract the main story from the web page and ensure that the cleaned text is greater than 1,100 characters long. We also check to make sure the document is not too large and does not contain an unusually large percentage of numbers, punctuation, or uppercase letters, all of which are signs that the documents will probably not be useful for comparability. We then send the document to Bing's language identification service and check that the language returned by the service is indeed the target language in with we were searching for the document (e.g. Croatian). The top five results returned for each search query that meet these minimum requirements are saved as likely comparable documents.

3.1.6 Results

Our retrieval method has a number of advantages over previous methods described in the related work. First, it does not require us to crawl all news sites in the Web. Instead, only a small set used as initial seeds is required; the rest of the documents are found using a search engine. Second, our method requires no specific lexical resources for named entity translation. We use all data which are publicly accessible, and therefore our methods can be implemented for any other language pairs.

By implementing this retrieval method, we managed to gather a large aligned comparable corpus. The size of comparable texts gathered for each language is shown in Table 2.

3.2 *Wikipedia*

Wikipedia is identified to be the world's largest multilingual encyclopaedia in the Web which, by September 2010, was estimated to contain over 16.5 million articles. Wikipedia covers all languages of the ACCURAT project and therefore is seen as a promising source of comparable documents. The number of Wikipedia documents has increased dramatically over the past few years and it covers a wide variety of topics. Even though Wikipedia documents can be edited by anyone, the contents

Table 2 Size of automatically retrieved comparable text

Language	Size of comparable text gathered (words)
Croatian	3,802,495
Estonian	11,409,322
German	27,422,578
Greek	13,438,848
Latvian	16,634,981
Lithuanian	28,750,162
Romanian	43,841,777
Slovenian	9,169,704
Total	154,469,867

are moderated and this has ensured Wikipedia documents to have reasonably higher quality than other Web documents.

An advantage of using Wikipedia as a source of comparable documents is the interlanguage links feature: a link connecting documents about the same topic but written in different languages. Several works were conducted to extract information from these interlanguage links, such as the creation of bilingual lexicon [1] by performing title extraction of the linked documents. Even though not all documents in Wikipedia have interlanguage links, i.e. documents which are only written in one language, the percentage of articles written in more than one language is considerably high and ranges between 46 % – 84 % as shown in Table 3. This data is based on Wikipedia dump in September 2010.

Wikipedia articles linked across languages are about the same entity, process, event, or topic, however there is a disagreement regarding how similar the information contained in these documents are. Mohammadi and GhasemAghee [14] found that many of the articles linked across languages contain very similar information and are thus ideal as a source for gathering comparable corpora. On the other hand, Adafre and Rijke [1] found that these connected documents do not necessarily talk about the same topics as they vary in length and may include additional, removal, or

Table 3 Percentage of documents with language links

Language	All pages	Page with language links	Average number of links
EN	3,110,586	1,425,938 (45.84 %)	4.84
DE	1,036,144	636,111 (61.39 %)	7.76
RO	141,284	106,321 (75.25 %)	17.24
LT	102,407	67,925 (66.33 %)	22.28
SL	85,709	58,489 (68.24 %)	21.02
HR	81,366	60,770 (74.69 %)	23.47
ET	72,231	49,440 (68.45 %)	25.47
EL	49,275	37,337 (75.77 %)	29.62
LV	26,227	22,095 (84.25 %)	33.36

completely different information. Attempting to align every pair of linked documents in Wikipedia to assess their comparability is a computationally expensive process and also unnecessarily wasteful if the articles are very different. Therefore, when gathering texts in Wikipedia using the link structure it is necessary to verify that the text itself is actually comparable.

3.2.1 Identifying Comparable Documents

In this section we describe our attempt to maximize the likelihood of finding comparable documents. We implement two simple filters, which are document's minimum size and length's difference. First filter, we eliminate pairs in which any of the documents is smaller than 2 KB. Afterwards, we compare the length between the two documents and disregard document pairs if the length difference is greater than 20 %. We implement this feature to focus the retrieval on documents of similar length, assuming that they will have similar structure and content, and therefore have higher probability to be comparable than other documents.

3.2.2 Results

An initial analysis find that most of the Wikipedia documents have a diverse size within different languages even though they describe the same topic. For example, the article about "Europe" in English has 12 main sections with the content reaching just under 10,000 words; the corresponding Latvian article, however, has just 6 main sections with a total word count of around 3,000 words. On average, articles from the corresponding under resourced languages have significantly smaller sizes compared to the English version. By finding articles of similar length only, we disregard over 80 % of the initial pairs, as shown in Table 4.

Table 4 Bilingual documents of similar size for each ACCURAT language pairs

Language pairs	All pairs	After size filtering	After doc length filtering
EL-EN	32,015 pairs	23,206 pairs	3,993 pairs
RO-EL	14,339 pairs	8,725 pairs	1,815 pairs
RO-EN	84,862 pairs	27,234 pairs	3,243 pairs
LT-EN	51,011 pairs	26,804 pairs	2,886 pairs
LV-EN	18,480 pairs	11,893 pairs	932 pairs
SL-EN	44,923 pairs	23,313 pairs	4,893 pairs
HR-EN	43,984 pairs	26,520 pairs	3,628 pairs
ET-EN	37,043 pairs	15,870 pairs	1,118 pairs
DE-EN	418,327 pairs	254,793 pairs	56,734 pairs
RO-DE	49,155 pairs	16,644 pairs	2,298 pairs
RO-LT	19,794 pairs	10,552 pairs	2,488 pairs
LT-LV	10,762 pairs	7,135 pairs	1,810 pairs

Furthermore, by performing manual assessment on some Wikipedia documents, we also found that some under resourced language documents were created by translating a main paragraph of English documents. This pair of documents may have a large difference in lengths, but these translated fragments are particularly useful for machine translation. These documents would not have been retrieved had a length difference filter been implemented as a filter. We therefore propose a different retrieval method as a future work as described in the Sect. 3.2.3.

3.2.3 Future Work in Wikipedia

As discussed in the previous section, the proposed filters disregard a large number of document pairs in Wikipedia. We therefore propose another method to filter out comparable documents by identifying comparability in a finer granularity, i.e. sentence level, which is defined below.

1. First, we crawl documents which have interlanguage links for all the ACCURAT language pairs and disregard those with a size smaller than our specified threshold.
2. Documents from the original source language were then translated into English using available MT system. In our case, we make use of Google Translate.
3. We then split the documents using a simple sentence splitter and filter out sentences which are not useful for the corpus, such as sentences which contain a large number of named entities or numbers.
4. Sentences which passed these filters were then paired to each sentence from the English document.
5. We count similarity score on each of these sentence pairs using Jaccard similarity measure and choose the highest scoring sentence for each of the sentences in the smaller documents.
6. If the Jaccard similarity score is higher than the defined threshold, we save this sentence pair and its score as a possible alignment. Otherwise, we disregard that sentence pair. We repeat step 5 and 6 until all sentences in the smaller documents have been paired to the highest scoring sentence of the other document.
7. We represent the comparability score of these documents as the average of these aligned sentences' scores.

By using these methods, we do not only filter documents based on the topic, but also whether there are any parts in the documents which can be aligned. This information is crucial for the next process: the phrase extraction. Even though the simple similarity scoring does not guarantee a perfect alignment of the sentences, we manage to filter out sentences which do not share any overlapping terminology. This feature works well in identifying similar sentences, unfortunately it also requires all documents to be translated to English using a fast and high quality machine translation, which is mostly not available for under resourced languages.

To avoid this problem, we implement a retrieval method which uses limited linguistic resources. This method is an adaptation of Adafre and Rijke [1] who find parallel sentences in bilingual Wikipedia documents by their anchor text information.

This method requires no linguistic resource apart from the information already available in Wikipedia. First, a bilingual lexicon is generated by extracting all Wikipedia titles which are connected by interlanguage links. We then translate all the anchor texts in source language into English using this lexicon. This lexicon will also be used to identify other parts in the documents existing in the lexicon, and translate them to English. Similarity scores are then calculated in the same manner between the English document and its translated source document. When tested in a Dutch-English environment, this method performs with high precision [1]. On the other hand, under resourced languages have considerably fewer documents which may limit the number of anchor texts inside the document. Furthermore, some words in Dutch and English share similar terms, which could assist the retrieval process, while this is not necessarily the case for documents from these under resourced languages.

3.3 Twitter

Another potentially interesting source of comparable documents is Twitter. Twitter allows users to post short text snippets (or tweets) which may contain news events, messages to other users, or comments about particular topics or links. Although the service only started in 2006, the use of Twitter has increased dramatically, with the number of users reaching 190 million in August 2010 [20]. As a resource of comparable documents, Twitter has several advantages. First, the messages are relatively short, which reduces issues surrounding alignment. Each tweet also contains the date of publication, enabling retrieval and alignment to be performed more accurately.

Around 10 % of tweets provide url links, which are sometimes accompanied with a short text message describing the link content or comments about it. An example of this is:

> *"Obama Wants Kids To Make Video Games [Politics]: US President Barack Obama ...http://bit.ly/cPrY4m."*

The link refers other users to the complete article of the news. Sometimes users also define their tweets by using # symbols, such as:

> *"#football HARGREAVES CLOSE TO RETURN - SURGEON: Owen Hargreaves' surgeon believes the Manchester United star is cl...http://bit.ly/dsf352"*

It is possible that the same links or topics are being tweeted in different languages. Our approach to use Twitter is to use URLs and topics as queries to retrieve comparable tweets in different language. We analysed two different retrieval methods: the first method uses English tweet as a seed to retrieve tweets from other languages, while the second method chooses a popular non English tweet and use it as a seed to find a comparable English tweet. The first method performed poorly due to the domination of English tweets in the search result, making it difficult for tweets from other languages to be retrieved. We manage to find interesting results by using the second method; an example of the retrieved tweets is shown in Table 5.

Table 5 Example of comparable tweets

Latvian tweet	
Original Text	Arkartigi interesants raksts par pasaule ietekmigaka tech portala Techcrunch raditaju un ipašnieku http://bit.ly/9nFsGa
Translated by Google	Extremely interesting article on the world's most powerful tech site Techcrunch creators and owners http://bit.ly/9nFsGa
Other retrieved English tweets	
Original Text	Interesting Inc article on Michael Arrington via @marshallk's podcast. "bust the door down and clean the mess up later" http://bit.ly/9nFsGa
Original Text	really interesting profile in Inc. magazine about Mr. TechCrunch himself - Michael Arrington http://bit.ly/9nFsGa
Original Text	Interesting article by @arrington on how he works. Except you can have more than 2 monitors with a Mac - http://bit.ly/9nFsGa

Our initial analysis found that there exist bilingual comparable tweets. However, there are several problems in retrieving tweets of under resourced languages. First of all, based on [13], over 50 % of tweets are in English, while the rest are in Japanese, Malay, Indonesian, and major European languages, such as Spanish, French and German. The use of ACCURAT languages in Twitter is relatively unpopular, which limits the available data retrieved from this source. Secondly, it was more difficult to identify language in Twitter as language identifier does not perform reliably on short texts. We also find that the same URL is mostly tweeted in that particular language only, which again limits the number of bilingual tweets found. Nevertheless, with the popularity of Twitter which is increasing rapidly, there is a possibility that tweets from these countries will also increase in the future. We plan to do further analysis once we have a bigger dataset for these language pairs.

4 Evaluation

At the end of the retrieval process we will have gathered pairs of documents, sentences or fragments, which were considered to be comparable by our retrieval methods. The next step will be to assess the accuracy of these methods to retrieve comparable texts. In this phase, we focus on the comparability level of the retrieved documents rather than the effect of these documents to an MT system. To decrease the judgment effort, we plan to evaluate only a sample of the retrieved text pairs. We describe the methods to choose the text pairs to be judged in Sect. 4.1. An online tool is under development and will be used by assessors to define comparability level of the chosen text pairs. The online judgment tool is described in Sect. 4.2 and both the classifier and assessment tool will be used iteratively as described in Sect. 4.3.

4.1 Classifier

As an evaluation method, we built a classifier which was trained using previously assembled comparable corpora. These Initial Comparable Corpora, later referred to as ICC, contains comparable documents of at least one million words of each language. The search was performed using semi automatic retrieval methods and each of the retrieved document pairs was annotated with its corresponding comparability level: parallel, strongly comparable or weakly comparable. Since the training data does not include non comparable document pairs, we created non comparable pairs by pairing documents from different domain or genre.

4.1.1 Features/Criteria of Comparability

We identified various features which are useful in identifying comparability level of a document pair as shown in Table 6. These features are divided into two categories: language-dependent features which need some translation methods or other linguistic knowledge, and language-independent features.

To test the performance of our classifier, we focused on Greek-English (EL-EN) corpora by using Google Translate to translate all the Greek documents into English. Google Translate is not expected to perform well on these under resourced languages and domains, nevertheless it can still be considered of better quality compared to bilingual dictionaries or MT based on our parallel corpora. Thus, the results of our classification can then be considered a realistic upper bound. The extracted language-dependent features include relative word overlap (the number of common unique words in the source and target documents divided by the number of unique words in the source document), relative stem overlap (the same as the relative word overlap apart from the fact that words were first stemmed by Porters Stemmer [21], and cosine similarity in the document vector space [18, 19]. In the vector space documents are represented as a t-dimensional vector where t is the number of words

Table 6 Features or criteria of comparability

Language-dependent features
1. Word/Stem overlap (relevant to source language)
2. Cosine similarities on word/stem occurrence
3. Cosine similarities on word/stem TF
4. Cosine similarities on stemmed word TF-IDF
5. Cosine similarities on stemmed bi-gram and tri-gram TF
Language-independent features
1. Out-link overlap
2. URL character overlap
3. URL number of slash difference
4. Image link word overlap
5. Image link filename overlap

(or stems) in the entire corpus of source (English) and translated target (Greek) documents. The vectors can be binary (1 if a word/stem is present in a document and 0 otherwise), include term (words or stems) frequencies (TF) or term frequencies weighted by the importance of a word in the corpus (TF-IDF). The cosine similarity of bi-gram and tri-gram TF vectors was also computed. The extracted language-independent features include out-link overlap (the number of common out-links in the two documents), image links overlap (the number of common image source URL in the two documents), URL level overlap (the difference of the number of slashes in the URL paths of both documents) and URL character overlap.

After extracting these features for all the document pairs, we trained a classifier using ECOC (Error Correcting Output Codes) [7] in order to predict the comparability level of newly retrieved document pairs. When evaluated using fivefold-cross-validation, our classifier showed promising results in identifying comparability levels of the document pairs, with precision scoring above 90 % for each comparability level.

We used our classifier to evaluate the comparability levels of the automatically retrieved news corpora mentioned in Sect. 3.1. The results are found to be promising with most of retrieved documents judged as strongly comparable as shown in Table 7.

4.2 Assessment Tool

To verify the classifier performance, we are currently preparing the assessment tool for the assessors. The retrieval methods described in Sect. 3 will result in pairs of segments of different granularity. For example, retrieval of news articles using named entities will find documents of the same topic. On the other hand, retrieval of Twitter or other pages using the anchor methods will find comparable fragments or sentences instead. For the sake of simplicity, we refer to these different granularities as segments. Given a pair of segments, assessors will be asked a series of questions to determine the comparability level. We will use this data to evaluate the classifier and improve its performance.

Table 7 Predicted comparability level of retrieved documents

Language	Non Comp. (%)	Weakly C. (%)	Strongly C. (%)	Parallel (%)
EL-EN	2.51	0.01	97.24	0.24
ET-EN	15.94	0	82.84	1.21
HR-EN	11.88	0	80.15	7.96
LV-EN	16.29	0	64.64	19.07
LT-EN	2.06	0	85.20	12.74
RO-EN	1.90	0	98.06	0.05
SL-EN	1.28	73.38	14.17	11.17

4.3 An Iterative Process

Having described the classification/ranking of document pairs by the retrieval techniques and the judging process that will be undertaken by the partners, here we illustrate an iterative process that could bootstrap both the quality of the classifier and the quality of the returned corpus.

- Ranking the documents: The aforementioned classifier, trained over the ICC can be used to classify document pairs returned by the retrieval methods described in previous sections. Each document pair can be then ranked based on the probability of being parallel, strongly comparable, weakly comparable or non comparable, respectively, with the documents with the highest probability of being parallel ranked on the top of the list and the ones with highest probability of being non-comparable ranked at the bottom.
- Sampling in a top-heavy manner: Given that assessing the comparability of all the returned document pairs requires extensive human effort, we propose to only assess the quality of a small sample of document pairs. The sampling method that could be used here is stratified sampling, with documents towards the top of the rankings (i.e. the ones that are more likely to be highly comparable) having higher probabilities of being sampled than the ones towards the bottom of the ranking. Measures, such as accuracy or precision at a certain cut-off along with information retrieval measures such as average precision, normalised discounted cumulative gain etc. can be statistically inferred.
- Re-train the classifier based on the results: Sampled document pairs are given to the human assessors to annotate them with respect to their comparability. After getting back the comparability grades of the deferent judged document pairs we can use these fresh data to re-train the classifier, re-classify/-rank document pairs - presumably with higher accuracy - and eventually bootstrap both the performance of the classifier and the quality of the produced comparable corpus.
- After the iteration process finished, we aim to have a fully trained classifier, which has high precision in predicting comparability levels between documents. This classifier will then be used as the criteria to evaluate all the retrieved comparable documents.

5 Conclusion

In this chapter, we discussed our work on developing retrieval methods to collect comparable documents from Web sources for under resourced languages. We identify three different Web sources and develop the appropriate retrieval methods to gather these comparable documents. News documents are retrieved using date and named entity which was translated using available online resources. We make use of the richness of interlanguage links in Wikipedia to retrieve comparable documents and perform several filters to filter out the irrelevant documents. Our method on retrieving tweets from Twitter involve the use of URL and topic as query. By using these methods, we managed to get a decent size of data from news and Wikipedia. Unfortunately, the use of under resourced languages in Twitter is less popular, and this has caused problems in retrieving high quality data.

We also developed methods to evaluate the retrieved documents by building a classifier to assess comparability levels of these automatically retrieved documents. Extraction of language dependent and language independent features was performed on the Initial Comparable Corpora (ICC) and these data were used as training data for the classifier. When evaluated using fivefold-cross-validation, the classifier managed to identify comparability levels of document pairs with a high accuracy. We used the classifier to evaluate the news corpora and found that over 85 % documents are found to be strongly comparable.

Our future work involves developing reliable independent features to extract comparable segments from Wikipedia and evaluate the results. We plan to compare the quality of retrieved documents found in different Web sources. We are also developing assessment tool to gather judgment from the assessors on a subset of document pairs. The judgment information will be used to evaluate the classifier's performance and improve the training data.

Acknowledgments The project has received funding from the ACCURAT Project, European Community's Seventh Framework Programme (FP7/2007-2013) under Grant Agreement no 248347.

References

1. Adafre, S.F., de Rijke, M.: Finding similar sentences across multiple languages in wikipedia. In: Proceedings of the 11th Conference of the European Chapter of the Association for Computational Linguistics, pp. 62–69 (2006)
2. Appelt, D.: An introduction to information extraction. Artif. Intell. Commun. **12**(3), 161–172 (1999)
3. Argaw, A.A., Asker, L.: Web mining for an Amharic-English bilingual corpus. In Proceedings of 1st International Conference on Web Information Systems and Technologies (WEBIST 2005), Miami, USA (May 2005)
4. Baroni, M., Bernardini, S.: Bootstrapping corpora and terms from the web. In Proceedings of LREC (2004)
5. Cavnar, W.B., Trenkle, J.M.: N-gram-based text categorization. In Proceedings of SDAIR-94, 3rd Annual Symposium on Document Analysis and Information Retrieval, pp. 161–175 (1994)
6. Chakrabarti, S.: Mining the Web: Discovering Knowledge from Hypertext Data, Science and Technology Books (2002)
7. Dietterich, T., Bakiri, G.: Solving multiclass learning problems via error-correcting output codes. J. Artif. Intell. Res. **2**(1), 263–286 (1995)
8. Do, T., Le, V., Bigi, B., Besacier, L., Castelli, E.: Mining a comparable text corpus for a Vietnamese-French statistical machine translation system. In: Proceedings of the Fourth Workshop on Statistical Machine Translation, pp. 165–172. Association for Computational Linguistics (2009)
9. Fung, P., Cheung, P.: Mining very non-parallel corpora: parallel sentence and lexicon extraction vie bootstrapping and EM. In: EMNLP, pp. 57–63 (2004)
10. Ghani, R., Jones, R., Mladenic, D.: Building minority language corpora by learning to generate web search queries. KAIS Knowl. Inform. Syst. **7**(1) (2005)

11. Grishman, R., Sundheim, B.: Message understanding conference-6: a brief history. In: Proceedings of the 16th International Conference on Computational Linguistics, Copenhagen (June 1996).
12. Hassan, A., Fahmy, H., Hassan, H.: Improving named entity translation by exploiting comparable and parallel corpora. In Proceedings of the 2007 Conference on Recent Advances in Natural Language Processing (RANLP), AMML Workshop (2007)
13. http://techcrunch.com/2010/02/24/twitter-languages/. Accessed 1 April 2011
14. Mohammadi, M., GhasemAghaee, N.: Building bilingual parallel corpora based on wikipedia. In: Proceedings of Second International Conference on Computer Engineering and Applications, vol. 2, pp. 264–268 (2010)
15. Munteanu, D., Marcu, D.: Improving machine translation performance by exploiting comparable corpora. Comput. Linguist. **31**(4), 477–504 (2005)
16. Munteanu, D. S., Fraser, A., Marcu, D.: Improved machine translation performance via parallel sentence extraction from comparable corpora. In: HLT-NAACL, pp. 265–272 (2004)
17. Resnik, P.: Mining the web for bilingual text. In: Proceedings of the 37th Annual Meeting of the Association for Computational Linguistics on Computational Linguistics, pp. 527–534, Morristown, NJ, USA. Association for Computational Linguistics (1999)
18. Salton, G., McGill, M.J.: Introduction to Modern Information Retrieval. McGraw-Hill, New York (1983)
19. Salton, G., Wong, A., Yang, C.S.: A vector space model for automatic indexing. Commun. ACM **18**(11), 613–620 (1975)
20. Schonfeld, E.: Costolo: Twitter now has 190 million users tweeting 65 million times a day. (2010). http://techcrunch.com/2010/06/08/twitter-190-million-users/ Accessed 1 September 2010
21. Sparck-Jones, K., Willet, P.: Readings in Information Retrieval. Morgan Kauffmann, San Francisco (1997)
22. Steinberger, R., Pouliquen, B., Ignat, C.: Navigating multilingual news collections using automatically extracted information. J. Comput. Inform. Technol. **13**(4), 257–264 (2005)
23. Talvensaari, T., Pirkola, A., Järvelin, K., Juhola, M., Laurikkala, J.: Focused web crawling in the acquisition of comparable corpora. Inform. Retr. **11**(5), 427–445 (2008)
24. Zhang, Y., Wu, K., Gao, J., Vines, P.: Automatic acquisition of Chinese-English parallel corpus from the web. In: Proceedings of 28th European Conference on Information Retrieval. ECIR '06 (2006)

Measuring the Distance Between Comparable Corpora Between Languages

Serge Sharoff

Abstract The notion of comparable corpora rests on our ability to assess the difference between corpora which are claimed to be comparable, but this activity is still art rather than proper science. Here I will discuss attempts at approximating the content of corpora collected from the Web using various methods, also in comparison to traditional corpora, such as the BNC. The procedure for estimating the corpus composition is based on selecting keywords, followed by hard clustering or by building topic models. This can apply to corpora within the same language, e.g., the BNC against ukWac as well as to corpora in different languages, e.g., webpages collected using the same procedure for English and Russian.

1 Introduction

The British National Corpus (BNC) has been collected in the beginning of the 1990s using various written and spoken sources from the 1970–1980s. It aims at representing modern English, but in many respects it is outdated and it does not cover many domains. The web is huge, it is easy to collect data from it either by using search engine queries [3, 16] or crawling respective websites [9]. However, once we have a corpus, we still do not know its composition, e.g., the proportion of webpages for medical doctors and patients in the corpus of [3], or to what extent the BNC is similar to a web corpus.

The problem of not knowing the content gets another dimension when we use comparable corpora, i.e., two corpora which are claimed to be similar in one aspect or the other. If comparable corpora have been collected in their own ways, e.g., from different websites or for different languages, they might be drawn from different

S. Sharoff (✉)
Centre for Translation Studies, School of Modern Languages,
University of Leeds, Woodhouse Lane, Leeds LS2 9JT, UK
e-mail: s.sharoff@leeds.ac.uk

S. Sharoff et al. (eds.), *Building and Using Comparable Corpora*,
DOI: 10.1007/978-3-642-20128-8_6, © Springer-Verlag Berlin Heidelberg 2013

distributions. For example, within the TTC project [5] our aim is to explore the possibility of mining terminological resources from specialised comparable corpora. For this task, we used parallel seeds to collect corpora on comparable topics by retrieving webpages returned in response to queries containing identical or nearly identical terms in several languages (below shown for English, Chinese and Russian):

fossil fuel	**化石燃料**	**ископаемое топливо**
power station	**发电厂**	**электростанция**
hydroelectricity	**水力发电**	**гидроэнергетика**
photovoltaics	**太阳能光伏**	**фотоэлектричество**

However, the crucial question is: do we get comparable pages by sending comparable queries? In a similar way, we would like to estimate the difference between web-derived corpora obtained by crawling respective domains like ukWac and itWac [4].

One approach to comparing corpora across languages is by translating the features obtained from their documents [1]. Li and Gaussier produced a metric for assessing the comparability of multilingual collections [13], see also their paper in this volume. Their similarity measure is based on the proportion of words for which a translation can be found in the opposite part of a comparable corpus. Their approach has been tested on a parallel corpus (Europarl) with added noise from other text types to decrease the degree of parallelism. Babych and Hartley also used parallel corpora from the TAUS Data Association[1] to determine the distance between corpora by measuring the difference in the frequencies of the translations of the top 500 words for each corpus [2].

The study reported in this paper is based on the same idea of a lexicon overlap measure via a dictionary, while the difference lies in (1) using the unsupervised methods for keyword selection, (2) clustering and topic modeling to assess and compare corpora with unknown composition, and (3) estimating the contents of really comparable rather than parallel corpora in realistic settings. The main contribution is that the procedure gives an answer to the question how similar large webcorpora such as ukWac or itWac [4] are to each other and to the BNC, as well as to the question about comparable corpora collected by using comparable queries, such as listed above.

2 Methodology

2.1 *Corpora Used for Analysis*

The BNC classifies its documents using a complex classification scheme [12], which includes such categories as

[1] http://www.tausdata.org/

Table 1 Corpora used in this study (sizes given in tokens and documents)

Corpus	BNC	ukWac	itWac	EN-Energy	RU-Energy	ZH-Energy
Tokens	111,246,939	2,119,891,296	179,512,658	7,505,765	7,766,462	12,431,752
Documents	4,054	2,541,926	175,646	5,762	5,126	3,287

- domains (eight labels in total: natsci, appsci, socsci, belief, imaginative, leisure, business, world affairs);
- genre (seventy labels in total, W.advert, W.newsp.brdsheet.national.arts, etc.)
- information about the audience, author, publication medium, etc.

In spite of the complexity of the classification system, it does not cover many substantial differences between the BNC texts, e.g., a text from socsci can be from the subdomain of linguistics or history, the domain of world affairs covers both local British and international politics. The composition of the BNC can be compared to ukWac and itWac [4], large corpora crawled from the .uk and .it domains to represent respectively British English and Italian (with subsequent language filtering). Another task is to compare the composition of specialised corpora collected using comparable seeds (Table 1).

In a somewhat similar study [17] the composition of I-En, I-Ru and ukWac was analysed in terms of their genres. Two separate sets of the genre categories of the BNC and RNC (Russian National Corpus) were mapped to metagenre labels, such as instruction or regulation, and these corpora were used to train supervised machine learning models for genre classification (using part-of-speech trigrams as the features for classification). The resulting classifiers were applied to the Web corpora. That study did show the similarities and differences between the metagenres of traditional corpora and those collected from the Web, but it was limited to a small set of eight metagenres. The composition of the topics on the Web remains to be investigated.

2.2 Keyword Selection

The procedure for selecting the keywords per each document was based on the Log-likelihood (LL) score [15]. Like the commonly used tf*idf score the procedure is language-independent, but unlike tf*idf it takes into account the relative frequency of a term in a document against the reference corpus as well as the absolute number of its occurrences as the evidence of its statistical significance. The LL-score also allows setting a straightforward threshold using the value of 10.83 ($p = 0.001$) or 15.13 ($p = 0.0001$), see [15] for more information on the rationale and the thresholds. An example of extracted keywords is shown below[2]:

[2] http://www.comp.leeds.ac.uk/biosystems/neuroscience.shtml from the time it was collected for ukWac.

LL-score	N	Keyword
126.07	7	Womble
101.47	9	neural
91.26	6	elegans
62.55	13	model
47.22	6	simulation
46.47	10	network
39.59	3	locomotion
36.66	3	biologically
26.71	3	constrain
21.95	3	nervous
21.78	3	cognitive
	...	

They are all useful for describing this individual webpage, but some of them (*Womble, elegans*) are too specific for the purpose of clustering webpages of a general-purpose corpus: *Womble* is a keyword for only 22 pages in ukWac, so it is less useful for finding its clusters. Also a corpus of the size of ukWac generates 89,394 unique keywords for clustering its 2,541,926 documents, causing dimensionality problems. Restricting the keyword lexicon down to the most common words[3] limits the ability to select pages with specific topics, so a simple approach used in this study was to reduce the amount of keywords down to the 10,000 words most often selected as keywords in the entire corpus. The use of the complete set of keywords was tested on the BNC and it was found to be marginally detrimental to the results of clustering. The clusters remained essentially the same, but a bit of noise was introduced, so a smaller keyword set was used in all experiments reported below.

2.3 Clustering Methods

For unsupervised detection of domains two scenarios were used, hard clustering and generation of topic models. Because of the need to cluster a fairly large number of webpages, the repeated bisection (RB) algorithm was used in Cluto [20], as it is quite efficient and tends to produce clearly interpretable results, cf also a study in [19]. A cluster is treated as a subcorpus and described in terms of its keywords against the entire corpus using the same LL-score as for obtaining the keywords of individual documents.

Generation of topic models is based on Latent Dirichlet Allocation (LDA), which estimates the distribution of probabilities of keywords belonging to different topics as well as the proportions of documents over the same set of topics (as in traditional clustering, the number of topics is set at the beginning of the experiment). This comes from an unsupervised procedure, in which the unknown distributions are derived in repeated Bayesian approximations from the distribution of hidden variables (like

3 As done, for example, in [10].

in Hidden Markov models). This setup helps in generalising inherent similarities between the keywords, see [6]. For each topic we get the degree of its association with documents and keywords. The advantage of topic models in comparison to RB clustering is that each document gets a proportion of its association with each topic, so that it can belong to several topics at once. However, this does not allow us to estimate the partitioning of the entire corpus into a set of identifiable segments in order to compare the relative size of each topic.

It is known from prior experience that there are more topics in the BNC than the set indicated in its classification scheme, so more than eight clusters need to be used. In all experiments with general-purpose corpora the option of 20 clusters was used. This number helped in producing interpretable models covering a diverse set of topics, while a larger number of clusters makes the comparison task inherently difficult. As discussed in [7], there is no correlation between human perception of the consistency of clusters and automatic measures (such as perplexity or I_2), so it is difficult to estimate the right amount of clusters automatically. Unfortunately, in the context of this study there was no scope for a proper human evaluation of the quality of each clustering solution for each corpus. CLUTO [20] uses I_2 as the standard objective function, which maximises the cosine similarity between each node and the centroid of the clusters it belongs to. I_2 on the BNC grew slowly when the number of clusters varied from 10 to 27 (reflecting the reduced size of each cluster), so no definite estimate towards the desired number of clusters can be made:

N	I_2	N	I_2	N	I_2	N	I_2	N	I_2	N	I_2
10	1.38	13	1.44	16	1.49	19	1.53	22	1.57	25	1.60
11	1.40	14	1.46	17	1.50	20	1.55	23	1.58	26	1.61
12	1.42	15	1.47	18	1.52	21	1.56	24	1.59	27	1.62

2.4 Cross-Linguistic Comparison

The procedure in Sect. 2.3 generates the clusters (or topics), which describe the composition of a corpus in a given language. The next step is to compare allegedly comparable corpora across languages. For this we need to match the clusters, either qualitatively by comparing their descriptive features or quantitatively by matching the translations of their features. In a similar study [2] the documents from the entire non-English parts of comparable corpora were translated into English using the Google MT server. For large corpora of the size of ukWac or I-RU, this approach is not feasible. So, the features need to be translated using a dictionary. However, there are also problems in using traditional bilingual dictionaries. First, comprehensive dictionaries are not available for many language pairs. Second, traditional dictionaries do not indicate the probabilities for choosing the translation equivalents. Finally, the procedure for mapping the features (keywords) of a language into another one runs into problems in two cases. First, there is possibility of natural variation in using terminology, so that some terms used in one document do not match its ostensibly parallel counterpart, e.g., very similar issues might be discussed under the labels of *copyright* or *IPR*, while we'd like to capture. Second, when there is a difference in the actual contents of the corpora in two languages, while the two corpora are close thematically, e.g., the recipes in English and Italian are likely to use different keywords.

The first two problems can be mitigated by using statistical dictionaries. The dictionaries used in this study were extracted from parallel corpora by Giza++ [14], with the MultiUN corpus [8] used for the English-Russian pair (230 MW, 8.2 million sentences), and Europarl [11] for the English-Italian pair (47 MW, 1.6 million sentences). The corpora have been lemmatised before extracting the equivalents in order to improve the chances of matching the variety of forms for morphologically rich languages (Italian and Russian). This results in a sparse translation matrix Tr, which for a word lists its equivalents with their translation probabilities:
Tr(**лопасть**) blade:0.778, vane:0.107, rotor:0.052
Tr(**возможность**) opportunity:0.290, possibility:0.100, capacity:0.068, possible:0.050, ability:0.023, able:0.020.[4]

The last problem can be mitigated by using the clusters, which provide a broader gist of the topics addressed and then mapping the gists rather than individual documents. The gist can be generated by either clustering or producing topic models. Even though the gists are based on the keywords and they are compared by translating them, their advantage over keyword mapping comes from the possibility of generalising the range of expressions used in individual text over the proximity to a more general field (such as property rights or food items). Formally speaking, the distance can be measured by the cosine similarity of the original English vectors E_i with the original feature vectors from other languages F_j which have been translated into English and weighted with the translation probabilities

$$\cos(E_i, F_j^t) = \frac{E_i \cdot F_j^t}{\|E_i\| \|F_j^t\|}$$

where F_j^t have been produced as

$$F_j^t = Tr \times F_j$$

3 Results

Even though corpus analysis via clustering is more important for web-derived corpora (because we do not know their exact composition), it makes sense to evaluate our methods on a corpus we know better, such as the BNC (Sect. 3.1). Then, in Sect. 3.2

[4] The Oxford Russian Dictionary, for example, translates возможность by *possibility, opportunity, means, resources*. Unlike the Giza++ dictionary, this does not cover the full range of probable translation equivalents and does not estimate their frequency.

I will compare I-RU and itWac, two general-purpose Internet corpora for Russian and Italian to ukWac [4, 16]. Finally, I will present a study of specialised comparable corpora across different languages (Sect. 3.3).

3.1 Clusters in the BNC

Table 2 lists the clusters and topic models with their keywords. The clusters produced by Cluto are numbered according to their internal consistency (the smallest number for the greatest consistency). Also, at the bottom of the list of clusters I indicate the most frequent genre categories from the BNC associated with this genre. In case the genre is not informative (W.misc), the most common domain category is given after a slash. Obviously not all cluster members share the same code, but the pattern is consistent. The following is the distribution of the BNC genre codes of documents belonging to Cluster 18 in Table 2:

N	BNC codes	N	BNC codes
424	W.fict.prose	5	W.letters.personal
49	W.misc	5	S.brdcast.discussion
43	S.oral.history	4	W.religion
29	W.fict.poetry	4	W.non.ac.medicine
26	W.biography	4	S.speech.unscripted
19	W.non.ac.soc	4	S.classroom
5	W.pop.lore	3	W.non.ac.humanities
5	W.essay.school	3	W.newsp.brdsht.social

The documents outside of W.fict.prose are still reasonably similar, as they include texts ADG (a book for teenage girls), ADM (accounts of travelling through Ireland), AP7 (experiences of old age), etc.[5]

The LDA procedure estimates the distribution of the topics for each document, so it is not possible to obtain information about the relative size of the topics and their constituent documents. In the end, they cannot be directly matched to the labels, thus the labels in Table 3 are indicative, selected on the basis of their keywords.

The two clustering methods agreed on a number of domains, such as fiction, local British and international affairs. However, for some domains (underlined in Table 2) the clusters and topic models disagreed. One considerable difference stems from the ability of topic models to differentiate between the everyday language and professional discourse (the topics labelled as business and health care). In addition to this, hard clustering did not produce clusters related to history and research, which constitute relatively small (but consistent) topics in the BNC, e.g., documents HHY, HJ0, HPN, etc, containing research project applications and texts on project management. Also, LDA generated an inconsistent topic, which combined references to sport clubs with other local British texts, possibly generalising on the membership of their keywords to place names in the UK.

[5] The ids are from the BNC index [12].

Table 2 Clusters in the BNC

0	3.3 %	Inc, Corp, software, user, Unix, system, lifespan, IBM, module, application, version, package, file
1	1.5 %	studio, video-tape, speaker, voice, read, over, report, say, yesterday, male, police, Oxford, Swindon
2	1.7%	pollution, environmental, conservation, nuclear, emission, waste, energy, forest, ozonosphere
3	7.6 %	Yeah, oh, get, yeah, well, know, yes, go, No, na, think, there, cos, gon, what, no, just, like, say
4	3.7 %	player, win, game, Cup, season, club, team, play, match, goal, championship, score, League, ball
5	1.7 %	aircraft, engine, railway, station, car, fly, train, pilot, squadron, locomotive, steam, line, crew
6	2.0 %	patient, disease, treatment, study, cell, infection, gastric, health, acid, clinical, concentration, care
7	2.8 %	God, Jesus, church, Christian, Christ, faith, king, bishop, prayer, gospel, spirit, pope, Holy
8	3.4 %	school, award, teacher, student, education, pupil, course, curriculum, research, study, ref, subject
9	6.6 %	okay, right, yeah, get, what, so, yes, if, just, well, go, think, know, actually, mean, there, oh
10	9.9 %	government, Minister, party, political, election, Soviet, labour, state, country, president, Prime
11	4.1 %	court, case, Act, law, defendant, plaintiff, contract, section, person, appeal, any, solicitor, under
12	4.9 %	music, film, guitar, band, play, song, album, Eliot, bass, movie, sound, musical, pop, actor
13	3.5 %	cell, gene, DNA, protein, energy, equation, sequence, molecule, temperature, surface, particle
14	6.5 %	company, market, rate, price, share, cost, profit, tax, business, firm, UK, investment, bank
15	4.6 %	Darlington, local, Council, council, housing, pension, councillor, authority, scheme, service, area
16	5.4 %	art, painting, century, artist, exhibition, Edward, building, William, museum, town, king, church
17	5.5 %	social, language, theory, word, may, text, behaviour, process, individual, information, meaning
18	15.5 %	her, him, me, my, say, look, eye, smile, back, go, feel, door, tell, hand, know, think, like, could
19	5.6 %	fish, water, your, plant, knit, bird, food, colour, breed, tank, can, horse, leaf, use, dry, egg, specie

0: W.non.ac.tech.engin; 1: W.news.script; 2: W.misc/W.app.science; 3: S.conv; 4: W.pop.lore/W.newsp.brdsht.nat.sports; 5: W.misc/W.leisure (texts for railway/aircraft enthusiasts); 6: W.ac.medicine; 7: W.religion; 8: W.misc/W.soc.science; 9: S.consult; 10: W.non.ac.polit.law.edu,W.newsp.brdsht.nat.reports; 11: W.ac.polit.law.edu; 12: W.newsp.brdsht.nat.arts,W.pop.lore; 13: W.non.ac.nat.science,W.ac.nat.science; 14: W.commerce; 15: S.meeting; 16: W.non.ac.humanities.arts,W.biography; 17: W.ac.soc.science; 18: W.fict.prose; 19: W.pop.lore,W.instructional/W.leisure

Table 3 Topic models in the BNC

news	police yesterday voice court read john hospital studio darlington speaker street road officer today claim britain crime council male murder charge
sport1	game club team player season match england league goal score race ball minute manager final john football championship sport test champion
fict1	door light smile across foot voice watch moment himself body sound stop dark walk mind black pull wall happen stare catch
pol1	political labour economic class society worker britain organisation population union individual community argue industrial economy process thus period influence production authority
med1	patient care health disease treatment hospital medical cell risk drug cause doctor test blood normal rate infection symptom factor parent cancer
ac	research education award student teacher course staff training project university department subject management pupil skill develop college experience date organisation library
biz1	price rate share bank firm sale profit financial investment account income product industry sell value customer capital rise management benefit director
rail	road building hotel town railway build village mile station north train walk street aircraft engine route centre site travel south park
lang	language sense experience kind human theory particular individual behaviour subject example knowledge english nature understand reason person process text itself suggest
hist	church century john lord christian england death himself english king royal father william jesus edward land roman bishop french amongst
leis1	music film artist record painting exhibition museum sound band american collection gallery song theatre picture press audience john director style television
comp	computer user software file application product corp datum technology module version unix window network machine package database design access available disk
leis2	colour food fish plant design garden light easy cover white flower size range available shape machine knit pattern piece wall grow
spok	yeah actually right hundred okay twenty pound sort alright nice thank fifty eight probably thirty nine anyway half seven thousand remember
legal	court person section order contract legal rule apply require authority appeal decision term claim property shall duty matter action provision application
pol3	council committee authority minister secretary labour community meeting office house national health scheme chairman county agree government scotland proposal matter district
fict2	mother love girl father friend herself remember wife door walk miss mind husband morning stay evening someone speak sister baby dress
envt	animal plant bird specie forest energy land environmental environment produce food million fish grow waste pollution soil scientist nuclear chemical site
med2	cell figure model datum value thus method function contain section sequence type produce structure test process surface solution occur position describe
pol2	minister president soviet national election foreign force political leader international military party union march economic europe april june official prime european

3.2 Comparing Internet Corpora to the BNC

The same procedure was applied to ukWac and generated the clusters and topics reported in Table 4. The clusters offer a quick way into comparing the composition of ukWac to that of the BNC. For instance, there is a cluster without specific keywords (Cluster 0 in ukWac). This closely corresponds to the fiction cluster in the BNC (Cluster 18 in Table 2). However, unlike the BNC the ukWac cluster contains little fiction and mostly consists of diary-like blogs, tabloid columns on everyday topics and chats.

Domains that reasonably match both the BNC and ukWac include political news, sports, health, business and religion. The differences concern a broader collection of topics in ukWac within each domain, as well as the presence of more modern texts. For example, clustering of ukWac detects two clusters with keywords related to computing, one of which (Cluster 7) is quite similar to Cluster 0 in the BNC. Another computing cluster of ukWac (15) belongs to the field of web-based communications (the longer list of its keywords also includes *HTML, Internet, click, Google*, etc) (Table 5).

RB clustering in Cluto seems to be considerably more efficient on a large corpus than LDA. Clustering of the entire ukWac took 1,494 s, while LDA was not able to deal with a selection of more than 500,000 documents from ukWac (on a computer with 4 GB memory), and producing topic models even for this subset took 4,896 s (clustering of the same subset with Cluto took 304 s).

Even though the clusters and the topic models in Table 4 are drawn from two slightly distributions, there is a broad similarity between their results. The clusters of music, movies, health, computing, communication, food and gardening, religion, politics, travel and business have compatible keywords.

In the case of topic modeling, it is also possible to estimate the distance of each of the topics generated for ukWac to the BNC topics in order to estimate the relative distance between the topics of a web corpus against an established reference one. The following is the list of topics in ukWac with their closest match in the BNC, sorted by the distance score:

Label	ukWac	BNC	cos
Legal	1	14	0.746
Politics	4	19	0.740
Locations	7	7	0.682
Diary/fict	0	16	0.599
Religion	9	9	0.598
	...		
Academic	13	18	0.201
Sports	3	7	0.179
Music	12	16	0.157
Online	18	18	0.115
Forum	19	19	0.061

Table 4 Keywords clusters and topic models in ukWac

0	12.2%	I, her, he, she, my, his, me, him, i, it, do, say, PM, post, man, go, have, love, think, but
1	7.5%	road, mile, town, Canal, park, Road, walk, route, fn, Park, Street, Museum, village, north
2	2.5%	game, poker, player, ball, play, Sudoku, puzzle, sudoku, score, Games, win, goal, casino, match, tournament
3	4.6%	school, teacher, education, pupil, learn, skill, learning, child, student, training, teaching, learner
4	5.8%	water, cleared, colour, dive, light, surface, diver, valve, boat, wheel, wall, battery, bike, engine
5	10.0%	sector, organisation, local, business, development, management, sustainable, environmental
6	5.6%	student, University, module, research, study, course, academic, Studies, degree, science, graduate
7	6.3%	datum, system, software, model, use, computer, network, data, technology, user, NUN, solution
8	1.4%	insurance, mortgage, loan, property, Insurance, auto, quot, lender, rate, insurer, Agents
9	2.1%	God, Jesus, Christ, Christian, church, Lord, he, Church, Bible, his, sin, faith, prayer, verse, Him
10	3.1%	patient, care, NHS, health, nurse, hospital, clinical, nursing, mental, medical, service, Trust, Nursing
11	3.5%	plant, fish, bird, garden, flower, tree, food, specie, fruit, wine, seed, vegetable, cook, sauce, soil, sugar
12	3.1%	disease, drug, treatment, cancer, patient, cell, blood, infection, symptom, therapy, animal, gene, risk
13	3.2%	club, race, season, win, League, Cup, team, player, championship, match, lap, football, Championship
14	7.0%	pension, government, Committee, Labour, union, that, political, shall, Minister, vote, member
15	5.4%	file, search, user, text, server, use, Windows, web, page, site, library, directory, Web, browser
16	5.9%	customer, company, your, you, payment, business, sale, product, card, any, charge, market, service, price
17	3.5%	hotel, bedroom, room, holiday, beach, accommodation, apartment, restaurant, bathroom, cottage
18	4.8%	music, song, band, album, guitar, sound, musical, dance, vocal, gig, jazz, bass, track, play, concert
19	2.4%	film, movie, cinema, camera, DVD, comedy, actor, star, Jolen, scene, Hollywood, character

For example, the highest cosine similarity score for Topic 1 in ukWac was Topic 14 in the BNC with a fairly large value of similarity between their keywords. On the other hand, the closest match for Topic 19 in ukWac was Topic 19 in the BNC, but the cosine similarity value in this case is still low.

The match between the majority of the BNC topics and ukWac is reassuring for the prospects of using ukWac as a reference corpus for linguistic research using larger

Table 5 Topic models in ukWac

fict	being something never always quite away sure though thought best enough night ever found rather anything left someone fact actually love
legal	shall court section legal person committee whether required office general agreement following authority notice decision apply being appropriate period personal contract
biz	management technology industry market business marketing network customer mobile quality working software manager training media corporate digital client leading commercial security
sports	game club season football race poker match league player sport ball final england best side goal racing playing world half city
pol1	international political british labour party state european united britain europe iraq national trade union states minister world military security american power
med1	care training health young education working social national advice safety trust learning service practice centre community police mental disabled hospital charity
envt	planning government environmental environment management national transport waste review energy sector funding strategy future economic sustainable impact consultation quality housing development
loc	road north water south river park west along species east left bridge station railway town land fish near route village found
comm	insurance credit property money rate financial income bank market price card loan pension cent mortgage value investment account house capital cash
relig	church jesus christ christian lord being human word faith upon love spirit shall bible nature true religious power holy truth death
design	water light power energy surface colour equipment fire control space speed size engine standard model quality front unit type temperature black
travel	hotel room garden house travel food holiday accommodation city centre airport wine parking bedroom kitchen town park property best floor restaurant
music	music film band theatre album sound series dance review story radio song video best festival love performance movie artist stage musical
acad	learning education ac teaching science research language college school academic knowledge social practice assessment degree english department studies analysis international understanding
med2	treatment disease food medical cancer risk clinical blood body patient animal drug health medicine pain hospital human control found skin heart
news	news june july march september october centre april november august meeting january february city december friday conference monday sunday saturday west
hist	john house century church royal king william history england james english thomas museum later george early british robert mary david henry
comp	data file software user computer server text windows program version code type value image screen click select control html object function
online	online website search email click html library internet guide information news download please address content index text resource section info send
forum	php forum id date message post asp view index member joined location author subject page thread from quote

Table 6 Keywords in clusters of itWac

0	2.59 %	film, cinema, regista, attore, personaggio, regia, scena, cinematografico, protagonista, pellicola, cast, storia
1	1.07 %	vino, olio, cucchiaio, uovo, burro, cuocere, pasta, piatto, latte, zucchero, formaggio, pepe, farina, cucina
2	11.53 %	mio, mi, che, suo, non, essere, dire, avere, io, uomo, tuo, fare, lui, quello, ti, ma, occhio, cosa, amore
3	13.99 %	articolo, numero, comma, decreto, legge, cui, previsto, lavoro, relativo, presente, servizio, lavoratore
4	2.37 %	musica, disco, canzone, album, musicale, brano, band, concerto, rock, suonare, chitarra, musicista, cd
5	3.19 %	scuola, scolastico, docente, insegnante, alunno, classe, formativo, didattico, formazione, istruzione, educativo
6	1.71 %	file, server, Windows, utente, software, Microsoft, versione, programma, web, browser, utilizzare, Linux
7	7.39 %	mi, ciao, io, non, messaggio, ti, mio, avere, fare, se, inviare, ma, me, scrivere, dire, sapere, cosa, tuo, Posted
8	3.86 %	chiesa, Dio, Ges, Cristo, cristiano, santo, papa, fede, suo, cattolico, vescovo, religioso, uomo, padre
9	6.79 %	articolo, emendamento, commissione, legge, esame, numero, relatore, corte, sentenza, comma, giudice
10	4.27 %	squadra, gioco, giocatore, gara, partita, giocare, atleta, campionato, calcio, vincere, gol, atletico, tifoso, sport
11	3.01 %	paziente, malattia, cellula, medico, farmaco, terapia, clinico, medicina, tumore, patologia, salute, sanitario
12	5.87 %	guerra, Iraq, americano, pace, Bush, contro, palestinese, militare, terrorismo, popolo, israeliano, iracheno
13	7.56 %	politico, governo, partito, europeo, paese, presidente, politica, Berlusconi, parlamento, maggioranza, voto
14	3.91 %	banca, euro, mercato, consumatore, milione, prezzo, aumento, miliardo, sindacato, paese, lavoratore, Cgil
15	3.89 %	teatro, arte, mostra, artista, opera, libro, spettacolo, romanzo, poesia, museo, autore, storia, scena, artistico
16	3.20 %	Internet, rete, utente, software, prodotto, tecnologia, sito, servizio, cliente, web, elettronico, azienda, digitale
17	6.67 %	universit?, ricerca, scientifico, studio, corso, universitario, laurea, professore, scienza, concorso, biblioteca
18	3.14 %	Hotel, mare, hotel, acqua, albergo, isola, zona, situare, vacanza, spiaggia, metro, antico, citt?, km, ristorante
19	3.96 %	Provincia, assessore, comune, sindaco, provinciale, parco, regione, comunale, territorio, area, cittadino

amount of more modern data. The two most dissimilar topics at the bottom (Online and Forum) reflect the genre changes which came with the Web. The changes in the musical cluster in ukWac against those in the BNC reflect the difference in the topics most discussed in the 1980s (BNC) vs 2000s (ukWac). There is also much greater variation in the amount of musical and sport subdomains covered in ukWac, which explains the divergence from the BNC clusters.

Table 6 lists the clusters generated for itWac, an Italian web-corpus collected using the same compilation procedure as ukWac. It again showed broad compatibility with ukWac. The domains of itWac pages include film (Cluster 0), food (1), blogs and chats (2 and 7), legal texts (3,9), music (4), education (5 and 17), computers (6), religion (8), sports (10), health (11), politics (12 and 13), business (14), culture (15), communication (16), travel (18) and local news (19). Some differences between the keywords in itWac and ukWac are related to the realities in individual countries (*Berlusconi* vs *Labour*). In ukWac there is a Do-It-Yourself cluster (4) and a cluster on NHS (10), while itWac contains two legal clusters (3 and 9), their longer keyword lists reveal that Cluster 9 primarily contains texts on criminal cases, while Cluster 3 is about commercial and civil legislation.

3.3 Comparing Comparable Corpora

Energy corpora for English, Chinese and Russian have been collected by making comparable queries. They are considerably smaller than ukWac (less than 10 million words). It was expected that we can find fewer subdomains in them. The I_2 value was again not indicative, so the experiment was set for 15 clusters. As in the previous experiments both clustering and topic models were used. However, on a smaller corpus hard clustering produced less clearly interpretable results (also probably because of the difficulty in clear separation between relatively similar topics), while LDA produced fairly reasonable models listed in Table 7. The analysis of topic models contradicted the original assessment made on the basis of term lists extracted from the corpora using the standard BootCat procedure [3][6]:

7467 renewable energy	**1508 возобновлять энергия**
3127 greenhouse gas	**1401 парниковый газ**
3049 natural gas	**2106 природный газ**
2320 solar energy	**2274 солнечный энергия**
1994 carbon dioxide	**1124 углекислый газ**
1920 solar cell	**2710 солнечный батарея**
1529 fuel cell	**862 топливный элемент**
	...

Given the overlap between the terms with similar frequencies and the fact that the corpora were collected using a parallel set of keywords, it was expected that their content is sufficiently similar, and overall the corpora are good candidates for extracting and aligning term lists. However, LDA identified that the Russian corpus was contaminated with documents relating to computers (because of links to *power supply*), general news and student essays, which contain low-level introductions into a range of topics in this field (Table 8). LDA also identified some topics present in both corpora, which are less ideal for term extraction, such as information for investors and news about utility companies, forums (we can expect less consistency

[6] Lemmatisation of term elements in Russian affects the syntactic pattern of the composite term [18].

Table 7 English renewable energy topics

1	Table, Equipment, Market, Consumption, Capacity, Production, Industry, Generation, Distribution, Transformers, Sector
2	earth, atmosphere, dioxide, surface, cool, cause, warming, fluid, radiation, methane, reservoir, human, rock, warm
3	reactor, Nuclear, uranium, radioactive, barrel, mine, Uranium, fission, cent, Petroleum, reserve, billion, mining, safety
4	ocean, wave, OTEC, Ocean, tide, Intel, Tidal, Wave, marine, Hawaii, offshore, Conversion, device, surface, conversion
5	Commission, Public, shall, bill, Utility, credit, Federal, contract, FERC, eligible, District, regulation, federal, County
6	distribute, consumer, distribution, network, peak, meter, period, investment, value, datum, average, sector, reliability, factor
7	speed, rotor, blade, field, magnetic, shaft, circuit, wire, engine, transformer, phase, connect, rotate, frequency, torque
8	cogeneration, hydrogen, ethanol, engine, wood, combustion, Biomass, boiler, burn, crop, convert, landfill, residue, gasoline
9	Green, News, Hydro, Business, India, Stock, Development, Sustainable, Geothermal, Alternative, Environmental
10	river, hydroelectric, hydro, reservoir, fish, River, head, blade, hydroelectricity, Hydro, Hydropower, height, stream
11	post, read, bill, want, look, article, problem, money, green, really, question, link, warming, news, storey, idea, What, talk
12	announce, billion, Tags, investment, megawatt, Kansas, expect, sign, April, News, green, California, feed-in, release, Farm
13	sustainable, policy, economic, sector, reduction, Development, management, national, international, community
14	module, light, silicon, inverter, watt, sunlight, hour, roof, saving, device, appliance, save, film, array, tower, house, electron
15	Program, National, Department, California, Center, Association, Resources, Public, Information, Efficiency, Service, page

in terminology used there) and legal texts. In general such lists of topics were useful in cleaning the corpus to achieve its greater consistency.

4 Conclusions

Three main outcomes of this study concern:

1. the possibility of rapid unsupervised assessment of the content of large corpora (clusters still need human interpretation, but this can be done quickly as long as they are consistent);
2. the possibility of comparing the content of corpora across languages and improving comparability;
3. the difference between hard clustering and LDA.

Table 8 Russian renewable energy topics

1	новость (news), ноутбук (laptop), процессор (processor), комментарий (comment), компьютер (computer), телефон (phone), ученый (scientist), рубрика (heading), память (memory), мобильный (mobile), intel, energy
2	финансовый (financial), государство (state), инвестиция (investment), мера (measures), доля (proportion), политика (politics/strategy), правительство (government), национальный (national), потребность (demand)
3	ооо (ltd), электронный (electronic), бесперебойный (uninterruptible), продажа (sale), дизельный (diesel), купить (buy), ремонт (repair), чертеж (drawing), ибп (UPS), фирма (firm), каталог (catalogue), товар (goods)
4	вот (this/yeah), кто (who), да (yes), сделать (do), сейчас (now), говорить (say), там (there), ни (neither), ли (would), надо (should), просто (simply), сам (-self), знать (know)
5	коммунальный (utilities), жилой (resident), федеральный (federal), водоснабжение (water supply), отопление (heating), оплата (payment), жкх (utilities), помещение (room), энергосбережение (energy saving)
6	рф (Russia), рао еэс (Unified Energy System), подробно (details), инвестиционный (investment), директор (director), правительство (government), президент (president), глава (chairman), новость (news), январь (January)
7	конференция (conference), выставка (exhibition), устойчивый (sustainable), научный (scientific), наука (science), охрана (protection), энергосбережение (energy saving), энергоэффективность (energy efficiency)
8	геотермальный (geothermal), биомасса (biofuel), водород (hydrogen), отходы (waste), возобновляемый (renewable), ветроэнергетика (wind generation), топливный (fuel), сжигание (burning), вэу (wind farm), солнце (sun)
9	море (sea), глобальный (global), океан (ocean), растение (plant), лес (forest), загрязнение (pollution), потепление (warming), природа (nature), парниковый (greenhouse), животное (animal), организм (organism), планета (planet)
10	плотина (dam), добыча (production), сооружение (installation), водохранилище (reservoir), месторождение (deposit), запас (resource), очистка (purification), сырье (raw materials), сточный (sewage)
11	аккумулятор (battery), воздушный (air), насос (pump), емкость (capacity), нагрузка (load), рисунок (drawing), ротор (rotor), конструкция (design), вал (shaft), корпус (body), вращение (rotation), мм (mm), охлаждение (cooling)
12	тэц (CHP), энергосистема (grid), газотурбинный (gas turbine), когенерация (cogeneration), киловольт (kV), нагрузка (load), котельная (boiler station), генерация (generation), теплоснабжение (heating supply), выработка (production)
13	паровой (steam), котел (boiler), силовой (power), трансформатор (transformer), гост (GOST), частота (frequency), пар (steam), линия (line), power, преобразователь (converter), передача (transmission), реактивный (reactive), переменный (alternating), провод (wires)
14	русский (Russian), улица (street), война (war), язык (language), московский (Moscow), республика (republic), школа (school), век (century), ребенок (child), франция (France), текст (text), германия (Germany), игра (game), километр, книга, транспорт, история, культура,
15	реферат (essay), реактор (reactor), движение (movement), ядро (nucleus), наука (science), поле (field), атом (atom), реакция (reaction), нейтрон (neutron), физика (physics), планета (planet), магнитный (magnetic)

With respect to (1), we can use clustering to reveal more information even about a well-annotated corpus, such as the presence of a considerable cluster of texts for railway and aircraft enthusiasts in the BNC (otherwise obscurely coded as W.misc or W.leisure). It also shows broad classes of texts in a corpus of unknown composition such as ukWac or itWac.

With respect to (2), the clusters can be used to compare the content of individual corpora, but the interpretation needs to be done manually again. Intersection between the keywords of clusters in two corpora can be also done automatically.

With respect to (3), clustering is a better approach to web-sized corpora, as LDA cannot reasonably handle ukWac and it takes considerably more time on the BNC or itWac. At the same time, on smaller corpora, LDA detects models which are easier to interpret in comparison to clustering solutions.

Acknowledgments The research leading to these results has received funding from the European Community's Seventh Framework Programme (FP7/2007-2013) under Grant Agreement No 248005, project TTC.

References

1. Adafre, S., de Rijke, M.: Finding Similar Sentences Across Multiple Languages in Wikipedia. In: Proceedings 11th EACL, pp. 62–69. Trento (2006)
2. Babych, B., Hartley, A.: Meta-Evaluation of Comparability Metrics using Parallel Corpora. In: Proceedings CICLING, (2011)
3. Baroni, M., Bernardini, S.: Bootcat: Bootstrapping Corpora and Terms from the Web. In: Proceedings of LREC2004. Lisbon (2004). http://sslmit.unibo.it/~baroni/publications/lrec2004/bootcat_lrec_2004.pdf
4. Baroni, M., Bernardini, S., Ferraresi, A., Zanchetta, E.: The WaCky wide web: a collection of very large linguistically processed web-crawled corpora. Lang. Resour. Eval. **43**(3), 209–226 (2009)
5. Blancafort, H., Daille, B., Gornostay, T., Heid, U., Mechoulam, C., Sharoff, S.: TTC: Terminology Extraction, Translation Tools and Comparable Corpora. In: Proceedings EURALEX2010. Leeuwarden (5–6 July 2010)
6. Blei, D.M., Ng, A.Y., Jordan, M.I.: Latent Dirichlet allocation. J. Mach. Learning Res. **3**, 993–1022 (2003)
7. Chang, J., Boyd-Graber, J., Wang, C., Gerrish, S., Blei, D.M.: Reading Tea Leaves: How Humans Interpret Topic Models. In: Proceedings Neural Information Processing Systems (2009)
8. Eisele, A., Chen, Y.: MultiUN: A multilingual Corpus from United Nations Documents. In: Proceedings of the Seventh Conference on International Language Resources and Evaluation (LREC'10). Valletta, Malta (2010). http://www.euromatrixplus.net/multi-un/
9. Joho, H., Sanderson, M.: The SPIRIT collection: an overview of a large web collection. SIGIR Forum **38**(2), 57–61 (2004)
10. Kilgarriff, A.: Comparing corpora. Int. J. Corpus Linguistics **6**(1), 1–37 (2001)
11. Koehn, P.: Europarl: A Parallel Corpus for Statistical Machine Translation. In: Proceedings MT Summit 2005 (2005). http://www.iccs.inf.ed.ac.uk/pkoehn/publications/europarl-mtsummit05.pdf
12. Lee, D.: Genres, registers, text types, domains, and styles: clarifying the concepts and navigating a path through the BNC jungle. Lang. Learning Technol. 5(3), 37–72 (2001). http://llt.msu.edu/vol5num3/pdf/lee.pdf

13. Li, B., Gaussier, E.: Improving Corpus Comparability for Bilingual Lexicon Extraction from Comparable Corpora. In: Proceedings COLING'10. Beijing, China (August 2010)
14. Och, F.J., Ney, H.: A systematic comparison of various statistical alignment models. Comput. Linguistics **29**(1), 19–51 (2003)
15. Rayson, P., Berridge, D., Francis, B.: Extending the Cochran Rule for the Comparison of Word Frequencies Between Corpora. In: Proceedings 7th International Conference on Statistical Analysis of Textual Data (JADT 2004), pp. 926–936. Louvain-la-Neuve (2004)
16. Sharoff, S.: Creating general-purpose corpora using automated search engine queries. In: Baroni, M., Bernardini, S. (eds.) WaCky! Working Papers on the Web as Corpus. Gedit, Bologna (2006). http://wackybook.sslmit.unibo.it
17. Sharoff, S.: In the garden and in the jungle: Comparing genres in the BNC and Internet. In: Mehler, A., Sharoff, S., Santini, M. (eds.) Genres on the Web: Computational Models and Empirical Studies, pp. 149–166. Springer, Berlin (2010)
18. Sharoff, S., Kopotev, M., Erjavec, T., Feldman, A., Divjak, D.: Designing and evaluating a Russian tagset. In: Proceedings of the Sixth Language Resources and Evaluation Conference, LREC 2008. Marrakech (2008)
19. Steinbach, M., Karypis, G., Kumar, V.: A Comparison of Document Clustering Techniques. In: KDD Workshop on Text Mining (2000)
20. Zhao, Y., Karypis, G.: Empirical and theoretical comparisons of selected criterion functions for document clustering. Machine Learning **55**(3), 311–331 (2004)

Exploiting Comparable Corpora for Lexicon Extraction: Measuring and Improving Corpus Quality

Bo Li and Eric Gaussier

Abstract We study in this chapter the problem of measuring the degree of comparability of bilingual corpora, with applications to bilingual lexicon extraction. We first develop a measure which can capture different comparability levels. This measure correlates very well with gold-standard comparability levels and is relatively robust to dictionary coverage. We then propose a well-founded algorithm to improve the quality, in terms of comparability scores, of existing comparable corpora, prior to showing that the bilingual lexicons extracted from corpora enhanced in this way are of better quality. All the experiments in this chapter are performed on French-English comparable corpora.

1 Introduction

Bilingual lexicons have been considered as an important resource in multilingual natural language processing tasks such as machine translation [14] or cross-language information retrieval (CLIR) [1]. Because building bilingual lexicons manually is a costly task, and because it is difficult to develop lexicons adapted to different domains and collections, researchers have tried to automatically extract bilingual lexicons from existing corpora. As most available bilingual corpora are comparable, i.e. corpora consisting of documents in different languages covering overlapping information without being the translation of each other, several studies have focused on the extraction of bilingual lexicons from comparable corpora [2–6, 12, 16, 18, 20, 22]. The basic assumption behind most studies is a distributional hypothesis, stating that

B. Li (✉) · E. Gaussier
Laboratoire d'Informatique de Grenoble (LIG), Université Joseph Fourier
Grenoble 1 / CNRS, Grenoble, France
e-mail: bo.li@imag.fr

E. Gaussier
e-mail: eric.gaussier@imag.fr

S. Sharoff et al. (eds.), *Building and Using Comparable Corpora*,
DOI: 10.1007/978-3-642-20128-8_7,

words which are translations of each other are likely to appear in similar contexts across languages. On top of this hypothesis, researchers have investigated the use of better representations for word contexts, as well as the use of different methods for matching words across languages, based on their context representation. These approaches seem to have reached a plateau in terms of performance, and the most recent improvements reported by recent studies in these directions occur when one deals with limited and constrained evaluation sets, e.g. [22], or with specific languages, e.g. [20].

More recently, and departing from such traditional approaches, we have proposed, in [10] and [11], two approaches based on improving the comparability of the corpus under consideration, prior to extracting bilingual lexicons. These approaches are interesting as there is no point in trying to extract bilingual lexicons from a corpus with a low degree of comparability, as the probability of finding translations of any given word is low in such cases. We give here a detailed presentation of these approaches and introduce new elements regarding both the comparability measure and the process to enhance corpus comparability, showing on this latter point that it is possible to guarantee a certain degree of *homogeneity* for the improved corpus, and that this homogeneity translates into a significant improvement of both the quality of the resulting corpus and the bilingual lexicons extracted.

The remainder of the chapter is organized as follows: All the materials used in the experiments are described in Sect. 2; In Sect. 3, we introduce the comparability measure and validate its reliability on the test collection with gold-standard comparability scores; Using this measure, we develop a strategy based on a clustering process to improve the quality of a bilingual corpus, a strategy which is then evaluated in the task of bilingual lexicon extraction in Sect. 4; A comparison with previous studies with similar goals is given in Sect. 5; Lastly, Sect. 6 concludes our work. For convenience, the following discussion will be made in the context of French-English comparable corpora.

2 Resources in the Experiments

The resources described here are used in the experiments presented in the following sections. First of all, we make use of several corpora: The parallel English-French *Europarl* corpus[1] [8], the TREC[2] *Associated Press* corpus (*AP*, English) and the corpora used in the multilingual track of CLEF[3] which includes the *Los Angeles Times* (*LAT94*, English), *Glasgow Herald* (*GH95*, English), *Le Monde* (*MON94*, French), *SDA French 94* (*SDA94*, French) and *SDA French 95* (*SDA95*, French). In addition to these existing corpora, two monolingual corpora from the Wikipedia

[1] http://www.statmt.org/europarl/

[2] http://trec.nist.gov/

[3] http://www.clef-campaign.org/

Table 1 The information of the corpora used in the experiments

Name	Language	Num of docs (k)	Num of words (m)
Europarl	English	–	51
	French	–	55
Associated Press	English	243	126
Los Angeles Times	English	113	71
Glasgow Herald	English	56	27
Le monde	French	44	24
SDA French 1994	French	43	13
SDA French 1995	French	43	13
Wiki-En	English	368	163
Wiki-Fr	French	378	169

We use the sentence-aligned version of the Europarl corpus and thus the number of documents is not available here. (k = 1000, m = 1000000)

dump[4] were built. For English, all the articles below the root category *Society* with a depth less than 4 were retained. For French, all the articles with a depth less than 7 below the category *Société* were extracted. The size of the corpora here are listed in Table 1.

The bilingual dictionary used in our experiments is constructed from an online dictionary. It consists of 33 k distinct English words and 28 k distinct French words, which constitutes 76 k translation pairs. Standard preprocessing steps like tokenization, POS-tagging and lemmatization are performed on all the linguistic resources. We directly work on lemmatized forms of content words (nouns, verbs, adjectives and adverbs).

3 Comparability Measure

The notion of comparability is a loose one, and comparable corpora range from lowly comparable ones to highly comparable and parallel ones. For data-driven NLP techniques, using better corpora often leads to better results, and a measure which can capture the different comparability levels is a necessary premise to evaluate the quality of a bilingual corpus. There have been some studies (e.g. [7, 17, 19]) investigating the comparison or characterization of two monolingual corpora. We know however of no such measure for bilingual corpora, and propose in this section a comparability measure based on the expectation of finding the translation of each word in the corpus vocabulary.

[4] The Wikipedia dump files can be downloaded at http://download.wikimedia.org.

3.1 Mathematical Formulation

The intuition behind the comparability measure we consider is based on the fact that it is easier to find the translation of a word present in a bilingual corpus if the corpus has a higher degree of comparability. Thus, for the comparable corpus $\mathcal{C}$, if we consider the translation process from the source language part $\mathcal{C}_s$ to the target language part $\mathcal{C}_t$, the comparability measure M_{st} can be defined as the expectation of finding, for each source word w_s in the vocabulary $\mathcal{C}_s^v$ of $\mathcal{C}_s$, its translation in the vocabulary $\mathcal{C}_t^v$ of $\mathcal{C}_t$, which directly reflects the above intuition. As one can note, this definition assumes that one knows in advance the translations of both source and target words. Such a knowledge is derived from existing, general bilingual dictionaries, which are usually independent from the corpora considered. Let σ be a function which indicates whether a translation from the translation set $\mathcal{T}_w$ of w is found in the vocabulary $\mathcal{C}^v$ of a corpus $\mathcal{C}$, i.e.:

$$\sigma(w, \mathcal{C}^v) = \begin{cases} 1 \text{ iff } \mathcal{T}_w \cap \mathcal{C}^v \neq \emptyset \\ 0 \text{ else} \end{cases} \tag{1}$$

M_{st} is then defined as the mathematical expectation of σ:

$$\begin{aligned} M_{st}(\mathcal{C}_s, \mathcal{C}_t) &= \mathbb{E}(\sigma(w, \mathcal{C}_t^v) | w \in \mathcal{C}_s^v) \\ &= \sum_{w \in \mathcal{C}_s^v} \underbrace{\sigma(w, \mathcal{C}_t^v) \cdot Pr(w \in \mathcal{C}_s^v)}_{A(w)} = \sum_{w \in \mathcal{C}_s^v \cap \mathcal{D}_s^v} A(w) + \sum_{w \in \mathcal{C}_s^v \backslash \mathcal{D}_s^v} A(w) \end{aligned} \tag{2}$$

where $\mathcal{D}_s^v$ is the source language vocabulary of a given bilingual dictionary $\mathcal{D}$ and $\mathcal{C}_s^v \backslash \mathcal{D}_s^v$ denotes set difference. As the comparable corpus and the bilingual dictionary are independent of one another, it is natural to assume that the probability of finding the translation in $\mathcal{C}_t^v$ of a word w is the same for w in $\mathcal{C}_s^v \cap \mathcal{D}_s^v$ and in $\mathcal{C}_s^v \backslash \mathcal{D}_s^v$. This assumption amounts to:

$$\forall w \in \mathcal{C}_s^v \backslash \mathcal{D}_s^v,\ A(w) = \frac{\sum_{w' \in \mathcal{C}_s^v \cap \mathcal{D}_s^v} A(w')}{|\mathcal{C}_s^v \cap \mathcal{D}_s^v|}$$

and leads to a simplification of M_{st} as:

$$M_{st}(\mathcal{C}_s, \mathcal{C}_t) = \frac{|\mathcal{C}_s^v|}{|\mathcal{C}_s^v \cap \mathcal{D}_s^v|} \sum_{w \in \mathcal{C}_s^v \cap \mathcal{D}_s^v} A(w)$$

There are several possibilities to estimate $Pr(w \in \mathcal{C}_s^v)$ in Eq. 2. However, the presence of common words suggests that one should rely on a presence/absence criterion rather than on the number of occurrences. For example, in the *Europarl* corpus, the English word *Europe* and the French word `Europe` are very common words. It means that even if one piece of English text and one piece of French text are randomly picked from *Europarl*, we can still expect to find many translation pairs *Europe*-`Europe`. Thus, to avoid the bias common words can introduce in the comparability measure, we make use of a direct estimate: $Pr(w \in \mathcal{C}_s^v) = \frac{1}{|\mathcal{C}_s^v|}$, which leads to:

$$M_{st}(\mathcal{C}_s, \mathcal{C}_t) = \frac{1}{|\mathcal{C}_s^v \cap \mathcal{D}_s^v|} \sum_{w \in \mathcal{C}_s^v \cap \mathcal{D}_s^v} \sigma(w, \mathcal{C}_t^v)$$

This formula shows that M_{st} is actually the proportion of source language words translated in the target language part of the comparable corpus. Similarly, the counterpart of M_{st}, M_{ts}, is defined as:

$$M_{ts}(\mathcal{C}_s, \mathcal{C}_t) = \frac{1}{|\mathcal{C}_t^v \cap \mathcal{D}_t^v|} \sum_{w \in \mathcal{C}_t^v \cap \mathcal{D}_t^v} \sigma(w, \mathcal{C}_s^v)$$

and measures the proportion of target language words in $\mathcal{C}_t^v$ translated in the source language part of the comparable corpus.

The two asymmetric measures M_{st} and M_{ts} above reflect the degree of comparability when considering the translation process in only one direction. They can be combined to form a comprehensive, robust and symmetric measure M, through a linear combination:

$$M(\mathcal{C}_s, \mathcal{C}_t) = \lambda \cdot M_{st}(\mathcal{C}_s, \mathcal{C}_t) + (1 - \lambda) \cdot M_{ts}(\mathcal{C}_s, \mathcal{C}_t)$$

where λ is a weight between 0 and 1 inclusive. Since M_{st} and M_{ts} can be interpreted as the proportion of words in one language (source or target) translated into the other language in the corpus, the weight λ can be defined as the ratio of source language words present in both the corpus and the dictionary to the whole vocabulary (again present in both the corpus and the bilingual dictionary), i.e.:

$$\lambda = \frac{|\mathcal{C}_s^v \cap \mathcal{D}_s^v|}{|\mathcal{C}_s^v \cap \mathcal{D}_s^v| + |\mathcal{C}_t^v \cap \mathcal{D}_t^v|}$$

This finally leads to the following measure for the degree of comparability of two corpora:

$$M(\mathcal{C}_s, \mathcal{C}_t) = \frac{\sum_{w \in \mathcal{C}_s^v \cap \mathcal{D}_s^v} \sigma(w, \mathcal{C}_t^v) + \sum_{w \in \mathcal{C}_t^v \cap \mathcal{D}_t^v} \sigma(w, \mathcal{C}_s^v)}{|\mathcal{C}_s^v \cap \mathcal{D}_s^v| + |\mathcal{C}_t^v \cap \mathcal{D}_t^v|} \qquad (3)$$

which corresponds to the overall proportion of the words for which a translation can be found in the comparable corpus. We now present a methodology to evaluate the above comparability measures.

3.2 A Methodology for the Evaluation of Comparability Measures

We present here a general methodology to assess comparability measures. This methodology is based on three steps:

1. Constructing gold-standard comparability levels from existing parallel corpora and monolingual corpora;
2. Defining an evaluation measure to compare the performance of different comparability measures w.r.t. the gold standards built in the previous step;
3. Assessing, from an experimental perspective, the robustness of the comparability measures w.r.t. the dictionary coverage.

We now detail all these steps.

Constructing Test Corpora

We will build several test corpora with different levels of comparability. To do so, we start from the parallel corpus of which the comparability level is the highest. Some noise, with content borrowed from the same and/or different domains, is then added to the parallel corpus to reduce its comparability. We will also start from a comparable corpus, and not a parallel one, consisting of content in the same domain and reduce its quality by importing noise from corpora in different domains.

Based on the above considerations, we developed gold-standard comparability scores from the *Europarl* and the *AP* corpora. We start from the parallel corpus, *Europarl*, of which we degrade the comparability by gradually importing some documents from *AP* which cover domains different from the ones covered in *Europarl*. Three groups (G_a, G_b, G_c) of comparable corpora are built in this fashion. Each group consists of test corpora with a gold-standard comparability ranging, arbitrarily, from 0 to 1 and corresponding to the proportion of documents in "parallel" translation. The comparability measures will be examined on each of the three groups of corpora.

The first group G_a is built from *Europarl* only. First, the *Europarl* corpus is split into 10 equal parts, leading to 10 parallel corpora ($\mathcal{P}_1, \mathcal{P}_2, \ldots, \mathcal{P}_{10}$) with a gold-standard comparability arbitrarily set to 1. Then for each parallel corpus, e.g. $\mathcal{P}_i$, we replace a certain proportion p of the English part with documents of the same size from another parallel corpus $\mathcal{P}_j (j \neq i)$, producing a new corpus $\mathcal{P}'_i$ less comparable than $\mathcal{P}_i$. For each $\mathcal{P}_i$, as p increases, we obtain several comparable corpora with decreasing comparability scores. All the $\mathcal{P}_i$ and their descendant corpora constitute the group G_a. The only difference between G_b and G_a is that, in G_b, the replacement in $\mathcal{P}_i$ is done with documents from the *AP* corpus and not from another parallel corpus from *Europarl*. In G_c, we start with 10 comparable corpora $\mathcal{P}'_i$ from G_a having a comparability score of 0. They thus contain documents from *Europarl* which are not translations of each other. Then each $\mathcal{P}'_i$ is further degraded by replacing certain portions with documents from the *AP* corpus. Because of the building process used, one can note that the gold-standard comparability scores in different groups, e.g. G_a and G_c, can not be compared with each other directly.

Correlation with Gold-standard Comparability Levels

The goal here is to assess whether the comparability measures we have introduced can capture the differences in comparability introduced in the three different groups G_a, G_b and G_c. In order to quantify this, we use the Pearson Correlation Coefficient to measure the correlation between the proposed measures and the comparability scores of different corpora:

$$r = \frac{\sum_i (X_i - \overline{X})(Y_i - \overline{Y})}{\sqrt{\sum_i (X_i - \overline{X})^2}\sqrt{\sum_i (Y_i - \overline{Y})^2}}$$

where X_i denotes the comparability score provided by one measure on a given bilingual corpus and Y_i is the arbitrary comparability score assigned to this corpus in the construction process. $\overline{X}$ represents the average of X_is over all the bilingual corpora considered in G_a, G_b or G_c (and similarly for $\overline{Y}$). The results obtained are listed in the columns M_{st}, M_{ts} and M of Table 2 where English is used as the source language and French as the target language. Each column, e.g. M_{st}, corresponds to the correlation scores between the specific comparability measure M_{st} and the gold-standard comparability levels on different test corpora. From these results, one can see that, on the three groups, M performs best and correlates very well with the gold standard as the Pearson coefficient is close to 1. M_{ts} performs worst among the three measures. Figure 1 plots the measures M and M_{ts} on ten comparable corpora and their descendants in G_c with respect to their gold-standard comparability scores. One can notice from Fig. 1b that the comparability scores from M_{ts} even decrease at a certain point as the gold standard scores increase. The reason for the different performance is that the asymmetric measures M_{st} and M_{ts} are sensitive to the length of the corpus. Given a single English document and a large French document collection, it is very likely that we can find the translations for most of the English words, although the two text sets are lowly comparable. In our case, since the average sentence length in *AP* is larger than that of *Europarl*, we increase the length of the English part of the test corpora remarkably when degrading the corpora in G_b and G_c, which leads to the poor performance of M_{ts}. The length related problem can be overcome by M and we will only consider the best measure M in the following experiments. As the results show here, M was able to capture all the differences in comparability artificially introduced in the degradation process we have considered. Meanwhile one can conclude that it

Table 2 The correlation scores between the comparability measures and the gold standard

	M_{st}	M_{ts}	M	M_c
G_a	0.897	0.770	0.936	0.972
G_b	0.955	0.190	0.979	0.978
G_c	0.940	−0.595	0.960	0.990

The column M_c is listed for the results in Sect. 3.3

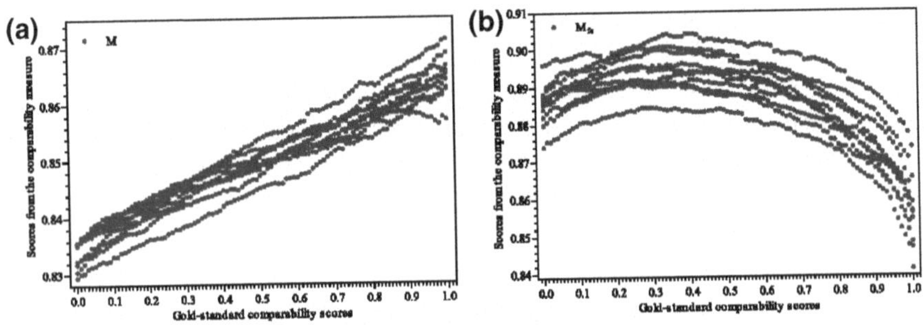

Fig. 1 Evolution of M and M_{ts} w.r.t. gold standard on the corpus group G_c (x-axis: gold-standard comparability scores; y-axis: comparability scores from M or M_{ts}. **a** M. **b** M_{ts})

is easier to capture the different comparability levels in G_b than in G_a and G_c, which coincides with our intuition, given the construction process considered.

Robustness of the Comparability Measure

As one can note from Eq. 3, M depends on the bilingual dictionary $\mathcal{D}$ used to bridge the language barrier. However, it is important that the comparability measure one retains remains consistent when the dictionary coverage of the corpus changes slightly, as this is a necessary condition to distinguish between different comparability levels. Indeed, if a slight change in the dictionary coverage entails an important change in the comparability score, then it becomes impossible in practice to compare different corpora, as they will likely have a different coverage with respect to the dictionary. In this part, several dictionaries of different sizes, corresponding to different coverages on the corpus vocabulary, are built by randomly choosing the subparts from the original dictionary. The coverage here is simply defined as the proportion of unique words in the corpus vocabulary that are covered by the dictionary. We choose to randomly pick certain proportions, from 50 to 99 % with a step of 1 %, of the original dictionary. For each proportion, 30 different dictionaries are built by randomly sampling the original dictionary 30 times at this proportion. These 1500 dictionaries are then used to compute M on different corpora with decreasing comparability scores. For clarity, we only take here the first parallel corpus $\mathcal{P}_1$ from $(\mathcal{P}_1, \mathcal{P}_2, \ldots, \mathcal{P}_{10})$ together with its descendant corpora which are built by setting the proportion p to different values in G_a. Lastly, for readability reasons, we only plot in Fig. 2 the comparability scores for some of the comparable corpora mentioned here, i.e. $\mathcal{P}_1$, $\mathcal{P}_1^{0.7}$, $\mathcal{P}_1^{0.4}$ and $\mathcal{P}_1^{0.1}$ (corresponding to the gold-standard comparability 1, 0.7, 0.4 and 0.1 respectively), w.r.t. the different coverages.

From Fig. 2 one can find that when the dictionary coverage lies above a certain threshold (inspected from the figure, roughly set to 0.62), the differences between

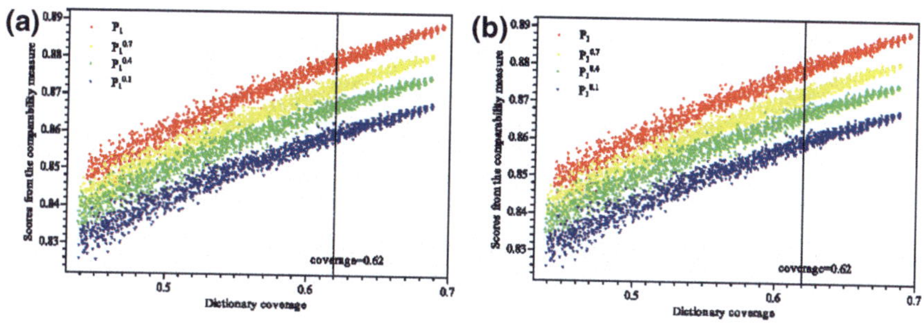

Fig. 2 Evolution of M w.r.t. different dictionary coverages on comparable corpora $\mathcal{P}_1$, $\mathcal{P}_1^{0.7}$, $\mathcal{P}_1^{0.4}$ and $\mathcal{P}_1^{0.1}$ in G_a (x-axis: different dictionary coverages; y-axis: comparability scores from M)

the 4 different comparability levels[5] can be captured very well, as the different data points are well separated. Similar conclusions can be made with different coverage thresholds on other groups of corpora.

The conclusion we can draw from the above observations is thus that, provided the coverage of the bilingual dictionary used is sufficient, the comparability measure M we have introduced is able to quantify the degree of comparability of a given bilingual corpus. Furthermore, in the case of two different bilingual corpora, if the coverage of the bilingual dictionary is roughly the same on these two corpora, then M can be used to assess whether a corpus is more comparable than the other. The dependence of M on the coverage of the bilingual dictionary is not surprising, and we know of no way to remove it. In practice however, relying on M with corpora displaying similar dictionary coverage will allow us to improve the comparability of a given corpus, as we will see in the next section.

3.3 A Step Further: Using Contextual Information

The comparability measure M defined by Eq. 3 and relying on the function in Eq. 1 "ignores" the fact that words may be ambiguous. We can make use of the assumption that words in parallel translation usually appear in similar contexts to disambiguate the translation candidates in the dictionary. This assumption has been used broadly in the task of bilingual lexicon extraction as detailed in Sect. 1. We have embedded this assumption in the function σ in Eq. 1.

Let us assume that the source (resp. target) language word w appears in the context word set S_s (resp. S_t) consisting of the words surrounding w_s (resp. w_t) in a certain window in the corpora. Then the similarity of the two *context sets* is measured by the overlap of the two sets which is directly the proportion of words of which the translation can be found in the counterpart set. Formally, the similarity of w_s and w_t,

[5] This is also true for all the comparability levels of comparable corpora although we only plot 4 levels in the figure.

based on their respective context information, can be written as:

$$s(w_s, w_t) = \frac{\sum_{w \in S_s \cap D_s^v} \sigma(w, S_t) + \sum_{w \in S_t \cap D_t^v} \sigma(w, S_s)}{|S_s \cap D_s^v| + |S_t \cap D_t^v|}$$

Let us recall the fact in σ that two words are treated as the translation of each other if the word pair appears in the bilingual dictionary. We make use of the context information and develop an enhanced version σ_c of σ, which is:

$$\sigma_c(w, \mathcal{C}^v) = \begin{cases} 1 \text{ if } \exists w' \in \mathcal{T}_w \cap \mathcal{C}^v, \quad s(w, w') > \zeta \\ 0 \text{ otherwise} \end{cases}$$

where ζ is the threshold for the similarity. According to σ_c, a word w is deemed to be translated, if at least one of its translations w' in the corpus, identified by the former function σ, is similar to w based on the context similarity measure $s(w, w')$. Replacing σ with σ_c in Eq. 3 will lead to the new comparability measure M_c as:

$$M_c(\mathcal{C}_s, \mathcal{C}_t) = \frac{\sum_{w \in \mathcal{C}_s^v \cap \mathcal{D}_s^v} \sigma_c(w, \mathcal{C}_t^v) + \sum_{w \in \mathcal{C}_t^v \cap \mathcal{D}_t^v} \sigma_c(w, \mathcal{C}_s^v)}{|\mathcal{C}_s^v \cap \mathcal{D}_s^v| + |\mathcal{C}_t^v \cap \mathcal{D}_t^v|}$$

Under the same experimental settings as above and setting the threshold ζ in σ_c to 0.3, we obtain for the correlation scores of M_c on G_a, G_b and G_c 0.972, 0.978 and 0.990 respectively, which are better than the ones for the original measure M as can be seen in Table 2. Although the value 0.3 set for ζ might not be the optimal one, it has already provided us with a best-performing measure M_c. However, the computational cost of the context-based approaches is much higher as it involves a comparison of the context of each word. For this reason, and because the difference between M and M_c is not important, we will make use of the measure M in the following experiments, as they require intensive calls to the comparability measure.

Having reviewed several comparability measures, we now turn to the problem of enhancing corpus comparability for bilingual lexicon extraction.

4 A Clustering Approach to Enhance Comparable Corpora

We introduce in this section the algorithm to improve the quality of a given comparable corpus. This algorithm relies on a clustering process to capture homogeneous documents. Our goal here is to construct, from a given comparable corpus, an enhanced version of it which displays a higher degree of comparability and preserves most of the original vocabulary.

4.1 Clustering Documents for High Quality Comparable Corpora

The distributional hypothesis underlying bilingual lexicon extraction method is more reliable when the documents in different languages describe the same or similar topics, since authors tend to use the same word combinations to describe similar topics (see [12] for a related analysis). In other words, if a corpus covers a limited set of topics, it is more likely to contain consistent information on the words used, leading to improved bilingual lexicons extracted. The term *homogeneity* directly refers to this fact, and we will say, in an informal manner, that a corpus is homogeneous if it covers a limited set of topics. The rationale for the algorithm we introduce here to enhance corpus comparability is precisely based on homogeneity. We conjecture here that if one can guarantee a certain degree of homogeneity in addition to a certain degree of comparability, then the bilingual lexicon extracted from the obtained corpus will be of higher quality. As we will see, this conjecture will be fully validated in the experimental section. In order to find document sets which are similar with each other (i.e. homogeneous), it is natural to resort to clustering techniques. Furthermore, since we need homogeneous corpora for bilingual lexicon extraction, it will be convenient to rely on techniques which allows one to easily prune less relevant clusters. To perform all this, we use in this work a standard hierarchical agglomerative clustering method but other clustering methods with associated pruning strategies can directly be used as well.

Bilingual Clustering Algorithm

The overall process retained to build high quality, homogeneous comparable corpora relies on the following steps:

1. Using the bilingual similarity measure defined in Equation 6 below, cluster source and target language documents so as to get bilingual dendrograms[6] from the original corpus $\mathcal{C}$ by grouping documents with related content;
2. Pick high quality sub-clusters by thresholding the obtained dendrograms according to the node depth;
3. Combine all these sub-clusters to form a new comparable corpus $\mathcal{C}_H$, which thus contains homogeneous, high-quality subparts;
4. Use again steps (1), (2) and (3) to enrich the remaining subpart of $\mathcal{C}$ (which will be denoted as $\mathcal{C}_L$) with external resources.

The first three steps are summarized in Algorithm 1. As one can note, only $\mathcal{C}$ is used in order to build $\mathcal{C}_H$, through clustering and pruning of documents. As such, Algorithm 1 aims at extracting the most comparable and homogeneous subpart of $\mathcal{C}$. Once this has been done, i.e. once $\mathcal{C}$ has been exploited, one needs to resort to

---

[6] As the low comparability scores of document pairs are filtered out to reduce the computational complexity, one usually obtains, instead of a fully connected dendrogram, several dendrograms from the clustering process.

new corpora if one wants to build an homogeneous corpus with a high degree of comparability from $\mathcal{C}_L$ (which is the part of $\mathcal{C}$ left after removing $\mathcal{C}_H$). To do so, we simply replace, in step (4), the input corpus $\mathcal{C}$ with two comparable corpora: The first one consists of the source language part of $\mathcal{C}_L$ and the target language part of an external corpus $\mathcal{C}_T$; The second one consists of the target language part of $\mathcal{C}_L$ and the source language part of $\mathcal{C}_T$. The two high quality subparts obtained from these two new comparable corpora in step (4) are then added to $\mathcal{C}_H$ to constitute the final comparable corpus of higher quality.

Algorithm 1 Bilingual Clustering Algorithm

Input:

Set $\mathcal{W}$ of all source and target language docs from $\mathcal{C}$

Positive real number θ (the depth threshold)

Output:

$\mathcal{C}_H$, high quality, homogeneous subpart of $\mathcal{C}$

```
Set C_H as the null set Ø;
Cluster W in order to get a set of dendrograms S;
for each dendrogram T in S do
    Set m as the maximal depth of T;
    Remove from T the low-quality sub-clusters of which the depth is lower than the depth
    threshold computed from m · θ;
    Add all the remaining documents in T to C_H;
end for
return C_H;
```

We now turn our attention to the similarity measure used in the clustering process.

Similarity Measure

Let us assume that we have two document sets (i.e. clusters) $\mathcal{R}_1$ and $\mathcal{R}_2$. In the task of bilingual lexicon extraction, two document sets are similar to each other and should be clustered if the combination of the two can complement the content of each single set, which relates to the notion of homogeneity introduced before. In other words, both the source language part $\mathcal{R}_1^s$ of $\mathcal{R}_1$ and the target language part $\mathcal{R}_1^t$ of $\mathcal{R}_1$should be comparable to their counterparts (respectively the target language part $\mathcal{R}_2^t$ of $\mathcal{R}_2$ and the source language part $\mathcal{R}_2^s$ of $\mathcal{R}_2$).[7] This leads to the following similarity measure for $\mathcal{R}_1$ and $\mathcal{R}_2$:

$$sim(\mathcal{R}_1, \mathcal{R}_2) = \beta M(\mathcal{R}_1^s, \mathcal{R}_2^t) + (1 - \beta) M(\mathcal{R}_2^s, \mathcal{R}_1^t) \quad (4)$$

where β $(0 \leq \beta \leq 1)$ is a weight controlling the importance of the two subparts $(\mathcal{R}_1^s, \mathcal{R}_2^t)$ and $(\mathcal{R}_2^s, \mathcal{R}_1^t)$. Intuitively, one would like to give more weight in the combination

[7] As $\mathcal{R}_1$ and $\mathcal{R}_2$ are clusters, then, by construction, their respective source and target language parts are comparable.

to the larger subpart, as it contains more information. We use here the number of document pairs to represent the amount of information contained in a comparable sub-corpus. Thus, the weight β can be defined as the proportion of possible document pairs in the current comparable corpus $(\mathcal{R}_1^s, \mathcal{R}_2^t)$ to all the possible document pairs, which is:

$$\beta = \frac{\#_d(\mathcal{R}_1^s) \cdot \#_d(\mathcal{R}_2^t)}{\#_d(\mathcal{R}_1^s) \cdot \#_d(\mathcal{R}_2^t) + \#_d(\mathcal{R}_2^s) \cdot \#_d(\mathcal{R}_1^t)}$$

where $\#_d(\mathcal{R})$ stands for the number of documents in $\mathcal{R}$. As the clusters are first formed from single documents, in the source and target languages, one can see that the similarity measure corresponds to a normalized comparability score between the source and target language sub-clusters making up the new cluster. However, the fact that imbalanced clusters (with say many more English than French documents) may arise in this process is only indirectly controlled through M. In order to have more control on this aspect, we introduce a penalty function ϕ aiming at penalizing corpora for which the number of documents in the source and target languages is too different. We expect the function to be between 0 and 1, with lower values identifying the stricter penalty. We propose a penalty function based on the relative difference between the corpus size in terms of document number, which is:

$$\phi(\mathcal{R}) = \frac{1}{\left(1 + log\left(1 + \frac{|\#_d(\mathcal{R}^s) - \#_d(\mathcal{R}^t)|}{\min(\#_d(\mathcal{R}^s), \#_d(\mathcal{R}^t))}\right)\right)} \tag{5}$$

The above penalty function leads us to a new similarity measure sim_l which is the one finally used in Algorithm 1:

$$sim_l(\mathcal{R}_1, \mathcal{R}_2) = sim(\mathcal{R}_1, \mathcal{R}_2) \cdot \phi(\mathcal{R}_1 \cup \mathcal{R}_2) \tag{6}$$

4.2 Computational Considerations

As comparable corpora usually consist of a large number of documents, the agglomerative clustering algorithm may cost a lot of memory space and computation time. We address this problem (a) by providing a lower bound of the comparability measure which can be computed efficiently, (b) by filtering out document pairs with comparability scores less than a predefined threshold η, and (c) by updating the similarity matrix iteratively in an efficient way during the clustering process. As the clustering process involves at each iteration the merging of the two closest clusters, relying on a lower bound ensures that the clusters to be merged have a high comparability score.

In our implementation, the measure $M(\mathcal{C}_s, \mathcal{C}_t)$ in Eq. 4 is replaced by a lower-bound[8] $\frac{1}{\#_d(\mathcal{C}_s) \cdot \#_d(\mathcal{C}_t)} \sum_{d_e \in \mathcal{C}_s, d_f \in \mathcal{C}_t} M(d_e, d_f)$ which yields a similarity measure

---

[8] For space constraints, we do not show here that the new measure we introduce is indeed a lower bound of M.

defined as the accumulative value of all the connections between two clusters. It is feasible, with this new measure, to update the similarity matrix iteratively in the clustering process. Assuming the clustering process merges, at some point, clusters $\mathcal{R}_1$ and $\mathcal{R}_2$ into $\mathcal{R}_{new}$, the similarity matrix between clusters should be updated and the similarity between $\mathcal{R}_{new}$ and any other cluster (e.g. $\mathcal{R}_3$) should be computed. According to Eq. 6 and the new similarity, the similarity between $\mathcal{R}_{new}$ and $\mathcal{R}_3$ can be written as:

$$sim_l(\mathcal{R}_{new}, \mathcal{R}_3) = \frac{(N_{\mathcal{R}_1} + N_{\mathcal{R}_2}) \cdot \phi(\mathcal{R}_1 \cup \mathcal{R}_2)}{\#_d(\mathcal{R}^s_{new}) \cdot \#_d(\mathcal{R}^t_3) + \#_d(\mathcal{R}^s_3) \cdot \#_d(\mathcal{R}^t_{new})}$$

where $(j = 1, 2)$ and:

$$N_{\mathcal{R}_j} = \frac{(\#_d(\mathcal{R}^s_j) \cdot \#_d(\mathcal{R}^t_3) + \#_d(\mathcal{R}^s_3) \cdot \#_d(\mathcal{R}^t_j)) \cdot sim_l(\mathcal{R}_j, \mathcal{R}_3)}{\phi(\mathcal{R}_j \cup \mathcal{R}_3)}$$

In the clustering process, since $sim_l(\mathcal{R}_1, \mathcal{R}_3)$ and $sim_l(\mathcal{R}_2, \mathcal{R}_3)$ are already known before the computation of $sim_l(\mathcal{R}_{new}, \mathcal{R}_3)$, one can directly update the similarity matrix at each iteration. Denoting N_c the number of clusters before a merge, the complexity of this update is $\mathcal{O}(N_c)$, whereas it reaches $\mathcal{O}(N_c \times \bar{C}^2)$ with the direct application of Eqs. 4 and 6 (with $\bar{C}$ the average number of documents per cluster).

4.3 Assessing the Comparability of the New Corpora

The experiments we have designed in this subsection aim at assessing whether the algorithm we have introduced yields corpora of higher quality in terms of comparability scores. In all our experiments, we use the method described in this chapter (a brief version also appears in [11]), as well as the one described in our previous work [10]. This latter method is the only alternative approach to enhancing corpus comparability we are aware of, hence its importance in the evaluation.

The clustering algorithm described in Sect. 4.1 is employed to improve the quality of the comparable corpus. The corpora *GH95* and *SDA95* are used as the original corpus $\mathcal{C}^0$. We consider two classes of external corpora to prove that the efficiency of our algorithm is not confined to a specific external resource. The first external corpus $\mathcal{C}^1_T$ consists of the corpora *LAT94*, *MON94* and *SDA94*. The second external corpus $\mathcal{C}^2_T$ consists of *Wiki-En* and *Wiki-Fr*. There are two empirical parameters to set for the corpora in consideration: the depth threshold θ in Algorithm 1 is set to 0.4, and the threshold η in Sect. 4.2 is set to 0.3. The two parameters are set from intuition in our experiments, and future work will be taken on to investigate different parameter settings.

After the clustering process, we obtain the corpora $\mathcal{C}^1$ (with the external corpus $\mathcal{C}^1_T$) and $\mathcal{C}^2$ (with the external corpus $\mathcal{C}^2_T$). As mentioned before, for comparison purposes, we also used the method described in [10] on the same data, leading to corpora $\mathcal{C}^{1'}$ (with $\mathcal{C}^1_T$) and $\mathcal{C}^{2'}$ (with $\mathcal{C}^2_T$) from $\mathcal{C}^0$. In terms of lexical coverage, $\mathcal{C}^1$

covers 97.9 % of the vocabulary of $\mathcal{C}^0$, while $\mathcal{C}^2$ covers 99.0 % of the vocabulary of $\mathcal{C}^0$. Hence, most of the vocabulary of the original corpus has been preserved, which was one of the requirements behind our approach. Concerning comparability scores, the comparability of $\mathcal{C}^1$ reaches 0.924 and that of $\mathcal{C}^2$ is 0.939. Both corpora are more comparable than the original corpus $\mathcal{C}^0$ of which the comparability is 0.881. Furthermore, both $\mathcal{C}^1$ and $\mathcal{C}^2$ are more comparable than $\mathcal{C}^{1'}$ (comparability 0.912) and $\mathcal{C}^{2'}$ (comparability 0.915), which shows that homogeneity is crucial for comparability.

4.4 Bilingual Lexicon Extraction Experiments

Following the experiments in Sect. 4.3, we will further show in this part that the bilingual lexicons extracted from the enhanced corpora are of higher quality. As previous studies on bilingual lexicon extraction from comparable corpora radically differ on resources used and technical choices, it is very difficult to compare them in a unified framework [9]. More importantly, our approach aims at enhancing corpus comparability, and can be coupled with any existing bilingual lexicon extraction method once the corpus has been enhanced. It is thus more interesting to directly asses whether such a coupling can lead to increased performance. To extract bilingual lexicons from comparable corpora, we directly use here the method proposed by Fung and Yee [4], and which has been referred to as the *standard approach* in more recent studies [2, 6, 22]. In this approach, each word is represented as a context vector consisting of its surrounding words. Source (or target) context vectors are then translated with an existing bilingual dictionaries. Finally, a translation score is given to any word pair based on the cosine of their respective context vectors.

Experimental Settings

As before, English is taken as the source language and French as the target language in the experiments. In order to measure the performance of the lexicons extracted, we divided the bilingual dictionary mentioned in Sect. 2 into 2 parts: 10 % of the English words together with their translations are randomly chosen and used as the evaluation set, the remaining words being used to compute context vectors similarity. English words not present in $\mathcal{C}_s$ or with no translation in $\mathcal{C}_t$ are excluded from the evaluation set. For each English word in the evaluation set, all the French words in $\mathcal{C}_t$ are then ranked according to their similarity with the English word. Precision, recall and the NMR measure are then computed on the first N translations.

The precision amounts in this case to the proportion of lists containing the correct translation (in case of multiple translations, a list is deemed to contain the correct translation as soon as one of the possible translations is present). The recall is the proportion of correct translations found in the lists to all the translations provided in the corpus. This evaluation procedure has been used in previous studies and is now standard. The precision or recall measure is not precise enough as it does not distinguish between candidate translations of different ranks. We thus use an additional measure NMR, previously discussed in [21, 22] to show the ability of the algorithm to precisely rank the selected translation candidates. Assuming the total number of English words in the evaluation set is m, NMR is defined as $\frac{1}{m}\sum_{i=1}^{m}\frac{1}{rank_i}$ where $rank_i$ is the rank of the first correct translation in the candidate translation list for the ith word in the evaluation set. If the correct translation does not appear in the top N candidates, $\frac{1}{rank_i}$ will be set to 0. In our experiments, N is set to 20.

Furthermore, several studies have shown that it is easier to find the correct translations for frequent words than for infrequent ones [15]. To take this fact into account, we distinguished different frequency ranges to assess the validity of our approach for all frequency ranges. Empirically, words with frequency less than 100 are defined as low-frequency words (W_l), whereas words with frequency larger than 400 are high-frequency words (W_h), and words with frequency in between are medium-frequency words (W_m).

Results and Analysis

In a first series of experiments, bilingual lexicons were extracted from the corpora obtained by our approach ($\mathcal{C}^1$ and $\mathcal{C}^2$), the corpora obtained by the approach described in [10] ($\mathcal{C}^{1'}$ and $\mathcal{C}^{2'}$) and the original corpus $\mathcal{C}^0$. Table 3 displays the results obtained. Each of the last two columns "$\mathcal{C}^1 > \mathcal{C}^0$" and "$\mathcal{C}^2 > \mathcal{C}^0$" contains the absolute and the relative difference (in %) w.r.t. $\mathcal{C}^0$. As one can note, the best results are obtained from the corpora built with the method we have described in this chapter. The lexicons extracted from the enhanced corpora are of higher quality, in terms of precision, recall and NMR, than the ones obtained from the original corpus and from the corpora built according to the methodology defined in [10]. The difference is more remarkable with $\mathcal{C}^2$, which is obtained from a large external corpus $\mathcal{C}_T^2$. Intuitively, one can expect to find, in larger corpora, more documents related to a given corpus, an intuition which seems to be confirmed by our results.

To assess the behavior of the methods w.r.t. word frequencies, we focus on the best results on $\mathcal{C}^{2'}$ from the previous approach and the best results on $\mathcal{C}^2$ from our approach. Table 4 summarizes the results obtained. As one can note, and not surprisingly, the results obtained with high frequency words are better than the ones obtained with low frequency words. Furthermore, our approach is superior for words in all the frequency ranges. The overall precision can be increased by 41.8 % relatively,

Table 3 Performance of the bilingual lexicon extraction from different corpora

	$\mathcal{C}^0$	$\mathcal{C}^{1'}$	$\mathcal{C}^{2'}$	$\mathcal{C}^1$	$\mathcal{C}^2$	$\mathcal{C}^1 > \mathcal{C}^0$ (%)	$\mathcal{C}^2 > \mathcal{C}^0$ (%)
Precision	0.226	0.277	0.325	0.295	0.461	0.069, 30.5	0.235, 104.0
Recall	0.103	0.122	0.145	0.133	0.212	0.030, 29.1	0.109, 105.8
NMR	0.119	0.150	0.175	0.150	0.257	0.031, 26.1	0.138, 116.0

Table 4 Comparison of the precision for words of different frequencies

	$\mathcal{C}^0$	$\mathcal{C}^{2'}$	$\mathcal{C}^2$	$\mathcal{C}^{2'} > \mathcal{C}^0$ (%)	$\mathcal{C}^2 > \mathcal{C}^0$ (%)	$\mathcal{C}^2 > \mathcal{C}^{2'}$ (%)
W_l	0.135	0.206	0.304	0.071, 52.6	0.169, 125.2	0.098, 47.6
W_m	0.256	0.390	0.564	0.134, 52.3	0.308, 120.3	0.174, 44.6
W_h	0.434	0.632	0.667	0.198, 45.6	0.233, 53.7	0.035, 5.5
All	**0.226**	**0.325**	**0.461**	**0.099, 43.8**	**0.235, 104.0**	**0.136, 41.8**

from 0.325 to 0.461. Comparing $\mathcal{C}^2$ with the original corpus $\mathcal{C}^0$, we note, for the overall precision, a relative increase of 104.0 %, from 0.226 to 0.461, which is very satisfactory in the context of general, large evaluation sets. Lastly, the improvement for the low-frequency and medium-frequency ranges is more significant in $\mathcal{C}^2$, which demonstrates that our approach behaves much better on what is generally considered to be a hard problem [15].

5 Related Work

We compare in this section our method with previous ones in the same vein (i.e. enhancing bilingual corpora prior to extracting information from them). Some researchers have tried to extract parallel subparts from comparable corpora, and thus obtain high quality corpus. For example, the work presented in [13] proposes methods to extract parallel fragments from comparable corpora, which can be adequately exploited in applications such as machine translation. However, their approach only focuses on a very small part of the original corpus, whereas our work aims at preserving most of the vocabulary of the original corpus. In many cases, this cannot be achieved if one only focuses on parallel fragments.

As we have mentioned before, we have followed here the general approaches proposed in [10] and [11] which consist in enhancing the quality of a comparable corpus prior to extracting information from it. We have shown here a method which ensures homogeneity of the obtained corpus, and which finally leads to comparable corpora of higher quality. In turn such corpora yield bilingual lexicons which are better in terms of their intrinsic quality (precision, recall and NMR). The algorithm we have detailed here is very general and can be coupled with different clustering techniques.

6 Conclusion

We have introduced in this chapter a comparability measure with which one can quantify and compare various comparability levels. We have furthermore introduced a methodology to assess the validity of comparability measures, in terms of both reliability and robustness. With this, we are able to show that the measure M proposed in this chapter is reliable and robust. We have then developed an approach to enhance corpus comparability in order to improve the quality of bilingual lexicons extracted from comparable corpora. We have demonstrated empirically that our approach guarantees a certain degree of comparability and homogeneity, while preserving most of the vocabulary of the original corpus. Our experiments furthermore show that the bilingual lexicons obtained are of better quality than the lexicons obtained with previous approaches.

Acknowledgments This work was supported by the French National Research Agency grant ANR-08-CORD-009.

References

1. Ballesteros, L., Croft, W.B.: Phrasal translation and query expansion techniques for cross-language information retrieval. In: Proceedings of the 20th ACM SIGIR, pp. 84–91, Philadelphia, Pennsylvania, USA (1997)
2. Déjean, H., Gaussier, E., Sadat, F.: An approach based on multilingual thesauri and model combination for bilingual lexicon extraction. In: Proceedings of the 19th International Conference on Computational Linguistics, pp. 1–7, Taipei, Taiwan (2002)
3. Fung, P., McKeown, K.: Finding terminology translations from non-parallel corpora. In: Proceedings of the 5th Annual Workshop on Very Large Corpora, pp. 192–202, Hong Kong (1997)
4. Fung, P., Yee, L.Y.: An IR approach for translating new words from nonparallel, comparable texts. In: Proceedings of the 17th International Conference on Computational linguistics, pp. 414–420, Montreal, Quebec, Canada (1998)
5. Garera, N., Callison-Burch, C., Yarowsky, D.: Improving translation lexicon induction from monolingual corpora via dependency contexts and part-of-speech equivalences. In: CoNLL 09: Proceedings of the Thirteenth Conference on Computational Natural Language Learning, pp. 129–137, Boulder, Colorado (2009)
6. Gaussier, E., Renders, J.M., Matveeva, I., Goutte, C., Déjean, H.: A geometric view on bilingual lexicon extraction from comparable corpora. In: Proceedings of the 42nd Annual Meeting of the Association for Computational Linguistics, pp. 526–533, Barcelona, Spain (2004)
7. Goeuriot, L., Grabar, N., Daille, B.: Characterization of scientific and popular science discourse in French, Japanese and Russian. In: LREC. Marrakech, Morocco (2008)
8. Koehn, P.: Europarl: a parallel corpus for statistical machine translation. In: Proceedings of MT Summit (2005)
9. Laroche, A., Langlais, P.: Revisiting context-based projection methods for term-translation spotting in comparable corpora. In: Proceedings of the 23rd International Conference on Computational Linguistics (Coling 2010), pp. 617–625, Beijing, China (Aug 2010)
10. Li, B., Gaussier, E.: Improving corpus comparability for bilingual lexicon extraction from comparable corpora. In: Proceedings of the 23rd International Conference on Computational Linguistics, pp. 644–652, Beijing, China (2010)
11. Li, B., Gaussier, E., Aizawa, A.: Clustering comparable corpora for bilingual lexicon extraction. In: Proceedings of the 49th Annual Meeting of the Association for Computational Linguistics: Human Language Technologies, pp. 473–478, Portland, Oregon, USA (June 2011)
12. Morin, E., Daille, B., Takeuchi, K., Kageura, K.: Bilingual terminology mining—using brain, not brawn comparable corpora. In: Proceedings of the 45th Annual Meeting of the Association for Computational Linguistics, pp. 664–671, Prague, Czech Republic (2007)
13. Munteanu, D.S., Marcu, D.: Extracting parallel sub-sentential fragments from non-parallel corpora. In: Proceedings of the 21st International Conference on Computational Linguistics and the 44th annual meeting of the Association for Computational Linguistics, pp. 81–88, Sydney, Australia (2006)

14. Och, F.J., Ney, H.: A systematic comparison of various statistical alignment models. Comput. Linguist. **29**(1), 19–51 (2003)
15. Pekar, V., Mitkov, R., Blagoev, D., Mulloni, A.: Finding translations for low-frequency words in comparable corpora. Mach. Transl. **20**(4), 247–266 (2006)
16. Rapp, R.: Automatic identification of word translations from unrelated English and German corpora. In: Proceedings of the 37th Annual Meeting of the Association for Computational Linguistics, pp. 519–526, College Park, Maryland, USA (1999)
17. Rayson, P., Garside, R.: Comparing corpora using frequency profiling. In: Proceedings of the Workshop on Comparing corpora, pp. 1–6, Hong Kong (2000)
18. Robitaille, X., Sasaki, Y., Tonoike, M., Sato, S., Utsuro, T.: Compiling French-Japanese terminologies from the web. In: Proceedings of the 11th Conference of the European Chapter of the Association for Computational Linguistics, pp. 225–232, Trento, Italy (2006)
19. Sharoff, S.: Comparing corpora using frequency profiling. In: Proceedings of Web as Corpus Workshop. Louvain-la-Neuve (2007)
20. Shezaf, D., Rappoport, A.: Bilingual lexicon generation using non-aligned signatures. In: Proceedings of the 48th Annual Meeting of the Association for Computational Linguistics, pp. 98–107, Uppsala, Sweden (2010)
21. Voorhees, E.M.: The trec-8 question answering track report. In: Proceedings of the 8th Text Retrieval Conference, pp. 77–82 (1999)
22. Yu, K., Tsujii, J.: Extracting bilingual dictionary from comparable corpora with dependency heterogeneity. In: Proceedings of HLT-NAACL 2009, pp. 121–124, Boulder, Colorado, USA (2009)

Statistical Corpus and Language Comparison on Comparable Corpora

Thomas Eckart and Uwe Quasthoff

Abstract With the wide availability of textual data in various languages, domains and registers it is easy to create text corpora for a variety of applications. These include, among many others, the field of Natural Language Processing. The Leipzig Corpora Collection creates and uses such corpora for more than fifteen years. However, the work on preprocessing distributed resources to ensure homogeneity and thus comparability is a steady process. As a result created corpora in identical formats allow the use of different statistical methods to generate various data for manual or automatic analysis. These are basis for applications in intra- and inter-language comparison or quality assurance of text stocks.

Keywords Corpus comparison · Language comparison · Corpus evaluation

1 Leipzig Corpora Collection

The project Leipzig Corpora Collection (LCC) continuously collects texts of various languages and genres for the creation of text corpora and their enrichment with different additional data (like POS tags or word base forms). At present, hundreds of corpora in more than 60 different languages are available. To cope with large amounts of gathered material (currently around 30 GB new text material per month) and to guarantee its fast processing a dedicated corpus creation toolchain was developed that is steadily improved [14].

T. Eckart (✉) · U. Quasthoff
University of Leipzig, Leipzig, Germany
e-mail: teckart@informatik.uni-leipzig.de
http://asv.informatik.uni-leipzig.de

U. Quasthoff
e-mail: quasthoff@informatik.uni-leipzig.de

S. Sharoff et al. (eds.), *Building and Using Comparable Corpora*,
DOI: 10.1007/978-3-642-20128-8_8, © Springer-Verlag Berlin Heidelberg 2013

Main field of application of the created corpora is statistical and pattern based language processing (like named entity recognition, clustering etc.) which are partially offered via different REST and SOAP based web services [4]. Additionally they are a valuable resource for teaching.

With a rising number of corpora, available text material and new usages of these resources, it became obvious that there is a strong need for a steady evaluation of corpus quality and a system that allows new ways of corpus-based work and analysis. This paper will concentrate on the questions of how to create comparable corpora that are based on heterogeneous input material and their comparative analysis for different applications. As it is impossible to give an in-depth view of this topic, the focus will be on overviewing used analysis types at the LCC.

2 Building Comparable Corpora

2.1 Introduction

In this paper, corpora in different languages are considered as comparable if they have the same genre and comparable sources. They are assumed to contain a considerable number of text pairs which deal with the same topic. One text might be the translation of another, but usually the relationship is weaker: The comparable texts usually have different length and are written independently. Examples are Wikipedias for different languages or newspaper articles dealing with the same topic.

As a variety of different input material is available, it is essential to use a work flow that is standardized as much as possible to avoid disturbances (by wrong selection of source material or variations in the corpus creation process) that can degrade statistical analysis. Therefore the used toolchain is supposed to be mainly language independent. However, to deal with different languages or text types some processing steps need additional resources (like lists of abbreviations, multi word units, tokenization rules, POS tagger etc.) that are specific for its kind of input material and therefore degrade comparability of created corpora. Because of the varying availability of tools, the amount of additional data for every corpus varies. A generic corpus consists of separated sentences, relevant source information together with additional statistical data like word frequencies or word co-occurrences. If available for the corresponding language, additional information like POS-tags, word base forms or semantic categories are included. Figure 1 shows a summary of the whole process.

2.2 Text Sources

The LCC uses a variety of different sources to aggregate text material and supplementary data. Text collection started in 1995. Since 2000 the central resource used is the World Wide Web. For this reason different crawlers and extractors for specific

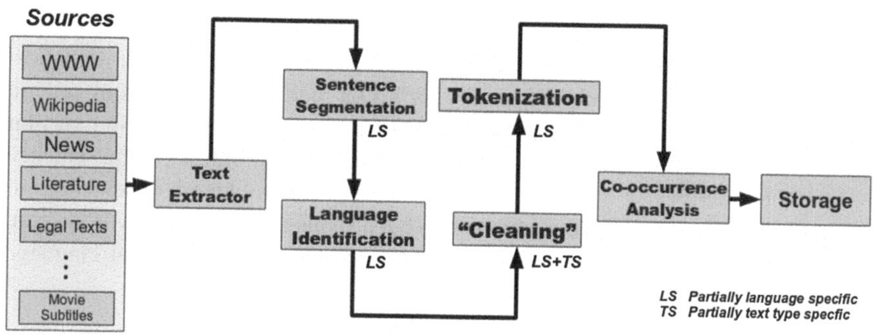

Fig. 1 Overview on generic corpora creation at the LCC

resources were created (including the distributed Web crawler Findlinks, cf. [1]). At present the major sources of text are:

- Online newspaper texts
- Randomly crawled text from the WWW
- Wikipedia

If needed, further sources are used as input material for new corpora. This includes corpora based on movie subtitles, chat room texts, Twitter messages or public available corpora of different time spans like texts provided by Project Gutenberg.

The key question when dealing with such a variety of formats is to convert them to a homogeneous format that can be a basis for all further preprocessing tasks. Therefore every text supply tool generates output in a specific XML format. As the focus of the LCC is on large data sets (vs. focus on the quality of single data entries) and text mining applications the loss of some specific annotations or meta data of the source material is tolerated.

2.3 Preprocessing

The whole preprocessing task is executed on the homogeneous format provided by the different text crawlers. This allows a mixture of different input material and generation of very large inter-domain (or inter-language) corpora. The preprocessing step can be seen as a generic (language independent) process, that can be adapted to reflect language or text type specifics if needed. Therefore it is a tradeoff between losing comparability of corpora and the need to reflect characteristics that otherwise would get lost.

As a first step the text is segmented into sentences. Language specific parameters used here are sets of sentence boundary marks and lists of abbreviations to enhance the sentence boundary detection process. If there is no language specific data available, generic data are used.

To remove all sorts of unwanted material three "cleaning" procedures are applied. First all sentence duplicates are removed to avoid problems with later statistical analysis. This turned out to be especially useful on material crawled from generic web pages which often contain a huge amount of boilerplates. As a next step regular expressions are used to remove all sorts of non-valid output. Similar to the sentence segmentation there exists a list of generic rules (like minimal length of sentences or unlikely ratio of letters to special characters) but also language and text type specific patterns. In the end the gained sentences are separated by a statistical language identification tool (cf. [7]).

2.4 Creating Text Databases

In a second step the sentences are analyzed by the statistical engine Medusa [3]. Processing steps include tokenization (partially language dependent), creation of inverse lists and word co-occurrence analysis using various significance measures. Further information about string and semantic word similarity are generated, stemming and POS tagging tools are applied, if available for the respective language.

2.5 Storage

All data are stored in relational databases to provide a generic access. This is especially important to permit new query types in the future, but also to allow users to directly work with the data by using their own SQL statements. However for time critical applications on huge datasets (like requests on very large word co-occurrences data) highly efficient binary data structures are used.

3 Measurements

To cover as many fields of interest as possible, more than 150 different measurement scripts were created and classified in the following six sections of analysis.

3.1 Corpus Meta Information

The following data are measured on the corpus creation process and contain information regarding the corpus and its creation: sources and their size, versions of preprocessing tools and duration of the processing tool chain etc. These information make the corpus production process transparent and repeatable.

3.2 Characters and Character N-Grams

Character statistics are well known and widely used (for example [13]). First of all, the alphabet used in the corpus can be identified by considering the most frequent characters up to a certain threshold. Single character probabilities can be used to investigate capitalization and the usage of foreign characters which are not considered as part of the alphabet. Moreover, frequencies of punctuation marks are of interest for analyzing the sentence structure.

High frequent character N-grams, especially at word beginnings or endings, show candidates for prefixes and suffixes. Character successor rates and character transition probabilities provide more insight into the boundaries of these affixes and other word formation rules. Figure 2 shows the most frequent letter bi- and trigrams of a German corpus. One might conclude that -en is a German suffix because of the many predecessors of this bigram. In the list of trigrams, -ung has a similar property, and hence is a candidate for a suffix.

If the letters are divided into vowels and consonants, we can investigate the vowel-consonant ratio or transition probabilities between vowels and consonants. While we cannot determine syllable boundaries without further knowledge, for some languages it is possible to count the (approximative) number of syllables per word and hence calculate the average syllable length. Here we approximate the number of syllables by the number of vocal blocks (consisting of one or more vocals) in the word. The quality of this approximation depends on the language, some language dependent additional rules are helpful.

3.3 Words and Multi-Words

Information about words is one of the most frequently used data to get an introduction to corpora. One of the reasons is its simplicity: word lists are almost always available and easy to generate (cf. [11]). Therefore a variety of different measures are used by several authors. Among others these include:

- Word length distribution (cf. [10])
- Text coverage

Rank	Frequency	Bigram
1	127758	-en
2	62558	-er
3	28815	-ng
4	25835	-te
5	19258	-es

Rank	Frequency	Trigram
1	29360	-ten
2	23406	-ung
3	20802	-gen
4	12387	-ter
5	12133	-hen

Fig. 2 Most frequent bi- and trigrams at words endings in a German corpus

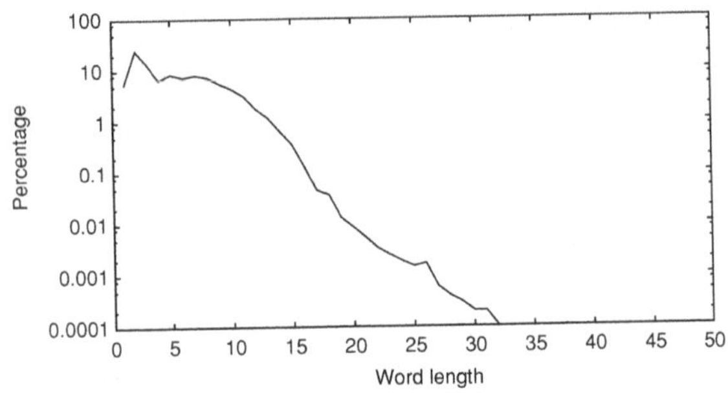

Fig. 3 Percentage of words for different word length in a Spanish newspaper corpus

- Sample words (with specific properties)
- Longest words in different frequency ranges
- Several variants of Zipf's law (cf. [17])
- Character transition probabilities
- Word string similarity (for example by using the Levenshtein distance)
- Average word length
- etc.

As a simple example Fig. 3 shows the distribution of word length in a Spanish newspaper corpus.

3.4 Sentences

The distribution of sentence lengths measured in words or characters should follow a typical pattern. Figure 4 shows the sentence length in characters for German newspaper texts and Ido Wikipedia texts. While the first distribution is as expected, the Ido Wikipedia shows remarkable irregularities. This is due to the fact that the Ido Wikipedia contains large portions of text generated by copy and paste with only minor editing.

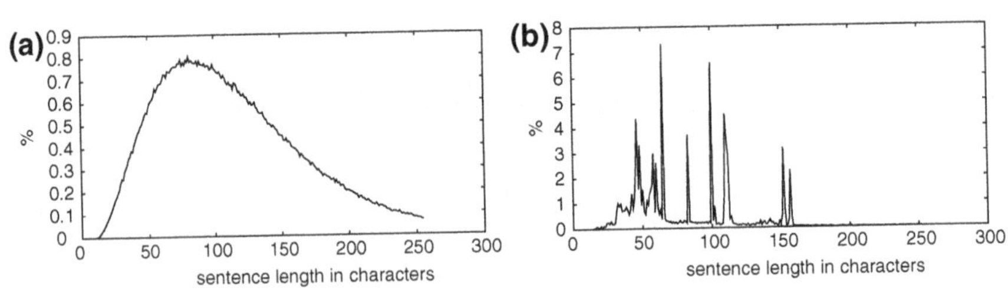

Fig. 4 Sentence length distribution in characters for a German and an Ido corpus

Typical sentence beginnings and endings of a certain number of words (N = 1, 2, 3, 4) give insights in typical phrases used in these positions. Figure 5 shows these values for an Australian English Web-based corpus.

Sample sentences are interesting for various reasons and can be chosen by using quite different methods. Here we will select sentences due to the average word rank (Fig. 6). Sentences with low average word rank will consist of mostly high frequent words. Those sentences are usually short and easy to understand. On the contrary, sentences with high average word rank do not contain the usual amount of stop words, have long average word length and are often very technical. In the case of poor corpus preprocessing there are some more possible reasons: The sentence may have another language or dialect, or the string under consideration does not contain meaningful words at all. Similar, but slightly different criteria to average logarithmic word rank are:

- sentences with minimal rank for highest ranked word
- sentences with maximal rank for lowest ranked word

N	Frequency	Sentence
1	429967	The
1	114828	This
2	36999	It is
2	22182	This is
3	4478	This is a
3	4030	If you are
4	1017	If you would like
4	968	It is important to

N	Frequency	Sentence
1	17480	it.
1	15830	Australia.
2	5359	in Australia.
2	4600	the world.
3	1493	in the world.
3	1287	New South Wales.
4	721	as soon as possible.
4	496	at the same time.

Fig. 5 Most frequent sentence beginnings/endings in an Australian web corpus for word lengths N = 1, 2, 3, 4

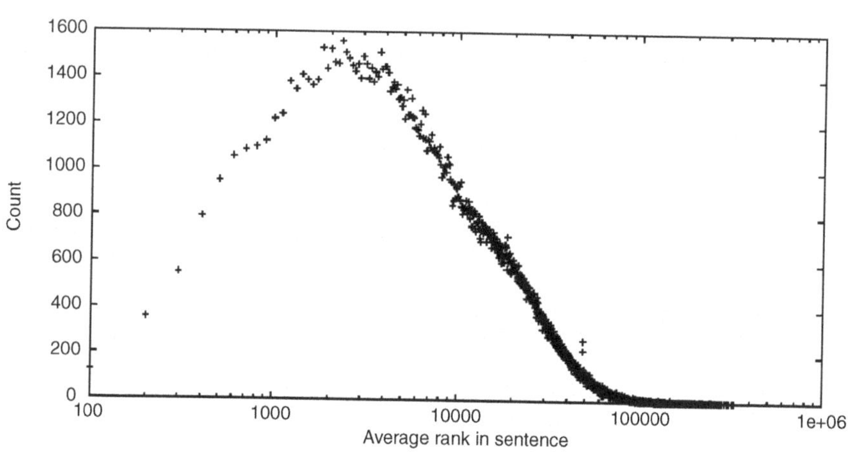

Fig. 6 Average word rank in sentences of a Danish web corpus

Table 1 Sentences with minimal rank of highest ranked word in a British Web corpus

Minimal rank	Sentence
55	This is who we are
74	We would not do that
80	If I had it to do over, I would
82	It's not for you to use

Table 2 Sentences with maximal rank of lowest ranked word in a British Web corpus

Maximal rank	Sentence
8111	Ave, ave, ave Maria!
7082	Happy Birthday Benny!
6923	Emma Watson Rocks!
5495	Gas Generator Manufacturer Expands!

Tables 1 and 2 show some examples for these two issues based on an English corpus.

3.5 Sentence Similarity

The treatment of duplicate sentences and nearly duplicate sentences depends on the purpose of the corpus. Especially in the case of a Web corpus, duplicates are usually unwanted. They are easy to remove by sorting all sentences. The treatment of similar sentences is more complicated. First, sentence similarity can be defined in several ways (see for example [12]). Some straightforward solutions are:

- using a hash function which respects similarity in some way, i.e.
  - ignores numbers and punctuation before hashing, or
  - uses only the word length instead of the words etc.
- comparing sentences as bag of words and ignoring all but the five (or so) least frequent words in a sentence.

For the measurement process, we can count the amount of sentences having similar sentences. If this amount is too high (either in general or for some of the sentences), one might want to remove all but one copy in a later corpus processing step.

3.6 Word Co-Occurrences

Calculating word co-occurrences (either as next neighbors or within a well defined window like a sentence) with a significance measure like Dunning's Log-likelihood (cf. [6]) allows several measurements:

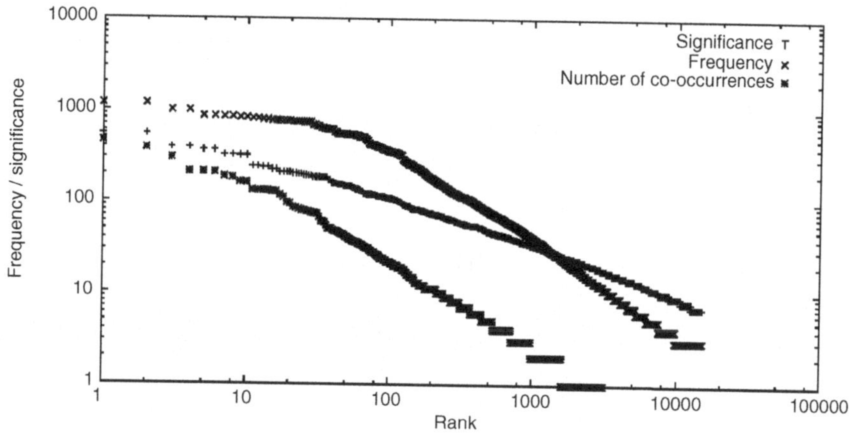

Fig. 7 Frequency resp. Significance resp. Number of co-occurrences per word versus rank in a Swedish newspaper corpus

- Choosing corpora of fixed size and the same lower threshold for the significance: How much word co-occurrences are found?
- What is the distribution of the significance? We can order all co-occurrence pairs by significance, starting with the pair having highest significance. Next define the rank of such a co-occurrence pair by its position according to this ordering. A Zipfian plot rank versus significance is shown in Fig. 7.
- One might measure semantic similarity of two words by the number of common co-occurring words. Results can be combined for different corpora or different languages.
- The co-occurrence relation generates a graph with words as nodes and weighted edges. Such graphs are known to have small world properties (cf. [16]). The corresponding small world parameters can be used for comparison.

3.7 Sources

In the case of newspaper corpora, usually there are different newspapers as sources. These sources may have different language characteristics resulting in different parameters. For a corpus, we are interested in the number of used sources, their typical size etc. Moreover, differences between the various sources measured in parameters as above and sources with extreme parameters are of interest. Table 3 shows the average sentence length in characters for some sources of a Russian corpus.

Table 3 Excerpt of Russian web sources list ordered by average sentences length in characters

Average sentence length	Source
74.60	http://www.bookap.by.ru/
65.68	http://militera.lib.ru/
56.54	http://lib.ru/
56.01	http://www.diary.ru/

3.8 *Variation of a Parameter Within the Corpus*

If a parameter is measured for many objects within a corpus (like for any word or for any sentence), both mean value and variance are of interest. Moreover sample objects with extreme parameter values are of interest, too, for various reasons. Such objects may be of special linguistic interest, may contain some entertainment value, or show some processing weakness (cf. Sect. 5.2).

3.9 *Variation of a Parameter with Corpus Growth*

Some parameters are related to the list of words which, of course, increase with corpus size. Some of them behave very regularly with the increase of the word list. For better comparison, we will consider a word list generated by a very large corpus. Instead of word lists generated by different corpora we will look at this given word list from the beginning up to a certain rank, usually 10^N for $N = 1, 2, \ldots$

Figure 8 shows two examples based on an Ukrainian corpus: the number of letter N-grams and the average word length for the 10^N most frequent words.

4 Linguistic Properties

From these purely statistical features, more linguistic properties can be derived:

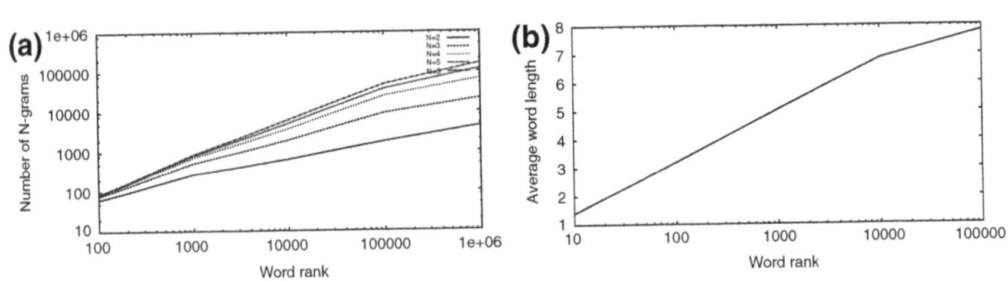

Fig. 8 Changing results for different word ranges (Ukrainian Wikipedia corpus). **a** Number of letter-N-grams for $N = 2, \ldots, 6$ for the most frequent words. **b** Average word length for the most frequent words

- Word formation hypotheses: Finding candidates for affixes, simple compound decomposition.
- Presence or absence of articles, prepositions and other kinds of high frequent function words
- Typical sentence beginnings and endings. Hence, candidates for constituents.
- Strength of word order in sentences.

In the case of language comparison, the parameters described above can be compared and aggregated result vectors can be visualized by using standard clustering algorithms. A deeper comparison of languages and semantic domains is possible by identifying cognates and describing typical variations.

5 Applications

5.1 Intra-Language and Inter-Language Comparisons

While language dependent parameters are expected to vary for different languages, their behavior for different genres within one language is difficult to predict. The Tables 4 and 5 compare three parameters (text coverage for the 20 most frequent words, the average word length in characters without multiplicity and the average sentence length in words) first for different text genres of German, and then the same parameters for newspaper corpora for different languages. The intra-language variation may help to decide whether differences between languages can be considered as significant. Moreover, for corpora of mixed or unknown genre such data help to decide whether more detailed information about the genres are necessary.

Table 4 Intra-language comparison for German corpora

Text type	Text coverage (20 top words) (%)	Average word length	Average sentence length
News	22.10	13.59	16.19
Web	21.57	14.06	16.03
Wikipedia	23.10	12.57	16.71
Movie subtitles	21.20	10.42	6.57

Table 5 Inter-language comparison

Language	Text coverage (20 top words) (%)	Average word length	Average sentence length
German	22.10	13.59	16.19
English	26.23	10.62	19.46
Czech	16.78	8.65	14.95
Vietnamese	12.44	4.97	23.64
Finnish	12.37	12.28	11.50

5.2 Using Statistical Anomalies for Corpus Cleaning

In many of the above mentioned analysis types a special value is measured for many objects like sentence length in characters for every sentence of a corpus. Looking at objects with extreme values (i.e. very small or very large), we often find effects of errors in the input material or poor preprocessing (cf. [9] or [5]). In the case of very short sentences we may find broken sentences. Moreover, sentences containing many very low frequent words are usually not well-formed. Table 6 shows sentences of an English Web corpus without high frequent words. Apparently some non-sentences were not removed by the cleaning component and the language identification failed in rejecting some non-English sentences.

Table 7 shows an excerpt of the distribution of semicolons in a 100,000 sentences Ukrainian corpus. These sentences that were accepted by the quality assurance procedures include among others:

- "Prossimo turno (22/5): Milan-Palermo (20/5, ore 20,30); Atalanta-Roma; Brescia-Messina; Cagliari-Udinese; Chievo-Bologna; Lazio-Fiorentina; Livorno-Juventus; Parma-Siena; Reggina-Lecce; Sampdoria-Inter." (Italian)
- "Machiaj: Divizia Make-up DUMAREX Parteneri media: EVENIMENTUL ZILEI; ZIUA; JURNALUL NATIONAL; CAPITALA; COTIDIANUL; METRO-BUS; AZI; BURDA ROMANIA; ANTENA 1 - Doina Levintza, "'Neata"; PRIMA TV - "Clubul de Duminica", "Stil"." (Romanian)
- "Siippainen kirjoitti lehtijuttujaan eri nimimerkeill kuten Iloinen, Petteri; Kaaleppi; Karho, Otto; Kimpinen; Kimpinen, Kalle; Miki, Urmas; O. S.; O. S-nen; Robin Hood; Saarto, Olavi; Svejk; Uolevi." (Finnish).

Table 6 Examples of sentences without high frequent words

Rank of most frequent word	Sentence
208	Fracture mechanics data—general—basic principles, example calculations, NDT
1773	Respiratory system: asthma, coughs, hay fever, helps sore throats
38817	Apa yang merupakan perzinahan dalam dunia monyet?
335857	Tetapi itupun tidak berkesan!

Table 7 Part of a semicolon distribution in Ukrainian sentences

# of semicolons	# of sentences
2	183
3	28
4	16
5	4
6	1
12	1

Another hint for problems in the corpus generation process is looking at extreme points of the distribution of specific characters.

This information provides a fast feedback and leads to more accurate data resources in the future. Statistical values that may indicate problems with input selection, inaccurate preprocessing tools or other issues are widely spread, ranging from character analysis to show character set problems to automated rating of the corpora sources based on their homogeneity of various statistical values. This is still to be evaluated.

6 Presentation

Central goal for the presentation of the created result pages was a web portal that should allow both researchers in the fields of natural language processing and linguistics an easy access and overview of existing corpora and a starting point for evaluating linguistic phenomena in the field of corpus, genre and language comparison. Each question, answered for a certain corpus, produces an HTML page containing the results. As described above, these result pages consist of a plot or of a (set of) table(s), or both. For comparisons, all corpora are assigned to three different categorization dimensions: language, text genre and corpus size. The Corpora and Language Statistics Website presented at http://www.cls.informatik.uni-leipzig.de supports this complex navigation. To achieve an easy access, despite the thousands of pages strongly related to each other, the ISO standard Topic Maps was used as underlying technology. Based on JRuby Topic Maps (cf. [2]) and tinyTIM, all existing resources were merged while allowing extensions to new fields and dimensions in the future (Fig. 9).

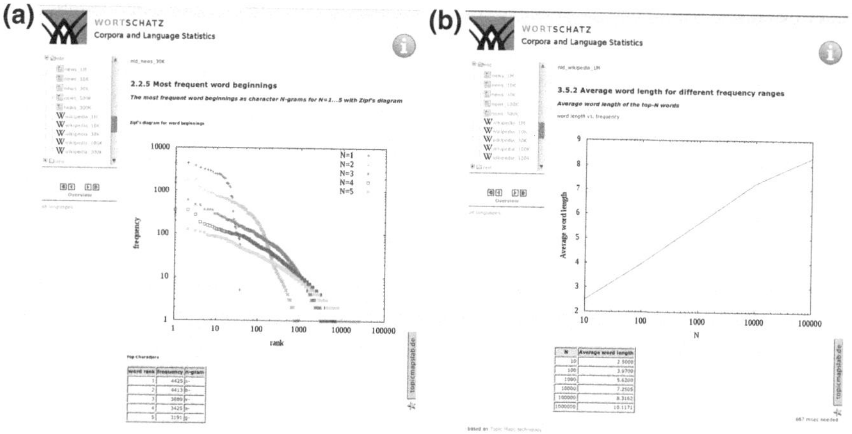

Fig. 9 Sample HTML pages. **a** Most frequent word beginnings. **b** Average word length for different frequency ranges

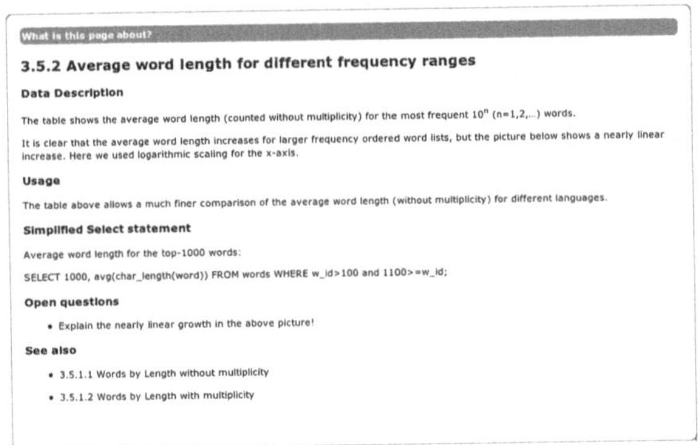

Fig. 10 Sample help screen

The user interface is designed to lower the entry barrier for the (possibly inexperienced) user: on the left side one can select between other languages, genres and corpus sizes. These links will show the corresponding page for the same question, but another corpus. The arrows allow linear scrolling through the different questions for one corpus.

An additional help screen gives detailed information about the data shown and the intentional background of the question (Fig. 10). A (sometimes slightly simplified) select-statement is provided. This can be used or modified for similar questions asked by the user. Some open problems and cross references complete this help screen. Those additional hints allow reproducibility and are basis for creation of new analysis scripts.

References

1. Biemann, C., Heyer, G., Quasthoff, U., Richter, M.: The Leipzig corpora collection—monolingual corpora of standard size. In: Proceedings of Corpus Linguistic, Birmingham, UK (2007)
2. Bleier, A., Bock, B., Schulze, U., Maicher, L.: JRuby topic maps. In: Proceedings of the 5th International Conference on Topic Maps Research and Applications (TMRA 2009), Leipzig, Germany (2009)
3. Buechler, M.: Medusa: Performante Textstatistiken auf grossen Textmengen - Kookkurrenz-analyse in Theorie und Anwendung. Vdm Verlag Dr. Müller, Saarbrücken (2008)
4. Buechler, M., Heyer, G.: Leipzig linguistic services—a 4 years summary of providing linguistic web services. In: Proceeding of TMS 2009 conference, Augustusplatz 10/11, 04109, Leipzig, Germany (2009)
5. Dickinson, M., Meurers, D.: Detecting annotation errors in spoken language corpora. In: Proceedings of the Special Session on Treebanks for Spoken Language and Discourse at the 15th Nordic Conference of Computational Linguistic (NODALIDA-05), Joensuu, Finland (2005)
6. Dunning, T.: Accurate methods for the statistics of surprise and coincidence. Comput. Linguist. **19**(1), 61–74 (1993)

7. Dunning, T.: Statistical identification of language. In: Technical report CRL MCCS-94-273. Computing Research Lab, New Mexico State University (1994)
8. Eckart, T., Quasthoff, U.: Statistical corpus and language comparison using comparable corpora. In: Workshop on Building and Using Comparable Corpora, LREC 2010, Malta (2010)
9. Eskin, E.: Automatic corpus correction with anomaly detection. In: Proceedings of the 1st Conference of the North American Chapter of the Association for Computational Linguistics (NAACL-00). Seattle, Washington, USA (2000)
10. Grzybek, P.: History and methodology of word length studies. The State of the Art. In: Grzybek, P. (Hrsg.) Contributions to the Theory of Text and Language. Word Length Studies and Related Issues, S. 15–90. Springer, Dordrecht (NL). ISBN 1-4020-4067-9 (HB) (2006)
11. Kilgarriff, A.: Using word frequency lists to measure corpus homogeneity and similarity between corpora In: Proceedings of 5th ACL SIGDAT Workshop on Very Large Corpora, pp. 231–245. Beijing and Hong Kong (1997)
12. Li, Y., McLean, D., Bandar Z., O'Shea, J., Crockett, K.: Sentence similarity based on semantic nets and corpus statistics. IEEE Trans. Knowl. Data Eng. **18**, 8 (2006)
13. Meier, H.: Deutsche Sprachstatistik. Olms, Hildesheim (1964)
14. Quasthoff, U., Eckart, T.: Corpus building process of the project "Deutscher Wortschatz". In: GSCL Workshop: Linguistic Processing Pipelines, Potsdam, Germany (2009)
15. Quasthoff, U., Richter, M., Biemann, C.: Corpus portal for search in monolingual corpora. In: Proceedings of the LREC 2006, Genoa, Italy (2006)
16. Watts, D.J., Strogatz, S.H.: Collective dynamics of 'small-world' networks. Nature **393**(6684), 440–442 (1998)
17. Zipf, G.K.: Human Behavior and the Principle of Least Effort: An Introduction to Human Ecology. Hafner reprint. New York, 1972, 1st edn. (Addison-Wesley, Cambridge, MA, 1949)

Comparable Multilingual Patents as Large-Scale Parallel Corpora

Bin Lu, Ka Po Chow and Benjamin K. Tsou

Abstract Parallel corpora are critical resources for building many NLP applications, ranging from machine translation (MT) to cross-lingual information retrieval. In this chapter, we explore a new but important area involving patents by investigating the potential of cultivating large-scale parallel corpora from comparable multilingual patents. Two major issues are investigated on multilingual patents: (1) How to build large-scale corpora of comparable patents involving many languages? (2) How to mine high-quality parallel sentences from these comparable patents? Four parallel corpora are presented as examples, and some preliminary SMT experiments are reported. We further investigate and show the considerable potential of cultivating large-scale parallel corpora from multilingual patents for a wide variety of languages, such as English, Chinese, Japanese, Korean, German, etc, which would to some extent reduce the parallel data acquisition bottleneck in multilingual information processing.

Keywords Multilingual patents · PCT patents · Parallel corpora · Machine translation · Sentence alignment

---

This chapter is based on the authors' previous work described in Lu et al. (2009, 2010a, 2010b, 2011).

B. Lu(✉) · B. K. Tsou
Department of Chinese, Translation and Linguistics, City University of Hong Kong, Kowloon, Hong Kong
e-mail: lubin2010@gmail.com; btsou99@gmail.com

B. Lu · K. P. Chow · B. K. Tsou
Research Centre on Linguistics and Language Information Sciences, Hong Kong Institute of Education, New Territories, Hong Kong

K. P. Chow
Hong Kong Institute of Education, New Territories, Hong Kong
e-mail: kpchow@ied.edu.hk

S. Sharoff et al. (eds.), *Building and Using Comparable Corpora*,
DOI: 10.1007/978-3-642-20128-8_9, © Springer-Verlag Berlin Heidelberg 2013

1 Introduction

Multilingual data are critical resources for building many applications, such as machine translation (MT) and cross-lingual information retrieval. Many parallel corpora have been built, such as the Canadian Hansards [9], the Europarl corpus [12], the Arabic-English and English-Chinese parallel corpora used in the NIST Open MT Evaluation.[1] However, few parallel corpora exist for many language pairs, such as Chinese-Japanese, Japanese-Korean, Chinese- French or Japanese-German. Even for language pairs where parallel corpora exist, such as Chinese-English and Arabic-English, the size of parallel corpora is still a major limitation for SMT systems to achieve higher performance.

In this chapter, we explore a new but important area involving patents by investigating the potential of comparable multilingual patents for building large-scale parallel corpora. Multilingual patents are not strictly parallel but considered comparable because of the similarity in content. We present a way to mine high-quality parallel sentences from these comparable patents by combining multiple clues for parallel sentence extraction, including length information, bilingual lexical overlap, and statistical translation probability. Some preliminary SMT experiments are also reported to estimate the extent to which increasing the size of the training corpus is useful to SMT.

Moreover, we investigate the potential to get large-scale parallel corpora for languages beyond the Canadian Hansards, Europarl and UN news used in NIST MT Evaluation. An estimation of the quantity of multilingual patents involving English, Chinese, Japanese, Korean, German, etc., is made. Since patents cover many technical domains (e.g. chemistry, biomedicine, electronics and vehicle), the large-scale parallel corpora mined from multilingual patents could be valuable resources not only in the patent domain but also in the related technical domains mentioned above.

One million high-quality Chinese-English sentence pairs have been provided to registered participants in the Patent MT evaluation at NTCIR-9,[2] and may be made publicly available to the research community with the hope to promote the research on Chinese-English machine translation in the patent domain.[3]

Related work is introduced in Sect. 2. Patents, PCT patents, multilingual patents are described in Sect. 3. Then harvesting in Sect. 4; alignment in Sect. 5; three parallel corpora are introduced in Sect. 6, the application to SMT is introduced in Sect. 7, followed by the estimation of the quantity of multilingual patents in Sect. 8. We discuss the results in Sect. 9 and conclude in Sect. 10.

---

[1] http://www.itl.nist.gov/iad/mig/tests/mt/.

[2] Anyone interested in the corpus are invited to contact the authors for more details.

[3] Retrieved March 2010, from http://www.collinslanguage.com/.

2 Related Work

Different approaches have been proposed to align sentences of bilingual documents which should ideally allow for an isomorphic relationship in content expressed through the different languages. Previous efforts have been based on the following information: (a) the sentence length in bilingual sentences [2, 9]; (b) lexical information in bilingual dictionaries [23]; (c) statistical translation model [5], or the composite of more than one approach [24, 30].

To overcome the lack of parallel documents, comparable corpora are also used to mine parallel sentences, which raises further challenges for finding parallel sentences since the bilingual contents are not strictly parallel. For instance, Zhao and Vogel [34] investigated the mining of parallel sentences for Web bilingual news collections which may contain much noise. Resnik and Smith [28] introduced the STRAND system for mining parallel text on the web for low-density language pairs. Munteanu and Marcu [25] presented a method for discovering parallel sentences in large Chinese, Arabic, and English comparable, non-parallel corpora based on a maximum entropy classifier. Wu and Fung [32] exploited Inversion Transduction Grammar to retrieve truly parallel sentence translations from large collections of highly non-parallel documents.

Another direction is to directly mine bilingual terms from the Web. Cao et al. [4] and Lin et al. [15] proposed two different methods utilizing the parenthesis pattern to extract term translations from bilingual web pages. Jiang et al. [16] presented an adaptive pattern-based bilingual data mining method which produced more than 7.5 million pairs of Chinese-English bilingual sentences and more than 5 million pairs of Chinese-English bilingual terms, both with over 80 % accuracy.

Smith et al. [29] investigated the viability of Wikipedia as a comparable corpus and extracted parallel sentences from it. In their experiments, they extracted more than 1 million sentences pairs for two language pairs, namely German and English and Spanish and English, as well as 140 thousand Bulgarian-English parallel sentences.

Only a few papers on related work in the patent domain. Higuchi et al. [11] used the titles and abstracts of 32,000 Japanese-English bilingual patents to extract bilingual terms. Utiyama and Isahara [31] mined about 2 million parallel sentences by using the "*Detailed Description of the Preferred Embodiments*" and "*Background of the Invention*" parts in the *description* section of Japanese-English comparable patents. The corpus was used for the NTCIR-7 Japanese-English patent machine translation task [7].

Multilingual Wikipedia has been a hot topic in the NLP field (e.g., [1, 29]). Multilingual patents have similar features ti wikipedia (e.g., large amounts of data and rapidly increasing new materials), but they also have different characteristics, which results in the difference of attention in the NLP field (see Sect. 9.1 for more discussion).

For statistical machine translation (SMT), tremendous strides have been made in the last two decades, including Brown et al. [3], Och and Ney [27] and Chiang [6]. For the MT evaluation, NIST has been organizing open evaluations for years, and the performance of the participants has been improved rapidly. The patent machine translation task at NTCIR-7 [7] and NTCIR-8 [8] has tested SMT performance on only the Japanese-English patent translation.

3 Patents and Multilingual Patents

In this section, patents, PCT patents and multilingual patents are introduced.

3.1 Patents

A patent is a legal document representing "*an official document granting the exclusive right to make, use, and sell an invention for a limited period*" (Collins English Dictionary). A patent application consists of different sections, namely, *bibliographical data (including title,abstract), drawings, claims, description*, etc. Since we focus on the text in the patent applications, only *title, abstract, claims and description* are used in the experiments discussed below.

Patent applications have been increasing very quickly in recent years as illustrated in Fig. 1. The application numbers filed in the leading patent offices from 1996 to 2008 are shown. We can observe that in about 12 years, China's patent applications have increased by 10 times, and USA and R. Korea have doubled in their patent applications, but USA and Japan are still the top two with most applications.

3.1.1 PCT Patents

Since the invention in a patent is only protected in the filing countries, a patent applicant who wishes to protect his invention outside the original country should file patents in other countries, which may involve other languages.

The Patent Cooperation Treaty (PCT) system offers inventors and industry an advantageous route for obtaining patent protection internationally. By filing one "*international*" patent application under the PCT via the World Intellectually Property Organization (WIPO), protection of an invention can be sought simultaneously (i.e. the priority date) in each of a large number of countries.

The number of PCT international applications filed is more than 1.8 million.[4] A PCT international application may be filed in any language accepted by the relevant receiving office, but must be published in one of the official publication languages (Arabic, Chinese, English, French, German, Japanese, Korean, Russian and Spanish). Other highly used languages for filing include Italian, Dutch, Finnish, Swedish, etc.

[4] Retrieved April, 2010 from http://www.wipo.int/pctdb/en/. The data below involving PCT patents comes from the website of WIPO.

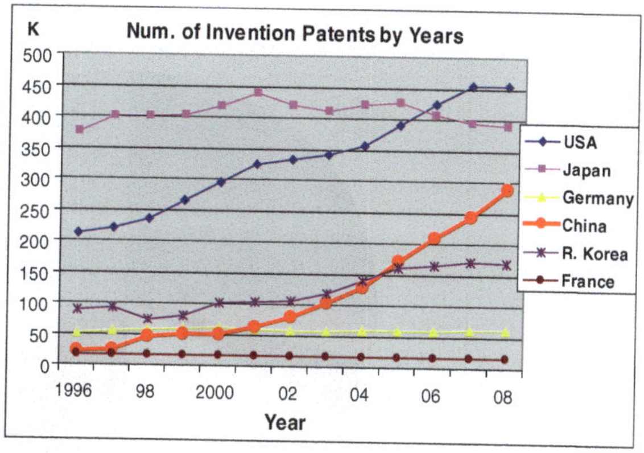

Fig. 1 Applications by the leading patent offices. Retrieved March 2010, from http://www.wipo.int/ipstats/en/statistics/patents/csv/wipo_pat_appl_from_1883_list.csv

Table 1 PCT application numbers for languages of publication and filing

	Lang. of filing	Share (%)	Lang. of publication	Share (%)
English	895K	52.1	943K	54.9
Japanese	198K	11.5	196K	11.4
German	185K	10.8	184K	10.7
French	55K	3.2	55K	3.2
Korean	24K	1.4	24K	1.4
Chinese	24K	1.4	24K	1.4
Other	336K	19.6	313K	18.2
Total	1.72M	100	1.72M	100

Table 1 shows the number of PCT applications for the most used languages of filing and publication.

From Table 1, we can observe that English, Japanese and German are the top 3 languages in terms of PCT applications, and English accounts for over 50% of applications in terms of language of both publication and filing. Figure 2 below shows the distribution of PCT applications for countries of origin in 2008, which is taken to be the country of residence of the first-named applicant in the PCT application. Figure 3 shows PCT application numbers for countries of origin from 2004 to 2009.

From Figs. 2 and 3, we can see that USA, Japan and Germany are the top 3 countries where the most PCT applications are originated. Although China increases its PCT applications quite rapidly in recent years, the number is still small. Compared to the patent numbers in Fig. 1, the gap of PCT application numbers between USA and China is much larger than that of application numbers, showing that China organizations are still in the process of learning to file PCT patents to protect their inventions outside China.

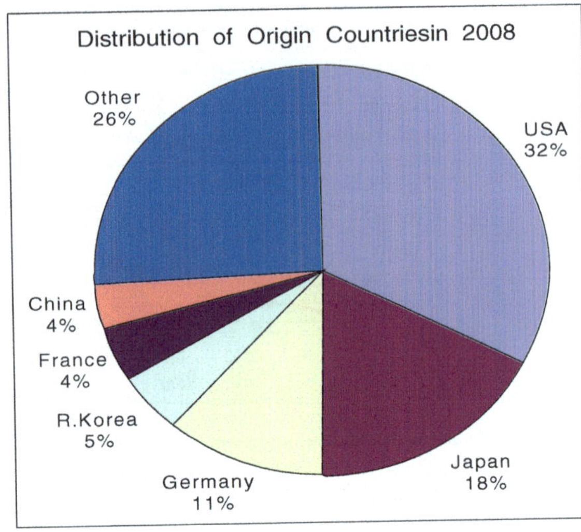

Fig. 2 Distribution of PCT applications for countries of origin in 2008

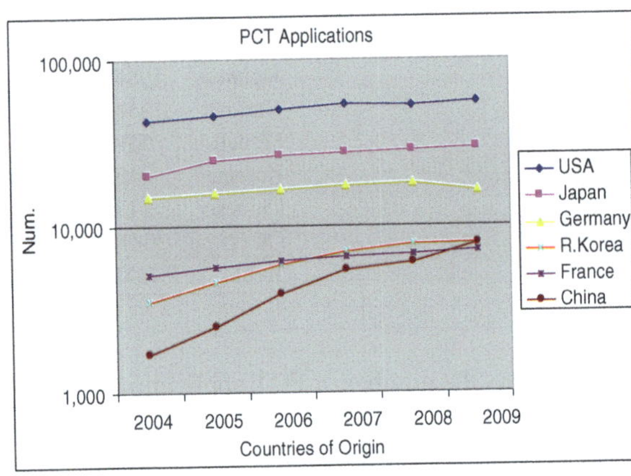

Fig. 3 PCT application numbers for the countries of origin by years

3.1.2 Multilingual Patents

A PCT application does not necessarily mean a multilingual patent. For example, a USA applicant who has submitted a PCT application through USA Patent Office may choose not to file it in other countries with other languages, and then that patent application would be only monolingual, instead of multilingual.

An applicant who has decided to proceed further with his PCT international application must fulfill the requirements for entry into the PCT national phase at the patent offices of countries where he seeks protection. For example, a Chinese company may first file a Chinese patent in China patent office and then file its international application also in Chinese under the PCT. Later on, it may have the patent translated into

English and file it in USA patent office, which means the patent becomes bilingual. If the applicant continues to file it in Japan with Japanese, it would be trilingual. Even more, it would be quadrilingual or involve more languages when it is filed in other countries.

Such multilingual patents are considered comparable (or noisy parallel) because they are not parallel in the strict sense but still closely related in terms of information conveyed [11, 17].

4 Harvesting Multilingual Patents from the Web

The official patent office in China is the State Intellectual Property Office (SIPO). In early 2009, by searching on its website,[5] we found about 200K Chinese patents previously filed as PCT applications in English and crawled their *bibliographical data, titles, abstracts* and *the major claim* from the Web, and then *other claims* and *descriptions* were also added.[6]

All PCT patent applications are filed through WIPO. With the Chinese patents mentioned above, the corresponding English patents were searched from the website of WIPO to obtain relevant sections of the English PCT applications,[7] including *bibliographical data, title, abstract, claims* and *description*. About 80 % (160K) out of the Chinese patents found their corresponding English ones.

Out of the above 160K bilingual patents, we managed to locate the corresponding Japanese version of about 130K of them from the Industrial Property Digital Library,[8] which is provided by Japan's National Center for Industrial Property Information and Training and is the public access portal of the Japanese Patent Office (JPO). The Japanese texts are downloaded and segmented using the chasen[9] utility.

5 Mining Parallel Sentences from Comparable Patents

The sentences in each section of Chinese patents were aligned with those in the corresponding section of the corresponding English patents to find parallel sentences after the Chinese sentences were segmented into words.

Since the comparable patents are not strictly parallel, the individual alignment methods mentioned in Sect. 2 would be not effective: (1) the length-based method is not accurate since it does not consider content similarity; (2) the bilingual dictionary-

---

[5] http://www.sipo.gov.cn/

[6] Some contents are in image format. Thus the images were OCRed and the characters recognized were manually verified.

[7] Some contents of the English patents were OCRed by WIPO.

[8] http://www.ipdl.inpit.go.jp/homepg.ipdl

[9] http://chasen.naist.jp/hiki/ChaSen/

based method cannot deal with new technical terms in the patents; (3) the translation model-based method would need training data to get a translation model. Thus, in this study we combine these three methods to mine high-quality parallel sentences from comparable patents.

We first use a bilingual dictionary to preliminarily align the sentences in each section of the comparable patents. The dictionary-based similarity score P_d of a sentence pair is computed based on a bilingual dictionary as follows [31]:

$$p_d(S_c, S_e) = \frac{\sum\limits_{w_c \in S_c} \sum\limits_{w_e \in S_e} \frac{\gamma(w_c, w_e)}{\deg(w_c)\deg(w_e)}}{(l_e + l_c)/2}$$

where w_c and w_e are respectively the word types in Chinese sentence S_c and English sentence S_e; l_c and l_e respectively denote the lengths of S_c and S_e in terms of the number of words; and $\gamma(w_c, w_e) = 1$ if w_c and w_e is a translation pair in the bilingual dictionary or are the same string, otherwise 0; and

$$\deg(w_c) = \sum_{w_e \in S_e} \gamma(w_c, w_e)$$

$$\deg(w_e) = \sum_{w_e \in S_c} \gamma(w_c, w_e).$$

For the Chinese-English bilingual dictionary, we combine three resources: namely, LDC_CE_DIC2.0[10] constructed by LDC, bilingual terms in HowNet and the bilingual lexicon in Champollion [23].

We then remove sentence pairs using length filtering and ratio filtering: (1) for length filtering, if a sentence pair has more than 100 words in the English sentence or more than 333 characters in the Chinese one, it is removed; (2) for length ratio filtering, we discard the sentence pairs with Chinese-English length ratio outside the range of 0.8 to 1.8. The parameters here are set empirically.

We further filter the parallel sentence candidates by learning an IBM Model-1 on the remaining aligned sentences and compute the translation similarity score P_t of sentence pairs by combining the translation probability value of both directions (i.e. Chinese->English and English->Chinese) based on the trained IBM-1 model [17, 24]. It is computed as follows:

$$p_t(S_c, S_e) = \frac{\log(P(S_e|S_c)) + \log(P(S_c|S_e))}{l_c + l_e}$$

where $P(S_e|S_c)$ denotes the probability that a translator will produce S_e in English when presented with S_c in Chinese, and vice versa for $P(S_c|S_e)$. Sentence pairs with similarity score P_t lower than a predefined threshold are filtered out as wrong aligned sentences.

[10] http://projects.ldc.upenn.edu/Chinese/LDC_ch.htm

Table 2 Statistics for each section

Sections	Chinese		English	
	#Chars	#Sents	#Words	#Sents
Title	89K	7K	49K	7K
Abstract	1.5M	29K	1.0M	32K
Claim	6.4M	145K	5.9M	201K
Description	27.7M	557K	20.9M	840K
Total	35.7M	738K	27.9M	1,080K

6 Parallel Corpora Mined from Multilingual Patents

In this section, we introduce the English-Chinese bilingual patents harvested from the Web and outline the methodology by which parallel sentences were mined from them. SMT experiments on the final parallel corpus are also described.

6.1 A Small Chinese-English Bilingual Corpus

This small corpus is extracted from about 7,000 Chinese-English comparable patents with about 6,000 of them containing full texts [17, 21]. The patents were first filed in the China Patent Office with Chinese as the original language. They were translated into English, and then filed in *USPTO* (United States Patent and Trademark Office).

We first get access to the English patents filed in USPTO, and then their Chinese corresponding patents were identified by using the priority information described in the *USPTO* patents. The priority information is actually the application number of the Chinese patent. Therefore we can get the sections for Chinese patents. In total, there are about 730K Chinese sentences and 1,080K English sentences in the comparable patents. The detailed statistics for each section are given in Table 2, and the statistics for the final parallel corpus in Table 3. The manual evaluation of the 160K parallel sentences shows that 92 % of the sentences are *correct* or *partially correct.*[11]

Table 3 Basic statistics of the small parallel corpus

#Patents	#Sentence Pairs	#Word Tokens		#Word Types	
		EN	CN	EN	CN
7K	160K	4,168K	4,130K	46K	44K

[11] *Correct* means the English sentence is exactly the literal translation of the Chinese one, or the content overlap between them are above 80 % with no need to consider phrasal reordering during the translation; partially correct means the Chinese sentence and the English one are not the literal translation of each other, but the content of each sentence can cover more than 50 % of the other; incorrect means the contents of the Chinese sentence and the English one are not related, or more than 50 % of the content of one sentence is not translated in the other. Please see [17] for more details.

Table 4 Selected percentages of different sections

Section	Title	Abstr.	Claims	Description	Total
#Candidates	7,029	15,755	56,667	275,737	352K
#Final Pairs	5,232	7,119	36,722	112,812	160K
Selected (%)	74.4	45.2	64.8	40.9	45.4

Table 5 Statistics of comparable patents

Sections	Chinese		English	
	#Char	#Sent	#Word	#Sent
Title	2.7M	157K	1.6M	157K
Abstract	33M	596K	20M	784K
Claim	367M	6.8M	217M	7.4M
Desc.	2,467M	48.8M	1,353M	54.0M
Total	2,870M	56.2M	1,591M	62.3M

We also compare the numbers of sentence pairs among different sections in the final corpus. The results in Table 4 show that the title and claims sections have two highest precisions: 74.4 and 64.8 % respectively; while the abstract and description sections show lower precisions: 45.2 and 40.9 % respectively. This shows that it is more difficult to find parallel sentences in the description or abstract section than in the title or claim sections, and that a large proportion of the patent titles are parallel.

6.2 A Large-Scale English-Chinese Bilingual Corpus

For the Chinese-English bilingual patents mentioned in Sect. 4, we automatically split them into individual sections according to the respective tags inside the patents, and segmented each section sentences according to punctuations. The statistics of each section for Chinese and English patents are shown in Table 5.

Based on the alignment approach mentioned in Sect. 5, we aligned the Chinese-English bilingual patents [19, 20]. Table 6 shows the sentence numbers and the percentages of sentences kept in each step above with respect to all sentence pairs. In the first row of Table 6, *1.DICT* denotes the first step of using the bilingual dictionary to align sentences; 2. *FL* denotes the length and ratio filtering; 3. *TM* refers to the final step of using translation models to filter sentence pairs.

Both the 31.5M parallel sentences after the second step *FL* and the final 14.3M after the third step *TM* are manually evaluated by randomly sampling 100 sentence pairs for each section. The evaluation metric follows the one in [17], which classifies each sentence pair into *Correct, Partially Correct* or*Wrong*. The results of manual evaluation are shown in Table 7.

From Table 7, we can see that: (1) In the final corpus, the percentages of *correct* parallel sentences are quite high, and the wrong percentages are no higher than 5 %; (2) Without the final step of TM, the accuracies of 31.5M sentence pairs are between 69 and 85 %, and the percentages of wrong pairs are between 7 and 15 %; 3) The

Table 6 Numbers of sentence pairs

	1. DICT	2. FL	3. TM (final)
Abstr.	503K	352K (70%)	166K (33%)
Claims	6.0M	4.3M (72.1%)	2.0M(33.4%)
Desc.	38.6M	26.8M (69.4%)	12.1M(31.3%)
Total	45.1M	31.5M (69.8%)	14.3M (31.7%)

Here the total number does not include the number of titles, which are directly treated as parallel

Table 7 Manual evaluation of the corpus

	Section	Correct (%)	Partially correct (%)	Wrong (%)
2. FL	*Abstr.*	*85*	*7*	*8*
	Claims	*83*	*10*	*7*
	Desc.	*69*	*15*	*15*
3. TM (final)	Abstr.	97	2	1
	Claims	92	3	5
	Desc.	89	8	3

Table 8 Statistics of comparable patents

Sections	English		Japanese	
	#Word	#Sent	#Word	#Sent
Title	1.6M	157K	1.4M	130K
Abstract	20M	784K	23M	460K
Claim	217M	7.4M	250M	3.8M
Desc.	1,353M	54.0M	854M	21.1M
Total	1,591M	62.3M	1128M	25.5M

abstract section shows the highest correct percentage, while the description section shows the lowest.

Thus, we could conclude that the mined 14M parallel sentences are of high quality with only 1–5 % wrong pairs, and our combination of bilingual dictionaries and translation models for mining parallel sentences are quite effective.

6.3 A Large-scale Japanese-English Bilingual Corpus

For the Japanese-English bilingual patents mentioned in Sect. 4, the similar pre-processing is conducted to get individual sections and sentences. The statistics of each section for Japanese and English patents are shown in Table 8.

We build the English-Japanese parallel corpus in several steps. In the absence of a good English-Japanese bilingual glossary suitable for alignment purpose, we perform first pass of alignment using sentence-length information, and obtain an automatic bilingual dictionary based on the sentence pairs with high scores using IBM model 1. In the second pass, we iteratively refine the alignment and learn the

Table 9 Numbers of English-Japanese sentence pairs

	RAW	1. HUNALIGN	2.FL	3. TM (Final)
Abstr.	460K	130K (28.2%)	126K (97%)	125K (99.4%)
Claims	3.8M	3.3M (86%)	2.4M (72.7%)	2.1M (87%)
Desc.	21.1M	6.7M (31.7%)	3M (44.7%)	2M (65%)
Total	25.36M	10.13M (39.9%)	5.52M (54.5%)	4.23M (76.6%)

automatically compiled dictionary, until it reaches a predefined iteration number. Next, we filter the sentences using a length ratio 0.5 to 2.0. Last, we do again the sentence alignment with IBM model 1 to obtain more accurate one-to-one translated sentence pairs.

Table 9 shows the sentence numbers and the percentages of sentences kept in each step above with respect to all sentence pairs. We will add the evaluation result in the camera-ready version.

6.4 A Chinese-English-Japanese Trilingual Corpus

With the trilingual patents harvested in Sect. 4, a trilingual sentence-aligned patent corpus can be established. As introduced in Sect. 6.2 and 6.3, the Chinese-English and English-Japanese corpora are first built; we then use the English sentences as the pivot to align Chinese-English and English-Japanese sentence pairs together as Chinese-English-Japanese sentence triplets.

This pivoting effort gives us 2.1M trilingual sentences with 1.4M from *Description*, 0.59M from *Claims* and 0.06M from *Abstract*. The preliminary manual evaluation of sampled sentences shows that about 70 % of the trilingual sentences are correctly aligned, but this accuracy is not satisfactory for the purpose as MT training data. Therefore, we further conduct the following filtering steps:

(1) First, automatically score each Chinese-Japanese sentence pair with a Chinese-Japanese bilingual dictionary, and keep only the highest scored 70 % pairs.
(2) Then, learn IBM Models [3, 26] on the remaining, sentences and compute the translation similarity score of sentence pairs by combining the translation probability value of both directions (i.e. Chinese->Japanese and Japanese->Chinese) based on the trained IBM models.
(3) Finally, sentences pairs with a translation similarity score lower than a predefined threshold are filtered out as wrong aligned sentences.

Finally, we arrive at about 1 million trilingual sentence triplets. The manual evaluation of 1,000 randomly sampled triplets show that 93 % of sentences are correctly aligned. Some triplet examples are shown in Appendix A. There may exist errors from the original patents in the sentences since some of them are OCR-ed, e.g., "either naturally occurring *of* artificially produced" should be "either naturally occurring *or* artificially produced" in the English sentence of the third triplet in Appendix A.

Table 10 SMT experimental results

System	BLEU (%)	#Sentence pairs for training
Model-A	17.94	300K
Model-B	19.96	750K
Model-C	20.09	1.5M
Model-D	20.98	3M
Model-E	22.60	6M

We also note that most sentences are not translated directly between Chinese and Japanese but via English, and thus the translations between Chinese sentences and Japanese ones sometimes seem incomplete or non-native.

6.5 Discussion on Augmenting on Trilingual Corpus

We had attempted to directly align Chinese and Japanese sentences in the Chinese-Japanese bilingual patents at the beginning. However, the results obtained were not satisfactory partly because of the lack of a good Chinese-Japanese bilingual lexicon suitable for alignment purpose. The publicly available bilingual lexical resources obtained tend to provide detailed definitions and explanations for each term. Such elaborate information does not represent the translated terms in actual usage in the patent texts. As a result, sentences cannot be well aligned based on such resources.

Another possible direction is to make use of the Chinese-Japanese sentence pairs in the current trilingual corpus established to compile a Chinese-Japanese bilingual lexicon via bilingual term extraction e.g. [10, 14, 17]. After the new lexicon is built, we can directly align Chinese and Japanese sentences from scratch again, anticipating more bilingual sentences to be mined and aligned. In return, the new set of increased bilingual sentences pairs can contribute to more trilingual sentences triplets through pivoting. This creates a cycle of value-adding stages: trilingual sentences after pivoting -> bilingual terms -> more bilingual sentences -> more trilingual sentences through pivoting. This iterative approach may also be generalized to multilingual corpora involving even more languages, but the complexity and computation cost may increase quickly.

7 Statistical Machine Translation on Patents

A Chinese-English SMT system is setup using Moses [13]. We train models based on different numbers of parallel sentences mined above. The test set contains 548 sentence pairs which are randomly selected and different from the training data. The sizes of the training data and BLEU scores for the models are shown in Table 10.

From Table 5, we can see that the BLEU scores are improving steadily when the training data increases. When the training data is enlarged by 20 times from

Table 11 Translations of a Chinese sample sentence

CN Sent.	电机 主轴 伸入 压缩机 壳体 内 的 工作 腔 中 ,
Ref.	the main shaft of the electric motor extends into the working cavity of the compressor shell
Model-A	the motor main shaft into the compressor the chamber
Model-B	motor shaft into the compressor housing: the working chamber
Model-C	motor shaft into the compressor housing: the working chamber
Model-D	motor spindle extends into the compressor housing: the working chamber
Model-E	motor spindle extends into the working chamber of the compressor housing

300K to 6M, the BLEU score increase to 22.60 from 17.94, which is a significant improvement. We show the translations of one Chinese sample sentence in Table 11.

From Table 11, we can see the translations given by Model-A to Model-C lack the main verb, the one given by Model-D has an ordering problem for the head noun and the modifier, and the one given by Model-E seems better than the others and its content closely resembles the reference despite the lexical difference.

8 Parallel Corpora for More Languages

In this section, we describe the potential of building large-scale parallel corpora for more languages, especially Asian languages by using the 1.8 million PCT patent applications and their national correspondents. By using PCT applications as the pivot, we can build multilingual parallel corpora from multilingual patents, which would greatly enlarge parallel data we could obtain.

The patent applications filed in one country should be in the official language(s) of the country, e.g. the applications filed in China should be in Chinese, those in Japan in Japanese, and so on. In Table 12, the second column shows the total numbers of patent applications in different countries which were previously filed as PCT ones; and the third column shows the total numbers of applications in different countries, which were previously filed as PCT ones with English as language of publication.

The number of the Chinese-English bilingual patents (CE) in Table 12 is about 188K, which is consistent with the number of 160K found in Sect. 4.1 since the latter contains only the applications up to early 2009. Based on Table 12, we estimate below the rough sizes of bilingual corpora, trilingual corpora, and even quadrilingual corpora for the different languages.

(1) Bilingual Corpora with English as one language

Compared to CE (188K), the Japanese-English bilingual corpus (269K) could be 50 % larger in terms of bilingual patents, the Korean-English one (134K) could be about 30 % smaller, and the German-English one (10K) would be much smaller.

Table 12 Estimated numbers of multilingual patents

National phase country[a]	All	English as lang. of publication
Japan	424K	269K
China	307K	188K
Germany	32K	10K
R. Korea	236K	134K
China and Japan	189K	130K
China and R. Korea	154K	91K
Japan and R. Korea	158K	103K
China and Japan & R. Korea	106K	73K

[a] For the national phase of the PCT System, the statistics are based on data supplied to WIPO by national and regional patent Offices, received at WIPO often 6 months or more after the end of the year concerned, meaning that the numbers are not up-to-date

(2) Bilingual Corpora for Asian Languages

The Japanese-Chinese bilingual corpus (189K) could be comparable to CE (188K) in terms of bilingual applications, the Chinese- Korean one (154K) could be about 20 % smaller, and the Japanese-Korean one (158K) is quite similar to the Chinese-Korean one.

(3) Trilingual Corpora and Beyond

In addition to bilingual corpora, we also build trilingual corpora from trilingual patents. It is quite interesting to note that the trilingual corpora could be quite large even compared to the bilingual corpora.

The trilingual corpora for Chinese, Japanese and English (130K) could be only 30 % smaller than CE in terms of patents. The trilingual corpus for Chinese, Korean and English (91K) and that for Japanese, Korean and English (103K) are also quite large. The number of the trilingual patents for the Asian languages of Chinese, Japanese and Korean (106K) is about 54 % of that of CE.

The number of the quadrilingual patents for Chinese, Japanese, Korean and English (73K) is about 38 % of that of CE. From these figures, we could say that a large proportion of the PCT applications published in English later have been filed in all the three Asian countries: China, Japan, and R. Korea.

Although we focus on a few specific languages in this section, there is potential for other languages as well, such as Arabic, French, Russian, Spanish, Italian, Dutch, Finnish, Swedish, etc., to mine parallel corpus from multilingual patents.

9 Discussion and Future Work

The websites from which the patents could be downloaded are not easy to access, and downtimes have been occasionally recorded during operation. To avoid excessive workload for the websites, the downloading speed had been limited. Some large patents would cost much time for the websites to respond and require considerably special attention. In short, considerable efforts were required to obtain these comparable patents.

In addition to our large-scale corpora mined in this study which is at least one order of magnitude larger, we can draw attention to give some other differences between ours and those introduced in Sect. 2 [11, 31]:

(1) Their bilingual patents were identified by the priority information in the US patents, and could not be easily extended to language pairs without English; while our method using PCT applications as the pivot could be easily extended to other language pairs as illustrated in Sect. 5.

(2) The translation processes involved are different: their patents were filed in USA Patent Office in English by translating from Japanese, while our patents were first filed in English as a PCT application, and later translated into Chinese. The different translation processes are likely to bias the linguistic characteristics.

9.1 Comparison of Wikipedia and Multilingual Patents

Here we compare multilingual patents with Wikipedia, which is a hot topic in the NLP fields (e.g. [1, 29]). They have similar features, such as large amounts of data and rapidly increasing new materials, but they have different characteristics in terms of the following aspects, which results in the difference of attention:

Accessibility: Wikipedia is open and easy to access, and the whole Wikipedia can be downloaded by a single point[12]; while multilingual patents are much harder to access because: (a) WIPO only has patent applications in the PCT version and other national versions in different languages have to be found in corresponding national patent offices, which are currently not so easy to obtain, (b) the national applications may not be freely distributed due to the copyright issue, and permission is need for distribution from the corresponding national patent offices. Few of them have made such permission.

File Format and Hyperlinks: Wikipedia is in the HTML format, which is easy to operate and has a large amount of hyperlinks and interwiki links among articles in different languages; while many national patent applications are in the PDF format or even image format, which requires more preprocessing to obtain the texts, and patents do not have hyperlinks except some references referring to related patents.

Domain Coverage: Wikipedia articles seem to cover more domains than patents, since they have not only technical articles but also articles on entertainment, history or news events.

Quality: Multilingual patents were usually manually translated by patent agents or professional, which means high-quality translations; while Wikipedia article pairs in different languages may range from almost completely parallel to containing almost no parallel sentences [29], and thus the quality of translations cannot be guaranteed.

[12] http://en.wikipedia.org/wiki/Wikipedia_database

The advantages of Wikipedia have made it a really hot topic. On the other hand, the disadvantages of multilingual patents have made them less attractive. However, the large amount, the high-quality translations, and fast expanding quantities of multilingual patents still draw attention to higher potential, which deserve closer attention in the NLP field.

9.2 *Future Work*

Since the PCT and multilingual patent applications increase rapidly in recent years as discussed in Sect. 3, we could expect more multilingual patents to enlarge the large-scale parallel corpora with the new applications and keep them up-to-date with new technical terms. On the other hand, patents are usually translated by patent agents or professionals. We could expect high quality translations from professionally produced multilingual patents. For example, compared to the parallel sentences mined from the Web in Jiang et al. [16], the 1–5 % wrong ratios of the parallel sentences mined here are quite low.

We note that patent corpora contain most sentences from technical domains and written in formal style. Thus, it would be interesting to know if the parallel sentences could improve the performance of SMT systems on NIST MT evaluation corpus containing news sentences and web sentences.

10 Conclusion

In this chapter, we show how large-scale high-quality parallel corpus can be mined from comparable multilingual patents harvested from the Web. Based on the mined parallel sentences, we show that adding more training data improves SMT performance significantly. One million high-quality Chinese-English sentence pairs have been provided to the participants in the Patent MT evaluation at NTCIR-9, and can be made publicly available to the research community.

With 1.8 million PCT patent applications and their corresponding national ones, there is considerable potential of constructing large-scale high-quality parallel corpora for a variety of language pairs. We give estimation on the sizes of multilingual parallel corpora which could be obtained from multilingual patents involving English, Chinese, Japanese, Korean, German, etc., which could reduce the parallel data acquisition bottleneck in multilingual information processing.

Acknowledgments We wish to thank our colleagues, Dr. Kataoka S. and Mr. Wrong B. and others, for their help in evaluating the sampled sentence pairs and triplets.

Appendix A. Chinese-English-Japanese Sentence Triplet Examples

Patent Publication Numbers			Sentence Examples		
EN	CN	JP	English	Chinese	Japanese
WO0000039	CN1310585	JP2000-556633	The present invention relates to a process and product for promoting weight loss in overweight dogs, and more particularly to a process for supplementing a canine diet with L-carnitine to promote weight loss improve body composition, and enhance satiety in the animal.	本发明涉及用于促进体重过重的狗减重的方法和产品，更具体地，其涉及补充带有L-卡尼汀的犬饮食以促进体重降低、改善身体构成并提高动物的饱满感。	本発明は、太り過ぎのイヌにおいて、体重減少を促進する方法、及び、産物、及び、より具体的には、体重減少を促進し、体の組成を改善し、また、動物において飽満を増強するために、イヌの食餌にL-カルニチンを補足するための方法に関連している。
WO0000039	CN1310585	JP2000-556633	In order that the invention may be more readily understood, reference is made to the following example which is intended to illustrate the invention, but not limit the scope thereof.	为了能够更容易地理解本发明，可以参考下列实施例，它们是为了说明本发明，但不限制其范围。	本発明をさらに容易に理解することができるために、本発明を説明することが意図されるがその範囲を限定しない以下の実施例に対して参照を示す。
WO0000107	CN1304298	JP2000-556693	The biocompatible material for the leaflets, rim strip and posts includes both biological material or synthetic polymers which could be either naturally occurring of artificially produced.	用于小叶、边缘磨带和支架的生物等效材料包括生物学材料或人造聚合体，它们是自然存在的或人工合成的。	葉状部、リムストリップ、およびポスト用の生体適合性材料は、自然に生成されるか、あるいは人工的に製造されるかのいずれかであってよい、生物学的材料または合成ポリマー（高分子）の両方を含む。

Patent Publication Numbers			Sentence Examples		
EN	CN	JP	English	Chinese	Japanese
WO0000132	CN1311653	JP2000-556718	Similarly the flange may be provided in any shape and preferably has a symmetrical shape preferably comprising a plurality of lobes (13).	同样，凸缘可以采取任何形状，较佳的是具有对称的形状，该形状最好包括多个突出部分(13)。	同様にフランジは任意の形で提供されるが、好ましくは複数のローブ（１３）を備えた対称形をとることが好ましい。
WO0000231	CN1335782	JP2000-556816	It is one object of this invention to provide a method for producing apertured film covers for use in a fluid absorbent material such as feminine care products, surgical drapes, fenestration reinforcement, absorbent pads and the like having aperture regions having a higher wettability than at least a portion of the top surface thereof.	本发明的一个目的是提供一种制造用在流体吸收材料，例如女性护理制品、手术被单，穿孔加强物、吸收衬垫和类似材料中的有孔薄膜外罩件的方法，该流体吸收材料具有孔区域，该孔区域具有比至少一部分顶表面高的可润湿性。	本発明の１つの目的は、女性用ケア製品、外科用ドレープ、開窓術用補強材、吸収体パッドなど、その上部表面の少なくとも一部分よりも高い濡れ性を持つ開口領域を備える、流体吸収体材料に使用される穿孔フィルム・カバーの製造方法を提供することである。

References

1. Adafre, S.F., de Rijke, M.: Finding similar sentences across multiple languages in wikipedia. In: Proceedings of EACL, pp. 62–69 (2006)
2. Brown, P.F., Lai, J.C., Mercer, R.L.: Aligning sentences in parallel corpora. In: Proceedings of ACL, pp. 169–176 (1991)
3. Brown, P.F., Della, S.A., Pietra, V.J., Pietra, D., Mercer, R.L.: Mathematics of statistical machine translation: parameter estimation. Comput. Linguist. **19**(2), 263–311 (1993)
4. Cao, G., Gao, J., Nie, J.: A system to mine large-scale bilingual dictionaries from monolingual web pages. In: Proceedings of MT Summit, pp. 57–64 (2007)
5. Chen, S.F.: Aligning sentences in bilingual corpora using lexical information. In: Proceedings of ACL, pp. 9–16 (1993)
6. Chiang, D.: Hierarchical phrase-based translation. Comput. Linguist. **33**(2), 201–228 (2007)
7. Fujii, A., Utiyama, M., Yamamoto, M., Utsuro, T.: Overview of the patent translation task at the NTCIR-7 workshop. In: Proceedings of the NTCIR-7 Workshop, pp. 389–400. Tokyo, Japan (2008)
8. Fujii, A., Utiyama, M., Yamamoto, M., Utsuro, T., Ehara, T., Echizen-ya, H., Shimohata, S.: Overview of the patent translation task at the NTCIR-8 workshop. In: Proceedings of the NTCIR-8 Workshop. Tokyo, Japan (2010)

9. Gale, W.A., Church, K.W.: A program for aligning sentences in bilingual corpora. In: Proceedings of ACL, pp. 79–85 (1991)
10. Ha, L.A., Fernandez, G., Mitkov, R., Corpas, G.: Mutual bilingual terminology extraction. In: Proceedings of the Sixth International Language Resources and Evaluation (LREC), pp. 28–30 (2008)
11. Higuchi, S., Fukui, M., Fujii, A., Ishikawa, T.: PRIME: a system for multi-lingual patent retrieval. In: Proceedings of MT Summit VIII, pp. 163–167 (2001)
12. Koehn, P.: Europarl: a parallel corpus for statistical machine translation. In: Proceedings of MT Summit X (2005)
13. Koehn, P., Hoang, H., Birch, A., Callison-Burch, C., et al.: Moses: open source toolkit for statistical machine translation. In: Proceedings of ACL Demo Session, pp. 177–180 (2007)
14. Kupiec, J.: An Algorithm for Finding Noun Phrase Correspondences in Bilingual Corpora. In: Proceedings of ACL-93, pp. 17–22 (1993)
15. Lin, D., Zhao, S., Durme, B.V., Pasca, M.: Mining parenthetical translations from the web by word alignment. In: Proceedings of ACL-08, pp. 994–1002 (2008)
16. Jiang, L., Yang, S., Zhou, M., Liu, X., Zhu, Q.: Mining bilingual data from the web with adaptively learnt patterns. In: Proceedings of ACL-IJCNLP, pp. 870–878 (2009)
17. Lu, B., Tsou, B.K., Zhu, J., Jiang, T., Kwong, O.Y.: The construction of an English-Chinese patent parallel corpus. In: Proceedings of MT Summit XII 3rd Workshop on Patent Translation (2009)
18. Lu, B., Tsou, B.K.: Towards bilingual term extraction in comparable patents. In: Proceedings of the 23rd Pacific Asia Conference on Language, Information and Computation (PACLIC'23), pp. 755–762 (2009)
19. Lu, B., Tsou, B.K., Jiang, T., Kwong, O.Y., Zhu, J.: Mining large-scale parallel corpora from multilingual patents: an English-Chinese example and its application to SMT. In: Proceedings of the 1st CIPS-SIGHAN Joint Conference on Chinese Language Processing (CLP-2010). Beijing, China. August, 2010 (2010a)
20. Lu, B., Jiang, T., Chow, K., Tsou, B.K.: Building a large English-Chinese parallel corpus from comparable patents and its experimental application to SMT. In: Proceedings of Workshop on Building and Using Comparable Corpora. Malta (2010b)
21. Lu, B., Chow, K.P., Tsou, B.K.: The cultivation of a trilingual Chinese-English-Japanese parallel corpus from comparable patents. In: Proceedings of Machine Translation Summit XIII (MT Summit-XIII). Xiamen (2011a)
22. Lu, B., Tsou, B.K., Jiang, T., Zhu, J., Kwong, O.: Mining parallel knowledge from comparable patents. In: Ontology Learning and Knowledge Discovery Using the Web: Challenges and Recent Advances. IGI Global (2011b)
23. Ma, X.: Champollion: A robust parallel text sentence aligner. In: Proceedings of the 5th International Conference on Language Resources and Evaluation (LREC). Genova, Italy (2006)
24. Moore, R.C.: Fast and accurate sentence alignment of bilingual corpora. In: Proceedings of AMTA, pp. 135–144 (2002)
25. Munteanu, D.S., Marcu, D.: Improving machine translation performance by exploiting non-parallel corpora. Comput. Linguist. **31**(4), 477–504 (2005)
26. Och, F.J, Ney, H.: A systematic comparison of various statistical alignment models. Comput. Linguist. **29**(1), 19–51 (2003)
27. Och, F.J., Ney, H.: The alignment template approach to machine translation. Comput. Linguist. **30**(4), 417–449 (2004)
28. Resnik, P., Smith, N.A.: The web as a parallel corpus. Comput. Linguist. **29**(3), 349–380 (2003)
29. Smith, J.R., Quirk, C., Toutanova, K.: Extracting parallel sentences from comparable corpora using document level alignment. In: Proceedings of NAACL-HLT, pp. 403–411 (2010)
30. Simard, M., Plamondon, P.: Bilingual sentence alignment: balancing robustness and accuracy. Mach. Transl. **13**(1), 59–80 (1998)
31. Utiyama, M., Isahara, H.: A Japanese-English patent parallel corpus. In: Proceeding of MT Summit XI, pp. 475–482 (2007)

32. Wu, D., Fung, P.: Inversion transduction grammar constraints for mining parallel sentences from quasi-comparable corpora. In: Proceedings of IJCNLP2005 (2005)
33. Wu, D., Xia, X.: Learning an English-Chinese lexicon from a parallel corpus, In: Proceedings of the First Conference of the Association for Machine Translation in the Americas (1994)
34. Zhao, B., Vogel, S.: Adaptive parallel sentences mining from web bilingual news collection. In: Proceedings of Second IEEE International Conference on Data Mining (ICDM-02) (2002)

Part II
Using Comparable Corpora

Extracting Parallel Phrases from Comparable Data

Sanjika Hewavitharana and Stephan Vogel

Abstract Mining parallel data from comparable corpora is a promising approach for overcoming the data sparseness in statistical machine translation and other NLP applications. Even if two comparable documents have few or no parallel sentence pairs, there is still potential for parallelism in the sub-sentential level. The ability to detect these phrases creates a valuable resource, especially for low-resource languages. In this chapter we explore three phrase alignment approaches to detect parallel phrase pairs embedded in comparable sentences: the standard phrase extraction algorithm, which relies on the Viterbi path; a phrase extraction approach that does not rely on the Viterbi path of word alignments, but uses only lexical features; and a binary classifier that detects parallel phrase pairs when presented with a large collection of phrase pair candidates. We evaluate the effectiveness of these approaches in detecting alignments for phrase pairs that have a known alignment in comparable sentence pairs. The results show that the non-Viterbi alignment approach outperforms the other two approaches in terms of F-measure.

1 Introduction

Statistical Machine Translation (SMT), like many natural language processing tasks, relies primarily on parallel corpora. The translation performance of SMT systems directly depends on the quantity and the quality of the available parallel data. However, such corpora are only available in large quantities for a handful of languages, including English, Arabic, Chinese and some European languages. Much of this data is derived from parliamentary proceedings (e.g. Europarl, Canadian Hansard, etc)

S. Hewavitharana (✉) · S. Vogel
Language Technologies Institute, Carnegie Mellon University, Pittsburgh, PA15213, USA
e-mail: sanjika@cs.cmu.edu

S. Vogel
e-mail: vogel+@cs.cmu.edu

S. Sharoff et al. (eds.), *Building and Using Comparable Corpora*,
DOI: 10.1007/978-3-642-20128-8_10, © Springer-Verlag Berlin Heidelberg 2013

and UN proceedings, though a limited amount of newswire text is also available. For most other languages, especially for less commonly used languages, parallel data is virtually non-existent.

Comparable corpora provide a possible solution to this data sparseness problem. Comparable documents are not strictly parallel, but are loose translations of each other, with overlapping information. A good example for comparable documents is the newswire text produced by multilingual news organizations such as AFP or Reuters. The degree of parallelism can vary greatly, ranging from *noisy parallel* documents that contain many parallel sentences, to *quasi parallel* documents that may cover different topics [3]. The Web is by far the largest source of comparable data. *Wikipedia* is becoming a more attractive source of comparable documents.

Comparable corpora may contain parallel data in different levels of granularity. This includes: parallel documents, parallel sentences, or parallel sub-sentential fragments. To simplify the process of extracting these different types of translational correspondences and to reduce the computational overhead, it can be divided into a cascade of tasks as shown in Fig. 1. First, a document level alignment is identified between the two collections of comparable corpora. This results in a set of comparable document pairs, some of which may already be parallel. In the next step, a sentence level alignment is detected within the identified document pairs. Even if two comparable documents have few or no parallel sentence pairs, there could still be partially parallel sentences, which contain parallel sub-sentential fragments, including word translation pairs, named entities, and long phrase pairs. These fragments are detected in the third step. The ability to identify these pairs would create a valuable resource for SMT, especially for low-resource languages.

In this chapter we explore several phrase alignment approaches to detect parallel phrase pairs embedded in comparable sentence pairs. We assume that a collection of comparable sentence pairs is already available at this stage (i.e. the first two steps in the framework in Fig. 1 is already applied). Our intention is to use the extracted phrases directly in the translation process, along with other phrase pairs extracted from parallel corpora. In particular, we study three alignment approaches:

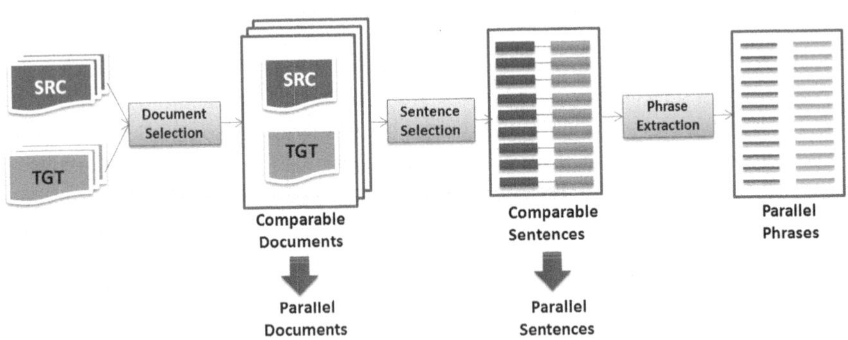

Fig. 1 Parallel data extraction framework

(1) the standard phrase extraction algorithm, which relies on the Viterbi path of the word alignment;
(2) a phrase extraction approach that does not rely on the Viterbi path, but only uses lexical features;
(3) and a binary classifier to detect parallel phrase pairs when presented with a large collection of phrase pair candidates.

We evaluate the effectiveness of these approaches in detecting alignments for phrase pairs that have a known translation a comparable sentence pair. The rest of the chapter is organized as follows: In Sect. 2 we discuss previous work on extracting parallel data from comparable corpora. Section 3 introduces the phrase alignment problem in comparable sentences and discusses some of the challenges involved. It also explains the different alignment approaches we explore. Section 4 presents the experimental setup and the results of the evaluation. We conclude, in Sect. 5, with an analysis of the results and some directions for future work.

2 Related Work

Many of the early attempts to exploit comparable corpora focused on learning bilingual word translations [4, 11, 12]. The underlying assumption in these efforts is that a word and its translation appear in similar context in the respective languages, and therefore co-occurrence statistics can be used to detect them. Several context similarity measures such as term frequency (*tf*) and inverse document frequency (*idf*) have been studied by Fung and Yee [4].

Cross-lingual information retrieval (CLIR) has been widely used to detect document level alignments in comparable corpora [8, 16, 19]. Here, a source language document is translated into the target language using a dictionary, and the resulting translation is used as a query to match against a collection of target documents. Resnik and Smith [13] find parallel article pairs from the Web by exploiting the similarities in the URL structure and the structure of meta information in the articles among other clues.

Detecting sentence alignment within comparable documents has been tried in different ways. Zhao and Vogel [18] and Utiyama and Isahara [16] extend parallel sentence alignment algorithms to identify parallel sentence pairs within comparable news corpora. Munteanu and Marcu [8] and Fung and Cheung [3] work on non-parallel documents, and do not assume that every sentence on source document has a corresponding translation on the target document. Each source sentence is matched against candidate target sentences and the sentence pairs above a certain similarity measure is picked as parallel. Fung and Cheung [3] use cosine similarity measure for this task, while Munteanu and Marcu [8] use a maximum entropy classifier. The former uses an iterative refinement where previously identified sentence pairs are added to the system in later iterations. Munteanu and Marcu [8] demonstrate that this method can be successfully used to improve the performance of a machine translation system. Tillmann and Xu [15] introduce a system that performs both the document selection and sentence alignment tasks in a single run without any document level pre-filtering. Such a system is useful when document level boundaries are not available in the comparable corpus.

The first attempt to detect sub-sentential fragments from comparable sentences was proposed by Munteanu and Marcu [9]. After identifying sentence pairs that have at least partial overlap, the parallel fragments are detected using an approach inspired by signal processing. Quirk et al. [10] later extended this work by introducing generative models that explicitly aligns comparable sentence pairs, including non-parallel sections. They showed improvements in translation performance when the method was applied on cross-domain test data. Kumano et al. [7] have proposed a phrasal alignment approach for comparable corpora using the joint probability SMT model. While this approach is appealing for low-resource scenarios as it does not require any seed parallel corpus, the high computational cost is a deterrent in its applicability to large corpora. Tillmann and Hewavitharana [14] present a single unified search algorithm that can be used flexibly on all the different extraction tasks handled by the various algorithms cited above. Translation pair extraction is handled as a chunk-alignment problem with no document-level pre-filtering, and data processed directly at the sentence level. The results indicate significant gains by using the extracted parallel data, especially when using the extracted fragments.

In all the approaches cited above, the extracted fragment data was used as additional training data to train alignment models. In contrast, this work focuses on extracting phrase pairs that can be directly used in the translation process, complementing a phrase table generated with existing parallel data.

3 Parallel Phrase Extraction

Figure 2 shows three sample sentences that were extracted from Gigaword Arabic and Gigaword English collections. For each comparable sentence pair, the Arabic sentence is shown first, followed by its correct English translation (in Italics). The English sentence is shown next. The parallel sections in each sentence are marked in boldface. In the first two sentences pairs, the English sentence contains the full translation of the Arabic sentence, but there are additional phrases on the English side that are not present on the Arabic sentence. These phrases appear at the beginning of sentence 1 and at the end of sentence 2. In sentence 3, there are parallel phrases as well as phrases that appear only on one side. The phrase "to Iraq" appears only on the Arabic sentence while the phrase "the former Egyptian foreign minister" appears only on the English side.

Standard word alignment and phrase alignment algorithms are formulated to work on parallel sentence pairs. Therefore, these standard algorithms are not well suited

1.

واضف انها تهدف لصرف انتباه الراي العلم عن الاعمال الوحشية المتزايدة التي يرتكبها النظام الاسرائيلي ضد الفلسطينيين في الاراضي المحتلة

[He] added that it aims to divert public attention from the growing atrocities committed by the Israeli regime against the Palestinians in the occupied territories.

" Iran considers these remarks as interference in its internal affairs , " Kharazi said , **adding that they are aimed at detracting public opinion from heightened atrocities committed by the Israeli regime against the Palestinians in occupied lands .**

2.

واضاف " لكن حتي الان لم نواجه مشكلات "

But "Until now we did not have problems"

" **but up to now , we didn't meet any problems** ; the afghan people are very kind to us , " he said.

3.

تعد هذه هي اول زيارة لموسي على العراق منذ توليه الامانة العامة للجامعة العربية في مايو الماضي

This is the first visit by Moussa** to Iraq, **since he became the General Secretary of the Arab League in last May.

This was also the first such visit by Moussa himself , the former Egyptian foreign minister , **since he assumed the post as AL chief in may last year .**

Fig. 2 Sample comparable sentences that contain parallel phrases

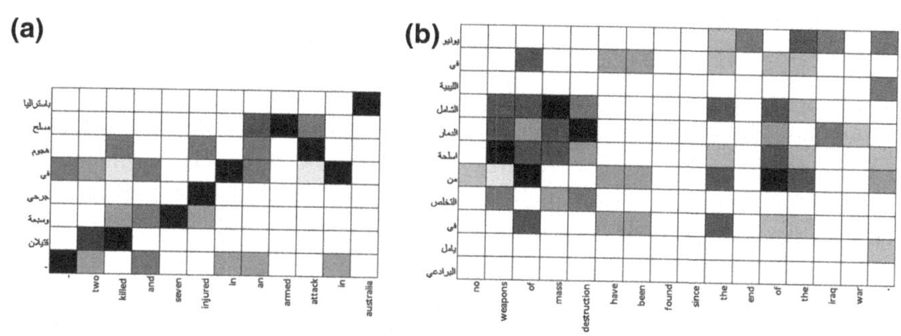

Fig. 3 Word-to-word alignment pattern for **a** a parallel sentence pair **b** a non-parallel sentence pair

to operate on partially parallel sentence pairs. Presence of non-parallel phrases may result in undesirable alignments.

Figure 3 illustrates this phenomenon. It compares a typical word alignment pattern in a parallel sentence pair (a) to one in a non-parallel sentence pair (b). The darkness of a square indicates the strength of the word alignment probability between the corresponding word pair. In Fig. 3a, we observe high probability word-to-word alignments (dark squares) over the entire length of the sentences. In Fig. 3b, we see one dark area above "weapons of mass destruction", corresponding to the parallel phrase pair, and some scattered dark spots, where high frequency English words pair with high frequency Arabic words. This spurious alignments pose problems to the phrase alignment, and indicate that word alignment probabilities alone might not be sufficient.

Our aim is to identify such parallel phrase pairs from comparable sentence pairs. In the following subsections we briefly explain the different phrase alignment approaches we use.

3.1 Viterbi Alignment

Here we use the typical phrase extraction approach used by Statistical Machine Translation systems: obtain word alignment models for both directions (source to target and target to source), combine the Viterbi paths using one of many heuristics, and extract phrase pairs from the combined alignment. We used Moses toolkit [6] for this task. To obtain the word alignments for comparable sentence pairs, we performed a forced alignment using the trained models.

3.2 Binary Classifier

We used a Maximum Entropy classifier as our second approach to extract parallel phrase pairs from comparable sentences. Such classifiers have been used in the past to detect parallel sentence pairs in large collections of comparable documents [8]. Our classifier is similar, but we apply it at phrase level rather than at sentence level. The classifier probability is defined as:

$$p(c|S,T) = \frac{exp\left(\sum_{i=1}^{n} \lambda_i f_i(c,S,T)\right)}{Z(S,T)}, \tag{1}$$

where $S = s_1^L$ is a source phrase of length L and $T = t_1^K$ is a target phrase of length K. $c \in \{0, 1\}$ is a binary variable representing the two classes of phrases: *parallel* and *not parallel*. $p(c|S,T) \in [0, 1]$ is the probability where a value $p(c = 1|S,T)$ close to 1.0 indicates that S and T are translations of each other. $f_i(c, S, T)$ are feature functions that are co-indexed with respect to the class variable c. The parameters λ_i are the weights for the feature functions obtained during training. $Z(S,T)$ is the normalization factor. In the feature vector for phrase pair (S,T), each feature appears twice, once for each class $c \in \{0, 1\}$.

The feature set we use is inspired by Munteanu and Marcu [8] who define the features based on IBM Model-1 [2] alignments for source and target pairs. However, in our experiments, the features are computed primarily on IBM Model-1 probabilities (i.e. lexicon). We do not explicitly compute IBM Model-1 alignments. To compute coverage features, we identify alignment points for which IBM Model-1 probability is above a threshold. We produce two sets of features based on IBM Model-1 probabilities obtained by training in both directions. All the features have been normalized with respect to the source phrase length L or the target phrase length K. We use the following 11 features:

1. Lexical probability (2): IBM Model-1 log probabilities $p(S|T)$ and $p(T|S)$.
2. Phrase length ratio (2): source length ratio K/L and target length ratio L/K.
3. Phrase length difference (1): source length minus target length, $L - K$.
4. Number of words covered (2): A source word s is said to be covered if there is a target word $t \in T$ such that $p(s|t) > \epsilon$, where $\epsilon = 0.5$. Target word coverage is defined accordingly.
5. Number of words not covered (2): This is computed similarly to 4. above, but this time counting the number of positions that are not covered.
6. Length of the longest covered sequence of words (2).

To train the classifier, we used parallel phrase pairs extracted from a manually word-aligned corpus. In selecting negative examples, we followed the same approach as in [8]: pairing all source phrases with all target phrases, but filter out the parallel pairs and those that have high length difference or a low lexical overlap, and then randomly select a subset of phrase pairs as the negative training set. The model parameters are estimated using the GIS algorithm.

3.3 Non-Viterbi (PESA) Alignment

A phrase alignment algorithm called "PESA" that does not rely on the Viterbi path is described in [17]. PESA identifies the boundaries of the target phrase by aligning words inside the source phrase with words inside the target phrase, and similarly for the words outside the boundaries of the phrase pair. It does not attempt to generate phrase alignments for the full sentence. Rather, it identifies the best target phrase that matches a given source phrase. PESA requires a statistical word-to-word lexicon. A seed parallel corpus is required to automatically build this lexicon.

This algorithm seems particularly well suited in extracting phrase pairs from comparable sentence pairs, as it is designed to not generate a complete word alignment for the entire sentences, but to find only the target side for a phrase embedded in the sentence. We briefly explain the PESA alignment approach below.

Instead of searching for all possible phrase alignments in a parallel sentence pair, this approach finds the alignment for a single source phrase $S = s_1 \dots s_l$. Assume that we have a parallel sentence pair (s_1^J, t_1^I) which contains the source phrase S in the source sentence s_1^J. Now we want to find the target phrase $T = t_1 \dots t_k$ in the target sentence t_1^I which is the translation of the source phrase.

A constrained IBM Model-1 alignment is now applied as follows:

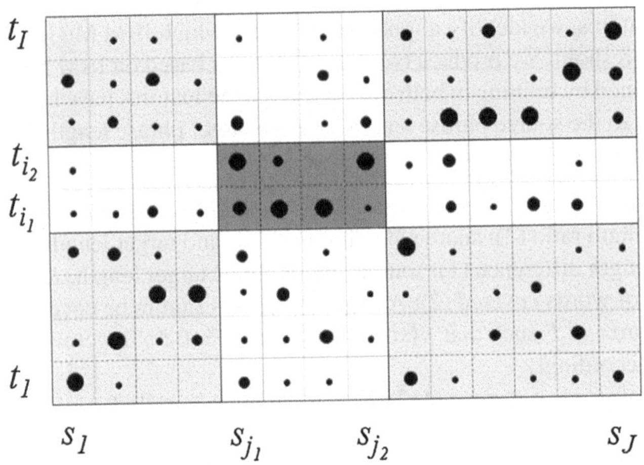

Fig. 4 PESA phrase alignment

- Source words inside phrase boundary are aligned only with the target words inside the phrase boundary. Source words outside the phrase boundary are only aligned with target words outside the phrase boundary.
- Position alignment probability for the sentence, which is $1/I$ in IBM Model-1, is modified to be $1/k$ inside the source phrase and to $1/(I-k)$ outside the phrase.

Figure 4 shows the different regions. Given the source sentence and the source phrase from position j_1 to j_2, we want to find the boundaries of the target phrase, i_1 and i_2. The dark area in the middle is the phrase we want to align. The size of the blobs in each box indicates the lexical strength of the word pair.
The constrained alignment probability is calculated as follows:

$$p(s|t) = \left(\prod_{j=1}^{j_1-1} \sum_{i \notin (i_1 \ldots i_2)} \frac{1}{I-k} p(s_j|t_i)\right) \times \left(\prod_{j=j_1}^{j_2} \sum_{i=i_1}^{i_2} \frac{1}{k} p(s_j|t_i)\right) \times \left(\prod_{j=j_2+1}^{J} \sum_{i \notin (i_1 \ldots i_2)} \frac{1}{I-k} p(s_j|t_i)\right) \tag{2}$$

$p(t|s)$ is similarly calculated by switching source and target sides in Eq. (2):

$$p(t|s) = \left(\prod_{i=1}^{i_1-1} \sum_{i \notin (j_1 \dots j_2)} \frac{1}{J-l} p(t_i|s_j) \right) \times \left(\prod_{i=i_1}^{i_2} \sum_{j=j_1}^{j_2} \frac{1}{l} p(t_i|s_j) \right) \times \left(\prod_{i=i_2+1}^{I} \sum_{j \notin (j_1 \dots j_2)} \frac{1}{J-l} p(t_i|s_j) \right) \quad (3)$$

To find the optimal target phrase boundaries, we interpolate the two probabilities in Eqs. (2) and (3) and select the boundary (i_1, i_2) that gives the highest probability, as in Eq. 4.

$$(i_1, i_2) = \underset{i_1, i_2}{\text{argmax}} \ \{(1-\lambda)\ log(p(s|t)) + \ \lambda\ log(p(t|s))\} \quad (4)$$

The value of λ is estimated using held-out data.

PESA can be used to identify all possible phrase pairs in a given parallel sentence pair by iterating over every source phrase. An important difference is that each phrase is found independently of any other phrase pair, whereas in the standard phrase extraction they are tied through the word alignment of the sentence pair. There are several ways we can adapt the non-Viterbi phrase extraction to comparable sentence.

- Apply the same approach assuming the sentence pair as parallel. The inside of the source phrase is aligned to the inside of the target phrase, and the outside, which can be non-parallel, is aligned the same way.
- Disregard the words that are outside the phrase we are interested in. Find the best target phrase by aligning only the inside of the phrase. This will considerably speed-up the alignment process.

4 Experimental Results

4.1 Evaluation Setup

We want to compare the performance of the different phrase alignment methods in identifying parallel phrases embedded in comparable sentence pairs. Using a manually aligned parallel corpus, and two monolingual corpora, we obtained a test corpus as follows: From the manually aligned corpus, we obtain parallel phrase pairs (S, T). Given a source language corpus $\mathcal{S}$ and a target language corpus $\mathcal{T}$, for each parallel phrase pair (S, T) we select a sentence s from $\mathcal{S}$ which contains S and a target sentence t from $\mathcal{T}$ which contains T. These sentence pairs are then non-parallel, but contain parallel phrases, and for each sentence pair the correct phrase pair is known. This makes it easy to evaluate different phrase alignment algorithms.

Ideally, we would like to see the correct target phrase T extracted for a source phrase S. However, even if the boundaries of the target phrase do not match exactly, and only a partially correct translation is generated, this could still be useful to improve translation quality. We therefore will evaluate the phrase pair extraction from non-parallel sentence pairs also in terms of partial matches.

To give credit to partial matches, we define precision and recall as follows: Let W and G denote the extracted target phrase and the correct reference phrase, respectively. Let M denote the tokens in W that are also found in the reference G. Then

$$Precision = \frac{|M|}{|W|} * 100 \tag{5}$$

$$Recall = \frac{|M|}{|G|} * 100 \tag{6}$$

These scores are computed for each extracted phrase pair, and are averaged to produce precision and recall for the complete test set. Finally, precision and recall are combined to generated the F score in the standard way:

$$F = \frac{2 \cdot Precision \cdot Recall}{Precision + Recall} \tag{7}$$

4.2 Evaluation

We conducted our experiments on Arabic–English language pair. We obtained manual alignments for 663 Arabic–English sentence pairs. From this, we selected 300 sentences, and extracted phrase pairs up to 10 words long that are consistent with the underlying word alignment. From the resulting list of phrase pairs, we removed the 50 most frequently occurring pairs as well as those only consisting of punctuations. Almost all high frequency phrases are function words, which are typically covered by the translation lexicon. Line 1 in Table 1 gives the n-gram type distribution for the source phrases.

Using the phrase pairs extracted from the manually aligned sentences, we constructed a comparable corpus as follows:

Table 1 N-gram type distribution of manually aligned phrases set

	1	2	3	4	5	6	7	8	9	10	All
Test set	2,826	3,665	3,447	3,048	2,718	2,414	2,076	1,759	1,527	1,378	24,858
Test set (found)	2,746	2,655	1,168	373	87	29	7	2	1	0	7,068

1. For each Arabic phrase, we search the Arabic Gigaword[1] corpus for sentences that contain the phrase and select up to 5 sentences. Similarly, for each corresponding English phrase we select up to 5 sentences from English Gigaword.[2]
2. For each phrase pair, we generate the Cartesian product of the sentences and produce a sentence pair collection. i.e. up to 25 comparable sentence pairs were constructed for each phrase pair.
3. We only select sentences up to 100 words long, resulting in a final comparable corpus consisting of 170 thousands sentence pairs.

Line 2 in Table 1 gives the n-gram type distribution for the phrase pairs for which we found both a source sentence and a target sentence in the monolingual corpora. As expected, the longer the phrases, the less likely it is to find them in even larger corpora.

We consider the resulting set as our comparable corpus which we will use to evaluate all alignment approaches. In most sentence pairs, except for the phrase pair that we are interested in, the rest of the sentence does not typically match the other side.

We obtained the Viterbi alignment using standard word alignment techniques: IBM4 word alignment for both directions, Viterbi path combination using heuristics ('grow-diag-final') and phrase extraction from two-sided training, as implemented in the Moses package [6]. Because the non-parallel segments will lead the word alignment astray, this may have a negative effect on the alignment in the parallel sections. Alignment models trained on parallel data are used to generate the Viterbi alignment for the comparable sentences. We then extract the target phrases that are aligned to the embedded source phrases. A phrase pair is extracted only when the alignment does not conflict with other word alignments in the sentence pair. The alignments are not constrained to produce contiguous phrases. We allow unaligned words to be present in the phrase pair. For each source phrase we selected the target phrase that has the least number of unaligned words.

The classifier is applied at the phrase level. We generate the phrase pair candidates as follows: For a given target sentence we generate all n-grams up to length 10. We pair each n-gram with the source phrase embedded in the corresponding source sentence to generate a phrase pair. From the 170 thousand sentence pairs, we obtained 15.6 million phrase pair candidates. The maximum entropy classifier is then applied to the phrase pairs. For each source phrase, we pick the target candidate for which $p(c = 1, S, T)$ has the highest value.

For the PESA alignment we used both inside and outside alignments, using only lexical probabilities. For each source phrase pair, we select the best scoring target phrase.

As our goal is to use these methods to extract parallel data for low resource situations, we tested each method with several lexica, trained on different amounts of initial parallel data. Starting from the full corpus with 127 million English tokens,

[1] Arabic Gigaword Fourth Edition (LDC2009T30).

[2] English Gigaword Fourth Edition (LDC2009T13).

Table 2 Results for alignment evaluation of test phrases

Lexicon	Viterbi				Classifier				PESA			
	Exact	P	R	F	Exact	P	R	F	Exact	P	R	F
Lex-Full	43.56	65.71	57.99	61.61	54.46	81.79	85.29	85.29	67.94	93.34	86.80	90.22
Lex-Large	42.95	65.68	56.69	60.85	53.57	81.32	88.34	84.69	67.28	93.23	86.17	89.56
Lex-Medium	41.10	63.60	51.15	56.70	52.38	80.30	86.64	83.35	65.81	91.95	84.73	88.19
Lex-Small	41.02	62.10	49.38	55.01	52.51	80.51	83.84	82.14	63.23	89.41	82.06	85.57
Lex-BTEC	19.10	26.94	23.63	25.18	18.76	45.90	36.17	40.46	17.45	46.70	36.28	40.83

we generated three additional parallel corpora, namely *Large*, *Medium* and *Small*, with 1/3, 1/9 and 1/27 of the original size, respectively. The three new corpora contain 40, 13 and 4 million English tokens. *Medium* and *Small* corpora are a better match to the resource levels for many languages. We also used data from the BTEC [5] corpus. This corpus contains conversational data from the travel domain, which is from a different genre than the document collections. Compared to other corpora, it is much smaller (about 190 thousand English tokens).

Table 2 gives the results for all three alignment approaches. Results are presented as percentages of: exact matches found (Exact), precision (P), recall (R) and F. The Viterbi alignment gives the lowest performance. This shows that the standard phrase extraction procedure, which works well for parallel sentence, is ill-suited for partially parallel sentences. Despite the fact that the classifier incorporates several features including the lexical features, the performance of the PESA alignment, which uses only the lexical features, has consistently higher precision and recall than the classifier. This demonstrates that computing both inside and outside probabilities for the sentence pair helps the phrase extraction. The classifier lacks this ability because the phrase pair is evaluated in isolation, without the context of the sentence.

Except for the BTEC corpus, the performance degradation is minimal as the lexicon size is reduced. This shows that the approaches are robust for smaller parallel amounts of parallel data.

Instead of using token precision, an alternative method of evaluating partial matches, is to give credit based on the length of the overlap between the extracted phrase and the reference. Precision and recall can then be defined based on the longest common contiguous subsequence, similar to [1]. Results obtained using this methods were similar to the results in Table 2.

5 Conclusion and Future Work

In this chapter we explored several phrase alignment approaches for extracting phrase pairs that are embedded inside comparable sentence pairs. We used the standard Viterbi phrase alignment, a maximum entropy classifier that works on phrase pairs, and a non-Viterbi PESA alignment in the evaluation process. The results show that PESA outperforms both the Viterbi approach and the classifier, in both precision and recall.

We plan to extend the PESA framework to use not only lexical features, but other features similar to the ones used in the classifier. We believe this will further improve the alignment accuracy.

While this work focuses on comparisons of different phrase alignment approaches in a realistic, yet controlled manner by selecting appropriate comparable sentence pairs for given phrase pairs, future experiments will focus on finding new phrase pairs from comparable corpora and evaluating the potential utility of the extracted data in the context of an end-to-end machine translation system.

References

1. Bourdaillet, J., Huet, S., Langlais, P., Lapalme, G.: TransSearch: from a bilingual concordancer to a translation finder. Mach. Transl. **24**(3–4), 241–271 (2010)
2. Brown, P.F., Pietra, S.A.D., Pietra, V.J.D., Mercer, R.L.: The mathematics of statistical machine translation: parameter estimation. Comput. Linguist. **19**(2), 263–311 (1993)
3. Fung, P., Cheung, P.: Mining very non-parallel corpora: parallel sentence and lexicon extraction via bootstrapping and EM. In: Proceedings of the Conference on Empirical Methods in Natural Language Processing, pp. 57–63, Barcelona, Spain (2004)
4. Fung, P., Yee, L.Y.: An IR approach for translating new words from nonparallel, comparable texts. In: Proceedings of the 36th Annual Meeting of the Association for Computational Linguistics, pp. 414–420, Montreal, Canada (1998)
5. Kikui, G., Sumita, E., Takezawa, T., Yamamoto, S.: Creating corpora for speech-to-speech translation. In: Proceedings of EUROSPEECH 2003, pp. 381–384, Geneva (2003)
6. Koehn, P., Hoang, H., Birch, A., Callison-Burch, C., Federico, M., Bertoldi, N., Cowan, B., Shen, W., Moran, C., Zens, R., Dyer, C., Bojar, O., Constantin, A., Herbst, E.: Moses: open source toolkit for statistical machine translation. In: Proceedings of the 45th Annual Meeting of the Association for Computational Linguistics, Prague, Czech Republic (2007)
7. Kumano, T., Tanaka, H., Tokunaga, T.: Extracting phrasal alignments from comparable corpora by using joint probability smt model. In: Proceedings of the International Conference on Theoretical and Methodological Issues in Machine Translation, Skvde, Sweden (2007)
8. Munteanu, D.S., Marcu, D.: Improving machine translation performance by exploiting non-parallel corpora. Comput. Linguist. **31**(4), 477–504 (2005)
9. Munteanu, D.S., Marcu, D.: Extracting parallel sub-sentential fragments from non-parallel corpora. In: Proceedings of the 21st International Conference on Computational Linguistics and the 44th Annual Meeting of the Association for Computational Linguistics, pp. 81–88, Sydney, Australia (2006)
10. Quirk, C., Udupa, R.U., Menezes, A.: Generative models of noisy translations with applications to parallel fragment extraction. In: Proceedings of the Machine Translation Summit XI, pp. 377–384, Copenhagen, Denmark (2007)
11. Rapp, R.: Identifying word translations in non-parallel texts. Proceedings of the 33rd Annual Meeting of the Association for Computational Linguistics, pp. 320–322, Cambridge, Massachusetts (1995)
12. Rapp, R.: Automatic identification of word translations from unrelated english and german corpora. Proceedings of the 37th Annual Meeting of the Association for Computational Linguistics, pp. 519–526, College Park, Maryland, USA (1999)
13. Resnik, P., Smith, N.: The web as a parallel corpus. Comput. Linguist. **29**(3), 349–380 (2003)
14. Tillmann, C., Hewavitharana, S.: A unified alignment algorithm for bilingual data. In: Proceedings of Interspeech 2011, Florence, Italy (2011)
15. Tillmann, C., Xu, J.-M.: A simple sentence-level extraction algorithm for comparable data. In: Companion Volume of NAACL HLT 09, Boulder, CA (2009)

16. Utiyama, M., Isahara, H.: Reliable measures for aligning Japanese–English news articles and sentences. In: Proceedings of the 41st Annual Meeting of the Association for Computational Linguistics, pp. 72–79, Sapporo, Japan (2003)
17. Vogel, S.: PESA: phrase pair extraction as sentence splitting. In: Proceedings of the Machine Translation Summit X, Phuket, Thailand (2005)
18. Zhao, B., Vogel, S.: Adaptive parallel sentence mining from web bilingual news collection. In: Proceedings of the IEEE International Conference on Data Mining, pp. 745–748, Maebashi City, Japan (2002)
19. Zhao, B., Vogel, S.: Full-text story alignment models for Chinese–English bilingual news corpora. In: Proceedings of the ICSLP '02 (2002)

Exploiting Comparable Corpora

Dragos Stefan Munteanu and Daniel Marcu

1 Introduction

Comparable corpora exhibit various degrees of parallelism. Fung and Cheung [3] describe corpora ranging from noisy parallel, to comparable, and finally to very non-parallel. The last category contains corpora composed of "… disparate, very non-parallel bilingual documents that could either be on the same topic (on-topic) or not". This is the type of corpora that out work is attempting to exploit.

Most existing methods for exploiting comparable corpora look for parallel data at the sentence level. However, we believe that very non-parallel corpora have none or few good sentence pairs; most of their parallel data exists ad the sub-sentential level. As an example, consider Fig. 1, which presents two news articles from the English and Romanian editions of the BBC. The articles report on the same event (the one-year anniversary of Ukraine's Orange Revolution), have been published within 25 min of each other, and express overlapping content.

Although they are "on-topic", these two documents are non-parallel. In particular, they contain no parallel sentence pairs; methods designed to extract full parallel sentences will not find any useful data in them. Still, as shown by the lines and boxes from the figure, some parallel fragments of data do exist; but they are present at the sub-sentential level.

In this work, we present a method for extracting such parallel fragments from comparable corpora. Figure 2 illustrates our goals. It shows two sentences taken from the articles in Fig. 1, and highlights and connects their parallel fragments.

Although the sentences share some common meaning, each of them has content which is not translated on the other side. The English phrase *reports the BBC's Helen Fawkes in Kiev*, as well as the Romanian one *De altfel, vorbind inaintea aniversarii* have no translation correspondent, either in the other sentence or anywhere in the

D. S. Munteanu (✉) · D. Marcu
SDL Language Weaver, Los Angeles, USA
e-mail: dragos@isi.edu

S. Sharoff et al. (eds.), *Building and Using Comparable Corpora*,
DOI: 10.1007/978-3-642-20128-8_11, © Springer-Verlag Berlin Heidelberg 2013

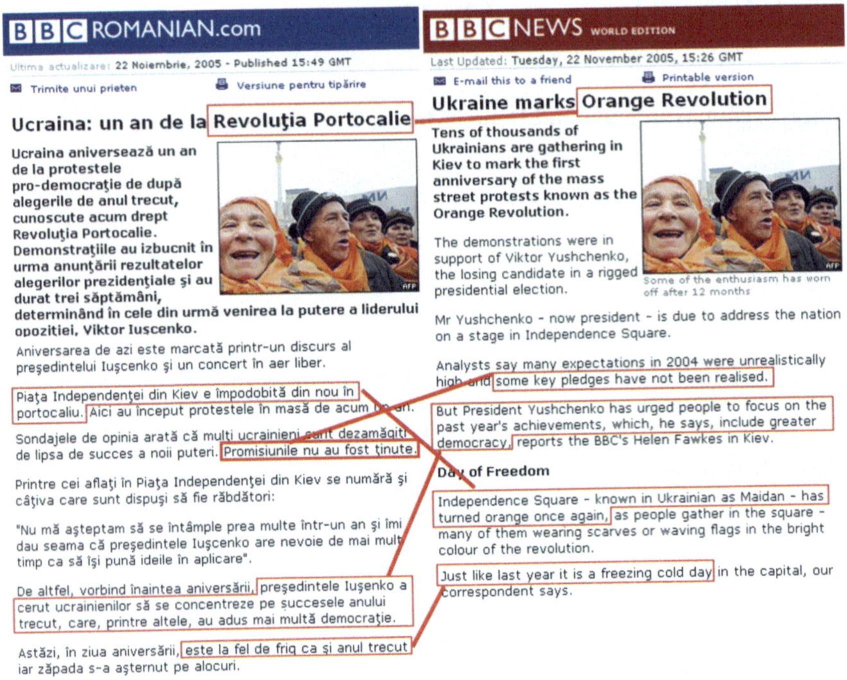

Fig. 1 A pair of comparable, non-parallel documents

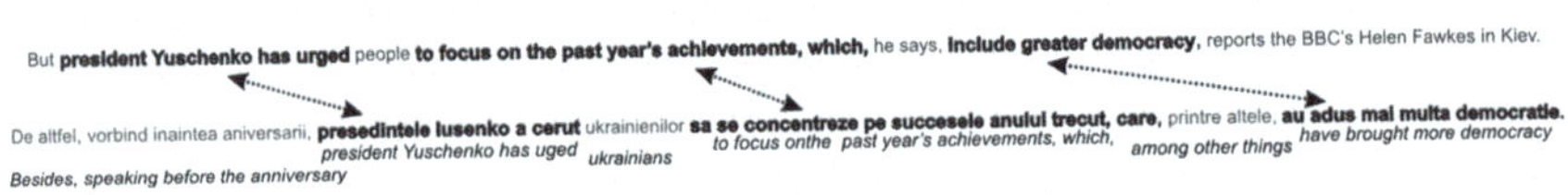

Fig. 2 A pair of comparable sentences

whole document. Since the sentence pair contains so much untranslated text, it is unlikely that any parallel sentence detection method would consider it useful. If the sentences would be used for another application, i.e. for training a Statistical Machine Translation (SMT) system, they might do more harm than good for the application's performance. The best way to make use of this sentence pair is to extract and use just the translated (highlighted) fragments. This is a difficult task, since it requires the ability to recognize translational equivalence in very noisy environments. However, a good solution to this problem would have a strong impact on parallel data acquisition efforts.

This chapter describes an approach for finding such parallel fragments. We evaluate the usefulness of the extracted data by showing it helps improve the performance of an SMT system. This work builds on our earlier results on parallel fragment extraction [13].

2 Finding Parallel Sub-Sentential Fragments in Comparable Corpora

2.1 Introduction

We start by describing the high-level architecture of our parallel fragment extraction system, shown in Fig. 3. The system takes two inputs. One is the comparable corpus, two collections of documents in the source and target language respectively. The other is a parallel corpus, a source-language text together with a parallel target-language translation.

The first step of the pipeline identifies document pairs that are similar, and therefore more likely to contain parallel data. Using a bilingual lexicon learned from the parallel data (ML-Lex), each source language document is translated wor-for-word in the target language. The translation is turned into a query which is run against the target language collection. The top 20 results are retrieved and paired with the source document.

From each document pair we generate all possible sentence pairs and run them through the second step in the pipeline, the candidate selection filter. This step discards pairs which have very few words that are translations of each other. The remaining sentence pairs are then presented to the fragment detection algorithm which, using a different lexicon (LLR-Lex), produces the system's output.

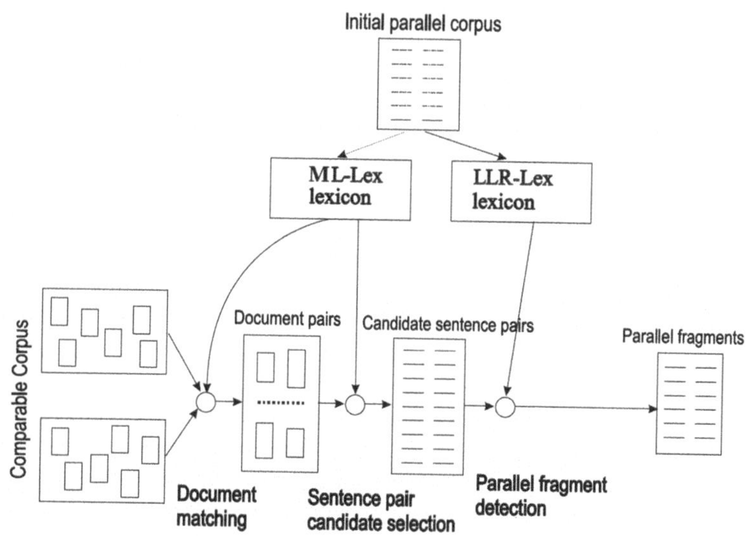

Fig. 3 A parallel fragment extraction system

The first two stages, document selection and candidate selection, are described in more detail in [12]. In the remainder of this section we will describe the fragment detection algorithm, as well as the LLR lexicon that it uses.

2.2 Using Log-Likelihood-Ratios to Estimate Word Translation Probabilities

Our system uses two probabilistic lexicons, learned automatically from the same initial parallel corpus. The first one, ML-Lex, is computed by aligning the parallel corpus at the word-level, and estimating translation probabilities using the maximum likelihood statistic on the word links. One of the characteristics of this lexicon is that each source word is associated with many possible translations. Although most of its high-probability entries are correct, there are a lot of entries (of non-negligible probability) where the two words are at most related. This characteristic is useful for the first two stages of our pipeline, which are not intended to be very precise. Their purpose is to accept most of the existing parallel data, and not too much of the non-parallel data. Using the ML-Lex lexicon helps achieve this purpose.

For the last stage, however, precision is paramount. We found empirically that when using ML-Lex, its incorrect entries have a serious impact on the quality of our results. We therefore developed a different approach for learning a bilingual lexicon, based on the Log-Likelihood-Ratio (LLR) statistic [2]. This approach has also been used by Moore [10, 11] and Melamed [9] as a measure of word association. Generally speaking, the LLR statistic gives a measure of the likelihood that two samples are not independent (i.e. generated by the same probability distribution). We use it to estimate the independence of pairs of words which cooccur in our parallel corpus.

If source word f and target word e are independent (i.e. they are not translations of each other), one can expect that $p(e|f) = p(e|\neg f) = p(e)$, i.e. the distribution of e given that f is present is the same as the distribution of e when f is not present. The LLR statistic gives a measure of the likelihood of this hypothesis. The LLR score of a word pair is low when these two distributions are very similar (i.e. the words are independent), and high otherwise (i.e. the words are strongly associated). However, high LLR scores can indicate either a positive association (i.e. $p(e|f) > p(e|\neg f)$) or a negative one; and one can distinguish between them by checking whether $p(e, f) > p(e)p(f)$ [11].

Thus, we can split the set of cooccuring word pairs into positively and negatively associated pairs, and obtain a measure for each of the two association types. The first type of association provides us with a cleaner lexicon, while the second allows us to estimate probabilities of words *not* being translations of each other.

Before describing the method more formally, we address the notion of word cooccurence. In the work of Moore [11] and Melamed [9], two words cooccur if they are present in a pair of aligned sentences in the parallel training corpus. However, most of the words from aligned sentences are actually unrelated; therefore, this is a rather weak notion of cooccurence. We follow Resnik et al. [20] and adopt a stronger definition, based not on sentence alignment but on word alignment: two words cooccur if they are linked together in the word-aligned parallel training corpus. We thus make use of the significant amount of knowledge brought in by the word alignment procedure.

We compute the LLR score using the following formula [2]:

$$LLR(s,t) = 2[logL(p_1,k_1,n_1) + logL(p_2,k_2,n_2) \\ - logL(p,k_1,n_1) - logL(p,k_2,n_2)]$$

where:

- $L(p,k,n) = p^k(1-p)^{n-k}$
- k_1, n_1 define the probability of observing target-language word t when source-language word s is present; that is, $k_1 = C(t,s)$ (i.e. the number of links between t and s in the aligned corpus) and $n_1 = C(s)$ (the number of links involving s).
- k_2, n_2 defined the probability of observing t when s is not present: $k_2 = C(t, \neg s)$ (links containing t but not s) and $n_2 = C(\neg s)$ (links not containing s).
- $p_1 = \frac{k_1}{n_1}$, $p_2 = \frac{k_2}{n_2}$, $p = \frac{k_1+k_2}{n_1+n_2}$

For a more detailed discussion of the relationship between this LLR formula and the frequencies and probabilities that arise in word-association computations, see [11].

Unfortunately, for this particular use of LLR scores, namely that of finding strongly associated word pairs, one of the terms in the formula raises problems. That term is n_2, the number of links that don't contain source word s; or the total number of links in the corpus minus those that contain s. The total number of links in a reasonably-sized corpus is very large compared to the number of links that involve any given word; therefore, this term has a very high value and renders all other terms in the LLR formula insignificant. The result is that all LLR scores will be relatively similar.

To solution we adopt is to pretend that the total number of links in the corpus is much smaller. Let us define W_s as the set of all target words linked to a source word s. For each s, we compute a word-specific total $T(s)$ as being the number of links that contain either s, or any word from W_s. Then, when computing $LLR(s,t)$, we use $n_2 = T(s) - C(\neg s)$. Note that LLR scores computed in this manner are no longer symmetrical: $LLR(s,t) \neq LLR(t,s)$

We use the LLR values to compute two conditional probability distributions: $P^+(t|s)$, the probability that source word s translates into target word t, and $P^-(t|s)$, the probability that s does not translate into t. We obtain these distributions by normalizing the LLR scores for each source word.

We describe below the lexicon estimation procedure:

- Word-align the parallel corpus. Following Och et al. [15], we run GIZA++ in both directions and then symmetrize the alignments

Table 1 Example of translation probabilities from the ML and LLR lexicons

Target translation	$ML(t\|s)$	$C(t)$	$LLR(s,t)$	$p(s,t) > p(s)p(t)$	$LLR^+(t\|s)$	$LLR^-(t\|s)$
Reserved	0.45	174	798.91	yes	0.82	
Restricted	0.07	398	68.17	yes	0.07	
Silent	0.02	76	18.02	yes	0.02	
*to	0.07	230178	16.61	no		0.08
*the	0.01	608638	175.45	no		0.85

Words marked with '*' are incorrect translations

- Compute all LLR scores. There will be an LLR score for each pair of words which are linked at least once in the word-aligned corpus
- Classify all $LLR(t,s)$ as either $LLR^+(t,s)$ (positive association) if $p(t,s) > p(t)p(s)$, or $LLR^-(t,s)$ (negative association) otherwise
- For each s compute the normalizing factors $\sum_s LLR^+(t,s)$ and $\sum_s LLR^-(t,s)$
- Divide all $LLR^+(t,s)$ terms by the corresponding normalizing factors to obtain $LLR^+(t|s)$
- Divide all $LLR^-(t,s)$ terms by the corresponding normalizing factors to obtain $LLR^-(t|s)$
- Reverse the source and target languages and repeat the steps outlined above (except for the first one, since the word alignment is symmetrical) to obtain $LLR^+(s|t)$ and $LLR^-(s|t)$

Table 1 illustrates the differences between the Maximum-Likelihood lexicon and the Log-Likelihood-Ratio lexicon. It lists several possible translations of Romanian source word *rezervata*, and presents the probabilities they have in lexicons computed from a Romanian-English parallel corpus. The entries from the last two rows are incorrect translations (and therefore marked with a star). They appear in ML-Lex because of alignment errors, caused by the high frequency of the respective English words. However, the LLR computation is able to determine that the association between these two words and the original source word is spurious, and to place them in the LLR^- lexicon.

2.3 Detecting Parallel Sub-Sentential Fragments

Intuitively speaking, our approach attempts to distinguish between source fragments that have a translation on the target side, and fragments that do not, using the information available in the lexicon. In Fig. 4 we show the sentence pair as Fig. 2, in which we underlined those words of each sentence which have a translation in the other sentence, according to LLR^+. Thus, boldface indicates true translations, while underline indicates translation according to the lexicon. The phrases "*to focus on the past year's achievements, which,*" and "*sa se concentreze pe succesele anului trecut, care,*" are mostly underlined. The lexicon is unaware of the fact that "achievements"

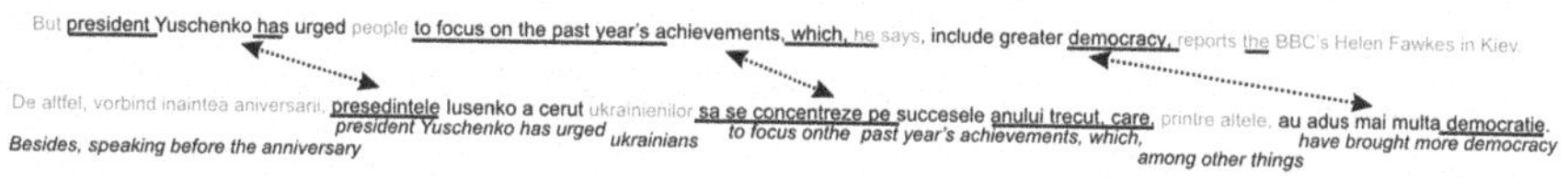

Fig. 4 Translated fragments, according to the lexicon

and "succesele" are in fact translations of each other (because "succesele" is a morphologically inflected form that does not cooccur with "achievements" in our initial parallel corpus). The rest of the sentences are mostly not underlined, although we do have occasional connections, some correct and some wrong. The best we can do in this case is to infer that these two phrases are parallel, and discard the rest. Doing this does gains a bit of new knowledge: the lexicon entry *(achievements, succesele)*.

We need to quantify more precisely the notions of "mostly translated" and "mostly not translated". Our approach is to consider the target sentence as a numeric *signal*, where translated words correspond to positive values (the translation probabilities from the LLR^+ lexicon), and the others to negative ones (coming from the LLR^- lexicon). We want to retain the parts of the sentence where the signal is mostly positive. This can be achieved by applying a smoothing filter to the signal, and selecting those fragments of the sentence for which the corresponding filtered values are positive.

The details of the procedure are presented below, and also illustrated in Fig. 5. Let the Romanian sentence be the source sentence S, and the English one be the target, T. We compute a word alignment $S \rightarrow T$ using the maximum-likelihood lexicon, and greedily linking each source word with its most likely target translation. For each of the linked target words, the corresponding signal value is the probability of the link (the alignment model is defined such that there can be at most one link for each target word). Thus, if target word t is linked to source word s, the signal value corresponding to t is $LLR^+(t|s)$ (the distribution described in Sect. 2.2), i.e. the probability that t is the translation of s.

For the remaining target words, the signal value should reflect the probability that they are not translated; for this, we employ the LLR^- distribution. Thus, for each non-linked target word t, I look for the source word least likely to be its non-translation: $s_0 = argmin_{s \in S} LLR^-(t|s)$. If s_0 exists, we set the signal value for t to $-LLR^-(t|s_0)$; otherwise, we set it to -1. This is the *initial signal*. We then compute the *filtered signal* by applying an averaging filter, which sets the value at each point to be the average of several values surrounding it. In our experiments, we use the surrounding five values, which produced good results on a development set. We then simply retain those fragments of T for which the corresponding filtered signal values are positive; these fragments are circled in the figure.

However, this approach will often produce incorrectly positive short fragments; examples of this are, in Fig. 5, the fragments "*president*", "*democracy*", and, "*reports*". In an attempt to avoid such errors, we discard all fragments with less than three words.

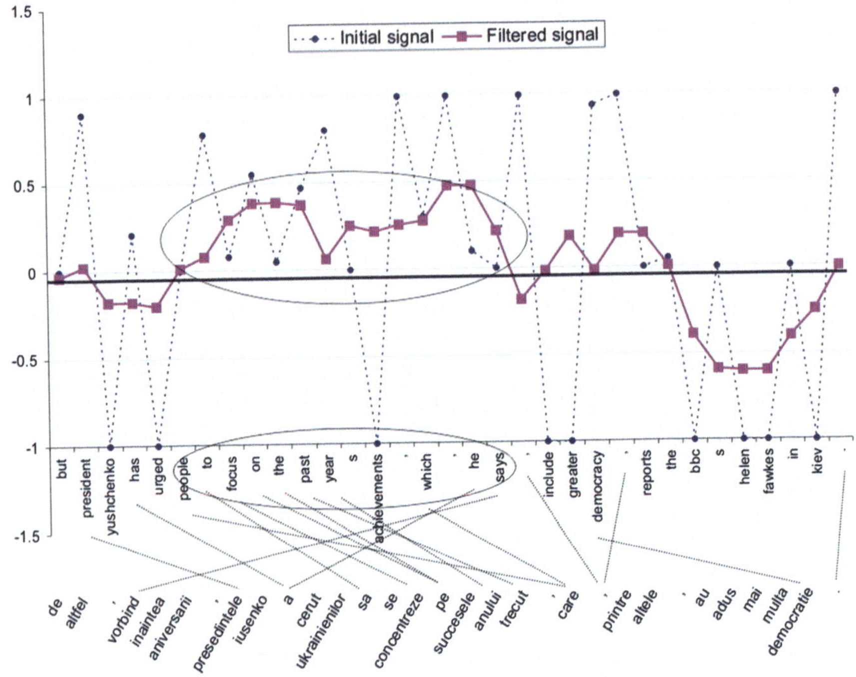

Fig. 5 A signal-filtering approach for detecting parallel fragments

We repeat the procedure for the other direction, $T \rightarrow S$ to obtain the fragments for s, and consider the resulting two text chunks as parallel. For the sentence pair from my example (Fig. 5), the system will output the pair:

people to focus on the past year's achievements, which, he says

sa se concentreze pe succesele anului trecut, care, printre

which is mostly correct.

2.4 Examples

Figure 6 presents a few fragment pairs extracted by our algorithm from the Romanian-English BBC Corpus (Sect. 3.1). For each example the figure contains the input, i.e. the initial sentence pair, and the output, namely the two fragments that were judged parallel. Each Romanian language sentence or fragment is followed by a human-produced translation, printed in bold and italic font. When printing the output, I used the "[...]" token to mark places where words from the original sentence were discarded. This token does not appear in the system output; it is only included in the figure to help make it more clear.

INPUT
- Dupa lansarea ofertei ar trebui sa urmeze o judecata severa facuta de alegatori.
A critical judgment passed by the electors should follow the launch of the offer.
- A critical judgment passed by the electors or their enthusiasm with the PDSR propositions should have followed the launch of the offer.

OUTPUT
- Dupa lansarea ofertei ar trebui sa urmeze o judecata severa facuta de alegatori.
A critical judgment passed by the electors should follow the launch of the offer.
- A critical judgment passed by the electors [...] have followed the launch of the offer.

INPUT
- Lista celor 23 de norocosi care si-au retras banii de la FNI Pitesti
The list of the 23 fortunates who whithdrew their money from FNI Pitesti
- Who withdrew money from the FNI shortly before that moment ?

OUTPUT
- [...] care si- au retras banii de la FNI Pitesti
[...] ***who withdrew their money from FNI Pitesti***
- Who withdrew money from the FNI shortly before [...]

INPUT
- Ieri, toate televiziunile nationale au solicitat ziarului Ziua caseta video, pentru a o difuza pe post.
Yesterday, all national television channels have requested the video tape from the Ziua newspaper , to broadcast it on the air.
- On the other hand, Chisinau has had a minimum reaction by placing news on the CEDO decision at the end of the news program on the national tv channel.

OUTPUT
- Ieri, toate televiziunile nationale au solicitat [...], pentru a o [...] pe post.
Yesterday, all national television channels have requested **[...]** ***, to it*** **[...]** ***on the air.***
- On the other [...] news on the [...] the end of [...] on the national tv channel.

Fig. 6 Examples of automatically extracted parallel sub-sententialfragments

The examples point out both positive and negative effects of the parallel fragment detection approach. In the first sentence pair from the figure, the model successfully discards the English phrase "*or their enthusiasm with the PDSR propositions*" which has no translation on the Romanian side. In the second pair, it manages to correctly extract translated fragments from two non-parallel sentences.

On the other hand, as shown by the last pair from the example, the method can also produce non-parallel fragment pairs. This is caused by the presence of incorrect lexicon entries, which wrongly give positive signal values to many function words. Thus, the approach might be improved by distinguishing content words from non-content words, and handling them differently in the model (maybe by giving less weight to the signal values corresponding to non-content words).

3 Experiments

3.1 Experimental Framework

We evaluate the data extracted with our algorithm in the context of Statistical Machine Translation. As explained in Sect. 2.1, our extraction system makes use of an initial parallel corpus. We can measure the impact of the extracted data on the performance of a translation system by training and comparing two SMT systems. The *Baseline* system is trained on the initial parallel corpus. The *PlusExtracted* system is trained on the initial corpus plus the extracted data. If the second system is better than the first, this provides evidence that our automatically extracted data is useful.

We perform experiments on three language pairs: Romanian-English, Arabic-English and Chinese-English. For each language pair we use a parallel corpus and a comparable corpus. The size of the initial parallel corpus clearly has a big influence on the size and quality of the lexicon used for extraction, and therefore on the quality of the extracted data. In order to measure this influence, we run, for each language, several extraction experiments. They all share the same comparable corpus, but use parallel corpora of various sizes.

3.2 Corpora and Translation Systems

Our Romanian-English parallel corpus contains data from two sources. One source is the training data made available through the Romanian-English word alignment task from the Workshop on Building and Using Parallel Corpora.[1] The remainder of the data comes from Romanian translations of the European Union's acquis communautaire, which we mined from the Web. The comparable data was downloaded from the Romanian and English editions of the online BBC news service. The test data used to measure the performance of the translation system consists of news articles from the TimeBank corpus which were translated into Romanian. It has 1000 sentences, and one reference translation.

For Arabic-English and Chinese-English we use data available through the Linguistic Data Consortium (LDC).[2] The parallel data, as well as the SMT test data, were compiled from corpora used in the NIST MT evaluations.[3] The test data has four reference translations. The comparable corpora are parts of the Arabic, Chinese and English Gigaword corpora available from the LDC.

Table 2 provides details about the comparable corpora used in the experiments.

The SMT systems used in my experiments are trained using a variant of the alignment template model described in [14]. All systems use two trigram language

Table 2 Size and composition of the comparable corpora

Language pair	News agency	Foreign		English	
		#articles	#tokens	#articles	#tokens
Romanian-English	BBC News	6k	2.5M	200k	118M
Arabic-English	AFP and Xinhua News	600k	115M	1.6M	330M
Chinese-English	CNA and Xinhua News	1.6M	456M	983k	247M

[1] http://www.statmt.org/wpt05

[2] http://www.ldc.upenn.edu

[3] http://www.nist.gov/speech/tests/mt

models: one trained on the English side of the parallel corpus, and one trained on the English side of the comparable corpus. Thus, any difference in performance between a baseline and a contrastive system are caused only by differences in their parallel training data. All systems trained on a particular language pair were tested on the same test set. We ensured that none of the documents from the comparable corpora (which are potentially training data) overlap with any of the test data. Translation performance was measured using the automatic BLEU [16] evaluation metric, against 4 reference translations. For readability purposes we report BLEU% numbers, i.e. we multiply the original scores by 100.

3.3 Experiments on the BBC Corpus

We ran two extractions on the Romanian-English BBC Corpus, using initial parallel corpora of sizes 1M words and 10M words respectively (on the English side). The results are presented in Fig. 7.

For each initial parallel corpus, the first two columns in the figure show the BLEU scores of the *Baseline* and *PlusExtract* systems described in Sect. 3.1. The first two rows in the table below the figure present the same information. The 95 % confidence intervals of these scores, computed by bootstrap resampling [7] indicate that a score

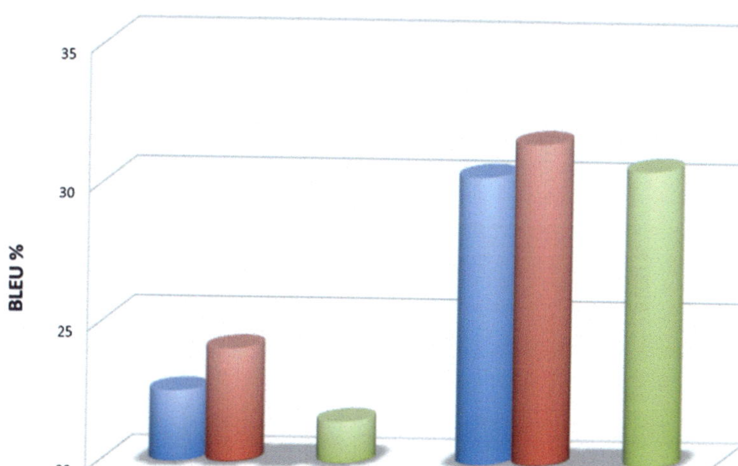

	1M	10M
Baseline	22.5	30.3
PlusExtract	24.04	31.51
Extracted data (Mw)	0.4	1.3
PlusExtract-noLLR	21.45	30.59

Fig. 7 SMT performance improvements using parallel fragments automatically extracted from the BBC corpus

increase of more than 1 BLEU% is significant. Thus, these scores show that the extracted parallel fragments yield significant translation performance improvements.

The third row in the table below the figure shows the amount of data produced by our extraction algorithm. The data size is measured in million words, on the English side. As expected, using a larger initial corpus enabled us to find more data, even though both extractions were performed on the same comparable data.

The last row in the table (as well as the last column in the figure) corresponds to an experiment intended to evaluate the LLR lexicons (Sect. 2.2). In order to verify that the LLR lexicons are indeed important to the success of our method, we ran the extraction algorithm using only the ML-Lex lexicon for all the stages of the pipeline. The resulting data fails to bring translation improvements, thus validating our claim.

3.4 Experiments on the Gigaword Corpus

For Arabic-English and Chinese-English we ran extractions with parallel corpora ranging from very small, 10k English words, to very large, 100M English words. The results are presented in Figs. 8 and 9.

For Arabic, the automatically extracted data brings significant improvements for all initial parallel corpora. For Chinese however, there are no improvements from

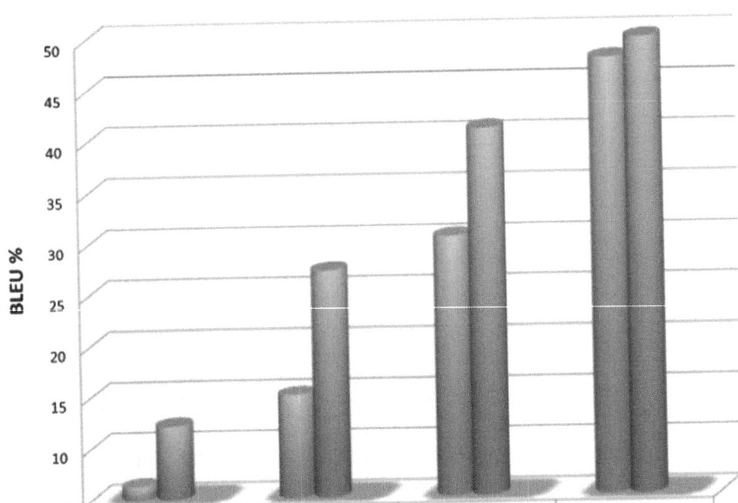

	10k	100k	1M	100M
Baseline	6.23	14.99	30.38	47.69
PlusExtract	12.1	27.14	40.89	49.66
Extracted data (Mw)	0.2	4	37	74

Fig. 8 Arabic-English MT performance improvements using parallel fragments automatically extracted from the Gigaword corpus

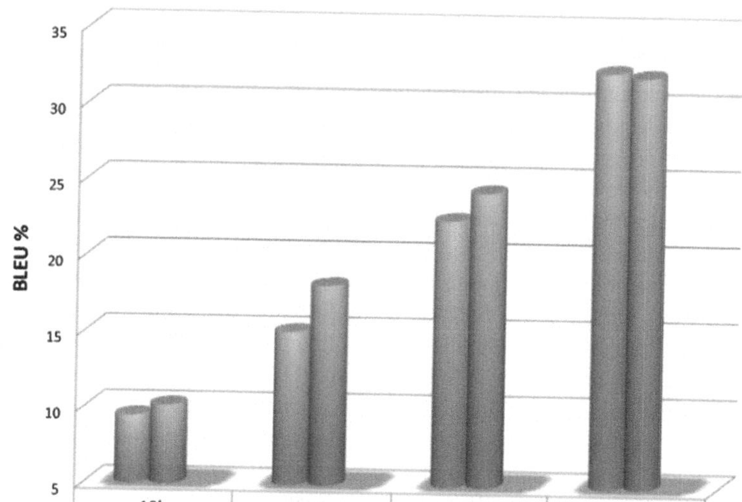

	10k	100k	1M	100M
Baseline	9.41	15	22.44	32.29
PlusExtract	10.1	18.07	24.29	32
Extracted data (Mw)	2.6	22	81	285

Fig. 9 Chinese-English MT performance improvements using parallel fragments automatically extracted from the Gigaword corpus

the extractions performed with the smallest and the largest parallel corpus. For the 100M-words corpus, the baseline system is too good and hard to beat. For the 10k-words system, the failure to improve on the baseline is due to the poor quality of the lexicon. The lexicon is computed from the automatically word-aligned parallel corpus; and for such a small corpus, the performance of the word aligner is quite poor. Having a noisy lexicon will significantly affect the performance of our fragment detection method.

We also performed *PlusExtracted-noLLR* experiments, running extractions without the LLR lexicons. All the resulting datasets failed to improve performance relative to the baseline, so we omitted them from the figure.

3.5 Bootstrapping

As can be seen from the experiments performed on the Gigaword corpora (i.e. Figs. 8 and 9), the amounts of data extracted with a small initial corpus (i.e. 10k tokens) are much smaller than the amounts extracted with larger initial resources. This is mostly due to the low coverage of the lexicon, which causes the candidate selection filter (Sect. 2.1) to discard many sentence pairs. The pairs are discarded because, according to the lexicon, they share too few words in common.

One possible solution is to make the filter less strict and allowing the generation of more candidate pairs. However, a lexicon computed from a small word-aligned parallel corpus has not just low coverage, but also low quality. Automatic word alignment algorithms perform poorly on small datasets, because they lack sufficient statistics for their parameters. Thus, lowering the precision of the filter, in conjunction with a noisy lexicon, would result in poor output quality.

A better solution is to use bootstrapping. After the initial extraction we can add the resulting data to the parallel corpus, compute a new lexicon, and run another extraction. The new lexicon should have much better coverage. Not only is it computed from more parallel data, but also from data obtained from the comparable corpus itself. Bootstrapping was first applied to this problem by Fung et al. [3].

Our results are presented in Figs. 10 and 11. We performed bootstrapping experiments on the Gigaword corpora, for Arabic-English and Chinese-English. We started from the smallest initial parallel corpora, of 10k English tokens. The first extraction, that starts from the initial parallel corpus, is called *Iteration 0* (since no bootstrapping is involved yet). Each subsequent iteration uses, as parallel corpus, the initial corpus plus the data extracted from the previous iteration (and nothing from any earlier iterations). All extraction runs use the same comparable corpus. Due to time and resource limitations we only performed two bootstrapping iterations.

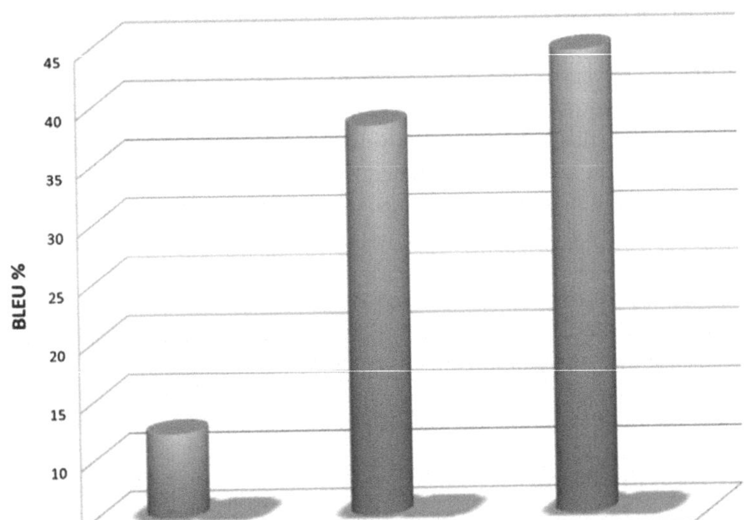

	Iteration 0	Iteration 1	Iteration 2
BLEU	12.1	38.2	44.4
Extracted data (Mw)	0.2	22	25.7

Fig. 10 Arabic-English bootstrapping iterations

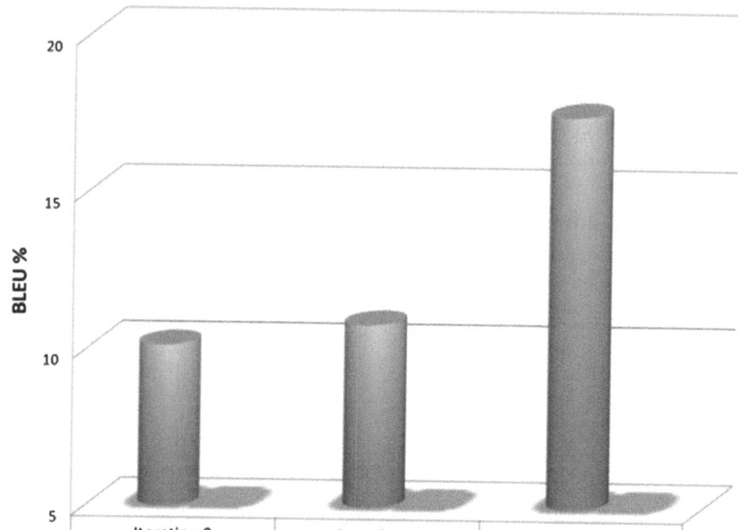

	Iteration 0	Iteration 1	Iteration 2
BLEU	10.1	10.8	17.5
Extracted data (Mw)	2.6	54	100

Fig. 11 Chinese-English bootstrapping iterations

4 Related Work

The research concerned with the exploitation of comparable corpora has advanced significantly over the years. The early efforts, which attempted to find word-for-word translations [6, 8, 18, 19] have evolved into algorithms for finding parallel documents [21, 22, 26], sentences [4, 5, 12, 23–25], and fragments [1, 13, 17]. Researchers increasingly evaluate their outputs in an extrinsic manner, by using the produced data in another NLP application (most often Machine Translation) and showing it brings performance improvements.

Another result of the field's maturity is that the algorithms are becoming ever more efficient, and are being applied to ever larger corpora. The largest experiment published to date is that of Uszkoreit et al. [22]. They describe a linear-time algorithm for detecting parallel documents, and apply it to a collection of 2.5 billion documents.

The earliest work concerned with the problem of parallel fragment detection is [13], which this chapter is based upon. It has been followed by two subsequent efforts that address several of our shortcomings.

Quirk et al. [17] present a more theoretically grounded approach to the problem of parallel fragment identification. They present generative models of loose translations, designed to explain the presence of large chunks of non-translated data. These models produce words by essentially switching between monolingual and bilingual generative models. The authors show how the model parameters can be estimated from data, and how the trained model can be used to extract parallel fragments from new sentence pairs.

Cettolo et al. [1] take an approach based on the availability of a repository of phrase pairs. The main contribution of their work is an algorithm for creating a phrase-level alignment between two texts, which achieves a good tradeoff between maximum coverage and minimum distortion. Once such an alignment is computed, the parallel fragments are simply the aligned pairs that are contiguous or in sufficiently close proximity. Interestingly, they obtain better results by aligning each target-language sentence not to source sentences, but to whole documents.

5 Discussion

We have presented a simple and effective method for extracting sub-sentential fragments from comparable corpora. We also presented a method for computing a better probabilistic lexicon using the LLR statistic, and showed that this lexicon helps improve the precision of our method. We demonstrated the quality of the data extracted with our algorithm by showing that it helps improve the performance of a statistical machine translation system.

It must be noted that the SMT improvements are partly due to the domain effect. For all our baseline systems the test data comes from a different domain, namely newswire articles, than the training data, which is mostly parliamentary proceedings. The automatically extracted data is also newswire. This means that the *PlusExtracted* systems have better domain overlap between their training and test data, which explains part of the improvements. However, the purpose of our experiments was simply to prove that the data produced by our system is of good quality. The domain effect does not invalidate this claim.

Our approach can be improved in several aspects. The signal filtering function is very simple. More advanced filters might work better, and eliminate the need for additional heuristics (such as our requirement that the extracted fragments have at least three words). The fact that the source and target signal are filtered separately is also a weakness; a joint analysis should produce better results. Despite the better lexicon, the greatest source of errors is still related to false word correspondences, generally involving punctuation and very common (closed-class) words. Giving special attention to such cases should help get rid of these errors and improve the precision of the method.

References

1. Cettolo, M., Federico, M., Bertoldi, N.: Mining parallel fragments from comparable texts. In: Proceedings of the 7th International Workshop on Spoken Language Translation, pp. 227–234 (2010)

2. Dunning, T.: Accurate methods for the statistics of surprise and coincidence. Comput. Linguist. **19**(1), 61–74 (1993)
3. Fung, P., Cheung, P.: Mining very non-parallel corpora: parallel sentence and lexicon extraction via bootstrapping and EM. In: Proceedings of the Conference on Empirical Methods in Natural Language Processing (EMNLP 2004), pp. 57–63 (2004)
4. Fung, P., Cheung, P.: Mining very non-parallel corpora: parallel sentence and lexicon extraction vie bootstrapping and em. In: Proceedings of the Conference on Empirical Methods in Natural Language Processing (EMNLP), pp. 57–63 (2004)
5. Fung, P., Cheung, P.: Multi-level bootstrapping for extracting parallel sentences from a quasi-comparable corpus. In: Proceedings of the 20th International Conference on Computational Linguistics (COLING), pp. 1051–1057 (2004)
6. Gaussier, E., Renders, J.M., Matveeva, I., Goutte, C., Dejean, H.: A geometric view on bilingual lexicon extraction from comparable corpora. In: Proceedings of the 42nd Annual Meeting of the Association for Computational Linguistics, pp. 527–534 (2004)
7. Koehn, P.: Statistical significance tests for machine translation evaluation. In: Proceedings of the Conference on Empirical Methods in Natural Language Processing, pp. 388–395 (2004)
8. Koehn, P., Knight, K.: Learning a translation lexicon from monolingual corpora. In: Proceedings of the Workshop of the ACL Special Interest Group on the Lexicon (SIGLEX), pp. 9–16 (2002)
9. Melamed, I.D.: Models of translational equivalence among words. Comput. Linguist. **26**(2), 221–249 (2000)
10. Moore, R.C.: Improving IBM word-alignment model 1. In: Proceedings of the 42nd Annual Meeting of the Association for Computational Linguistics, pp. 519–526 (2004)
11. Moore, R.C.: On log-likelihood-ratios and the significance of rare events. In: Proceedings of the 2004 Conference on Empirical Methods in Natural Language Processing, pp. 333–340 (2004)
12. Munteanu, D.S., Marcu, D.: Improving machine translation performance by exploiting non-parallel corpora. Comput. Linguist. **31**(4), 477–504 (2005)
13. Munteanu, D.S., Marcu, D.: Extracting parallel sub-sentential fragments from non-parallel corpora. In: Proceedings of the 21st International Conference on Computational Linguistics and 44th Annual Meeting of the ACL, pp. 81–88 (2006)
14. Och, F.J., Ney, H.: The alignment template approach to statistical machine translation. Comput. Linguist. **30**(4), 417–450 (2003)
15. Och, F.J., Ney, H.: A systematic comparison of various statistical alignment models. Comput. Linguist. **29**(1), 19–51 (2003)
16. Papineni, K., Roukos, S., Ward, T., Zhu, W.J.: Bleu: a method for automatic evaluation of machine translation. In: Proceedings of the 40th Anniversary Meeting of the Association for Computational Linguistics, pp. 311–318 (2002)
17. Quick, C., Udupa, R.U., Menezes, A.: Generative models of noisy translations with applications to fragment extraction. In: Proceedings of MT Summit XI (2007)
18. Rapp, R.: Identifying word translation in non-parallel texts. In: Proceedings of the Conference of the Association for Computational Linguistics, pp. 320–322 (1995)
19. Rapp, R.: Automatic identification of word translations from unrelated English and German corpora. In: Proceedings of the 27th Annual Meeting of the Association for Computational Linguistics, pp. 519–526 (1999)
20. Resnik, P., Oard, D., Levow, G.: Improved cross-language retrieval using backoff translation. In: Proceedings of the 1st International Conference on Human Language Technology Research (2001)
21. Resnik, P., Smith, N.A.: The web as a parallel corpus. Comput. Linguist. **29**(3), 349–380 (2003)
22. Uszkoreit, J., Ponte, J., Popat, A., Dubiner, M.: Large scale parallel document mining for machine translation. In: Proceedings of the 23rd International Conference on Computational Linguistics (COLING), pp. 1101–1109 (2010)
23. Utiyama, M., Isahara, H.: Reliable measures for aligning Japanese-English news articles and sentences. In: Proceedings of the 41st Annual Meeting of the Association for Computational Linguistics, pp. 72–79 (2003)

24. Wu, D., Fung, P.: Inversion transduction grammar constraints for mining parallel sentences from quasi-comparable corpora. In: Proceedings of 2nd International Joint Conference on Natural Language Processing (IJCNLP), pp. 257–268 (2005)
25. Zhao, B., Vogel, S.: Adaptive parallel sentences mining from web bilingual news collection. In: 2002 IEEE International Conference on Data Mining, pp. 745–748 (2002)
26. Zhao, B., Vogel, S.: Full-text story alignment models for Chinese-English bilingual news corpora. In: Proceedings of the International Conference on Spoken Language Processing (2002)

Paraphrase Detection in Monolingual Specialized/Lay Comparable Corpora

Louise Deléger, Bruno Cartoni and Pierre Zweigenbaum

Abstract Paraphrases are a key feature in many natural language processing applications, and their extraction and generation are important tasks to tackle. Given two comparable corpora in the same language and the same domain, but displaying two different discourse types (lay and specialized), specific paraphrases can be spotted which provide a dimension along which these discourse types can be contrasted. Detecting such paraphrases in comparable corpora is the goal of the present work. Generally, paraphrases are identified by means of lexical and/or structural patterns. In this chapter, we present two methods to extract paraphrases across lay and specialized French monolingual comparable corpora. The first method uses lexical patterns designed according to intuition and linguistic studies, while the second is empirical, based on n-gram matching. The two methods appear to be complementary: the n-gram method confirms the initial lexical patterns and identifies other patterns. Besides, differences in the direction of application of paraphrase patterns highlight differences between specialized and lay discourse.

Keywords Paraphrase identification · Lexico-syntactic patterns · Pattern discovery · Discourse distinction · Medical domain · Monolingual comparable corpora

1 Introduction

Comparable corpora refer to text corpora sharing common characteristics. They consist of texts in two languages that are collected by applying the same sampling frame

L. Deléger (✉) · P. Zweigenbaum
LIMSI-CNRS, Orsay F-91403, France
e-mail: louise.deleger@limsi.fr

P. Zweigenbaum
e-mail: pz@limsi.fr

B. Cartoni
Département de Linguistique, Université de Genève, 1211 Genève 4, Switzerland
e-mail: bruno.cartoni@unige.ch

S. Sharoff et al. (eds.), *Building and Using Comparable Corpora*,
DOI: 10.1007/978-3-642-20128-8_12, © Springer-Verlag Berlin Heidelberg 2013

without being translations of each other [20]. As a matter of fact, this notion also applies to monolingual texts. In a monolingual context, comparable corpora can be texts from different sources (such as articles from various newspapers) or from different genres (such as specialized and lay texts) but dealing with the same general topic. Comparable corpora have been used to perform several Natural Language Processing tasks, such as extraction of word translations [6, 12, 21, 25] in a multilingual context or acquisition of paraphrases [4, 9, 26] in a monolingual context. In this work, we are interested in using comparable corpora belonging to two discourse types (specialized and lay) but in the same domain (medicine) and language (French).

We therefore present a study that relies on monolingual comparable corpora of two different discourse types (lay vs. specialized) from the medical domain to empirically explore paraphrase phenomena between lay and specialized languages. Understanding paraphrasing mechanisms is a key issue for many NLP tasks such as text simplification or authoring medical documents dedicated to the general public.

This chapter describes two approaches. The first approach, which summarizes previous experiments [8], is a linguistic method based on a priori lexical knowledge which extracts specific, pre-defined types of paraphrases and compares their use across texts belonging to lay and specialized discourse types. The second approach tries to generalize this comparison of the two discourse types in order to discover new relevant paraphrases and paraphrasing patterns as well as confirm the relevance of the paraphrases identified in the first approach.

In the first approach, we looked at three types of paraphrases that are often said to be potentially relevant to the contrast between lay and specialized languages. The relevance of these types of paraphrases was defined in more introspective linguistic studies, and the empirical pieces of research performed here are also meant to verify such statements. All these paraphrases involve some kind of morphological variation: noun to verb variations, noun to adjective variations and variations between a neoclassical form and its modern-language equivalent. The general method to identify these paraphrases was to define lexico-syntactic patterns and to rely on morphological resources (pairs of nouns/verbs, adjectives/nouns) or on a morphosemantic analyzer.

The detection of paraphrases also implies the application of patterns that are frequent enough, and the relevance of a pattern is sometimes hard to predict in advance. In the second part of this chapter, we attempt to extend the extraction of paraphrases to a more data-driven approach that can catch frequent, yet undiscovered patterns. In that approach, no pre-established patterns are used. Paraphrase extraction is based on a comparison of stemmed lexical n-grams. Once new paraphrase candidates are found, a generalization step is performed (mainly based on POS tags), to identify supplementary paraphrasing patterns. We evaluate these patterns and examine them to determine which ones might be typical of the lay/specialized contrast. These empirically gathered patterns also help to evaluate the actual coverage of the patterns designed in the first method.

This chapter is structured as follows. Section 2 presents a short review of previous work on this topic. Section 3 describes the comparable corpora used in this study. In Sect. 4, we introduce the two extraction methods that we experimented: lexical-based patterns, and n-grams. Section 5 presents the results of the two methods. Finally, Sect. 6 discusses the complementarity of the two methods.

2 Background

Paraphrases are expressions that convey similar meanings. Detecting them can be useful in various applications, including information retrieval [14], information extraction [26], document summarization [3] and text simplification [9]. Several methods have been designed to extract paraphrases, many of them dealing with comparable text corpora. A few paraphrase acquisition approaches used plain monolingual corpora to detect paraphrases, such as [15] who detects term variants or [24] who extract paraphrases from random Web documents.

Such a type of corpus does not insure the actual existence of paraphrases, hence a majority of methods have relied on corpora with a stronger similarity between the documents, thus more likely to provide a greater amount of paraphrases. Some of them used monolingual parallel corpora, i.e., multiple translations or versions of the same texts. For instance Barzilay and McKeown [5] detected paraphrases in a corpus of English translations of literary novels, and Pang et al. [23] did so with a corpus of English translations of Mandarin Chinese news texts. Another, quite different method consists in observing revisions in a wiki such as Wikipedia [19]. However such corpora are not always available and approaches which rely instead on other types of corpora are actively investigated.

Bilingual parallel corpora have been exploited to acquire paraphrases in English [2] and French [18], using foreign phrases as pivot expressions to identify equivalent source phrases. Comparable corpora are another useful source of paraphrases. In this regard, only closely related corpora have been used, especially and almost exclusively corpora of news sources reporting the same events, also called 'synchronous' corpora. Barzilay and Lee [4] generated paraphrase sentences from news articles using finite state automata. Shinyama and Sekine [26] extracted paraphrases through the detection of named entity anchors in a corpus of Japanese news articles. In the medical domain, Elhadad and Sutaria [9] worked with a comparable, almost parallel, corpus of medical scientific articles and their lay versions from the ReutersHealth E-line newsfeed to associate specialized and lay terms of the UMLS Metathesaurus [17].

In this chapter, we explore more open methods to collect pairs of specialized and lay expressions in the medical domain. First, the comparable corpora we use generally have no parallel documents. Second, we do not restrain our search to medical terms found in pre-existing terminological bases such as the UMLS. Third, we do not only focus on noun phrases, but also include verbal and adverbial expressions.

3 A Collection of Comparable Medical Corpora

For this study, we built comparable corpora of French-language medical specialized and lay texts from the Web, in the domains of diabetes, nicotine addiction and cancer. For cancer, documents were directly downloaded from a website providing comparable texts: the SOR (*Standards Options Recommendations*) website[1] which publishes guidelines for physicians and their lay versions for the general public. Since such ready-made comparable corpora are scarce resources, we also gathered documents on the nicotine addiction and diabetes topics from various sources through a guided search: we thus identified and queried a number of relevant websites. This includes health search engines (the health Web portals CISMeF[2] and HON[3]); French governmental websites (the HAS[4] website which issues guidelines for health professionals, and the INPES[5] website which provides educational material for the general public); and health websites dedicated to the general public (Doctissimo,[6] Tabac Info Service,[7] Stoptabac[8] and Diabète Québec[9]). This resulted in three corpora, each consisting of two parts: a specialized part and a lay part. To balance the corpora, we sampled sentences to obtain approximately the same number of words in the two parts of each corpus. We then gathered the three corpora into one large corpus, with a specialized part and a lay part.

Table 1 shows the sizes of the total corpus and of its sub-corpora. The corpus was tagged with parts-of-speech and lemmatized using Treetagger[10] and the French lemmatizer Flemm.[11]

Table 1 Sizes of the corpora

	Diabetes		Nicotine addiction		Cancer		Total corpus	
	S	L	S	L	S	L	S	L
Sentences	29,692	44,799	25,460	40,840	10,838	13,389	65,990	99,028
Words	581,100	581,712	604,206	604,309	228,784	228,793	**1,414,090**	**1,414,805**

S Specialized, *L* Lay

1 http://www.sor-cancer.fr/
2 http://www.cismef.org/
3 http://www.hon.ch/
4 http://www.has-sante.fr/
5 http://www.inpes.sante.fr/
6 http://www.doctissimo.fr/
7 http://www.tabac-info-service.fr/
8 http://www.stop-tabac.ch/
9 http://www.diabete.qc.ca/
10 http://www.ims.uni-stuttgart.de/projekte/corplex/TreeTagger
11 http://www.univ-nancy2.fr/pers/namer/Telecharger_Flemm.htm

4 Extraction Methods

This section describes the two methods used to extract paraphrases: first, the focused, pattern-based method of [8] (Sect. 4.1) and second, a more open n-gram-based method (Sect. 4.2). The results of the two methods are presented in Sect. 5.

4.1 Pre-defined Lexical Paraphrasing Patterns

Local paraphrases are similar to term variants: they build on morphologically and semantically related words and on syntactic variation [15]. Our extraction method based on lexical patterns hypothesizes and tests specific kinds of lexical variation. For every lexically based variation, we built paraphrase patterns that link the two alternating variants. Such patterns rely on existing lists of alternating variants. In the following, we first describe the alternating linguistic phenomena that may give rise to paraphrases (Sect. 4.1.1), then the methods used to collect data to be exploited by the lexical patterns (4.1.2), and finally the various patterns used in the extraction phase (Sect. 4.1.3).

4.1.1 Linguistic Phenomena Under Study

Deverbal nouns are nouns that are morphologically derived from verbs. They denote the action expressed by the verb (e.g., *treat/treatment*), hence noun phrases with a deverbal head can often be paraphrased with a verb phrase (e.g., *treatment of the patient/treat the patient*). These nominalizations are often said to be preferred by specialized language as opposed to verbs in lay language [10].

Neoclassical compounds are words formed from Greek and Latin roots. A common observation (e.g., [27]) is that technical language and especially medical language use many neo-classical compounds and that such compounds constitute a great proportion of the neologisms. Neo-classical compounds are often 'compositional' in the sense that the meaning of a compound is often the combination of that of its parts [e.g., $gastritis = stomach\ (gastr) + inflammation\ (itis)$]. It is this property that we use to look for paraphrases of compounds.

Relational adjectives are said to be very frequent and typical of specialized domains [16] and are very often used to catch morphological variants in term extraction tasks [7]. Relational adjectives are derived from nouns; they designate a relation between (i) the entity denoted by the noun they are derived from and (ii) the entity denoted by the noun they modify [11]. In a noun phrase such as *muscle abdominal* (EN: *abdominal muscle*), the adjective *abdominal* designates the relation between the head noun (*muscle*) and the base noun of the adjective: *abdomen*. From this relation comes also the fact that the same concept can be expressed by an alternating prepositional phrase [*muscle de l'abdomen* (EN: *muscle of the abdomen*)], which is often the case in terminological variation.

Another related phenomenon is the prefixation of relational adjectives. In a prefixed relational adjective [e.g., *anticancéreux* (EN: *anticancerous*)], the formal base of the adjective is the relational adjective [*cancéreux* (EN: *cancerous*)], but on the semantic side, the prefixation rule applies to the nominal base (*cancer*) [11]. Such a prefixation can be paraphrased as *against cancer*.

In the same line, we include in this investigation another alternating phenomenon that can be compared with the previous one: deverbal adjectives coined out of verbs with suffixes *-if*, *-eur* and *-oire*, that are alternatively used in place of a deverbal noun, both sharing the same verbal base. In this respect, they also have a relational meaning. For example, *troubles dépressifs* [EN: *depressive disorders*] and *troubles de la dépression* [EN: *disorders of depression*] are equivalent terms.

4.1.2 Lexical Resources Used for Each Paraphrasing Phenomenon

The above-mentioned study emphasizes morphological phenomena that can occur in paraphrases. We describe here the sources of morphological knowledge we used to detect these phenomena.

Deverbal noun lexical resource: We relied on an existing lexicon of French deverbal nouns paired with their base verbs: the Verbaction lexicon [13].

Neo-classical compound resource: We compiled a list of all neo-classical compounds used in the corpus by analyzing nouns and adjectives with the French morphological analyzer Dérif [22]. This analyzer allows us to segment compounds into their different components and to link each component to its modern-language equivalent: $gastritis \rightarrow gastr + itis \rightarrow stomach + inflammation$.

Relational adjective resources: For this phenomenon, three resources were gathered based on the lists of all the adjectives found in the entire corpus.

- The *relational adjective* list gathers all the relational adjectives analyzed as such in the corpus, with their base nouns (873 pairs).
- The *prefixed relational adjective* list is similar to the previous one. Each prefixed relational adjective found in the corpus is listed with its nominal base (303 pairs).
- The *deverbal adjective* list contains deverbal adjectives linked with their corresponding deverbal nouns, both being coined on the same verb (214 pairs).

These lists (a total of 1,390 pairs) were built by (1) gathering all adjectives from both parts of the corpus; (2) analyzing them through the French morphological analyzer Dérif [22]; and (3) manually checking each analyzed pair. Unanalyzed adjectives were also carefully checked to extend the coverage of the resources.

4.1.3 Lexical Patterns

Three sets of paraphrasing patterns have been set up to extract the three paraphrase types under study: noun to verb variations, noun to adjective variations and variations between a neo-classical form and its modern-language equivalent. They each rely on the corresponding lexical resources described above. All of these patterns are 'oriented', which means that, following linguistic description and intuition, the left side of the pattern is searched in the specialized part of the corpus, and the right side in the lay part of the corpus.

To identify nominalization paraphrases, we built a set of lexico-syntactic patterns using the noun/verb pairs as anchors. The general method is to look for phrases with corresponding content words in the context of the anchors.

For instance:

$$N_1 p\,(d)\,N_2 \rightarrow V_1\,(d)\,N_2$$
$$N_1 p\,(d)\,N_2 A_3 \rightarrow V_1\,(d)\,N_2 A_3$$
$$N_1 p\,(d)\,N_2 \rightarrow N_2 (V)\,V_1$$

where N is a noun, A an adjective, V a verb, p a preposition and d a determiner (in these patterns equivalent words share the same index; for instance, the noun/verb pair shares index 1).

To detect paraphrases of neo-classical compounds, we looked for the modern-language equivalents of the word components (in the case of *gastritis* this means searching for *inflammation* and *stomach*) close to each other within a syntactic phrase. We empirically set a threshold of 4 words as the maximum distance between the modern-language translations of the components. A pattern used to search those paraphrases is for instance:

$$C \rightarrow C_1 W_{0-4} C_2 (A \mid N) N_1$$

where C is a neo-classical compound in a specialized text segment, C_1 and C_2 are the modern-language components of C and W_i is an arbitrary word.

To identify paraphrases with relational adjectives, a single lexico-syntactic pattern was used:

$$N_2\,A_1 \rightarrow N_2 p\,(d)\,N_1$$

which relies on the previously described adjective/noun pairs (in this pattern equivalent words share the same index; for instance, the adjective/noun pair shares index 1).

We applied the left side of the patterns to the specialized part of the corpus and the right side to the lay part of the corpus. This gave us sets of candidate specialized/lay paraphrases.

4.2 N-Gram-Based Paraphrasing Patterns

In this second approach, we attempt to discover paraphrases in a more scalable way. Instead of relying on pre-established, manually defined patterns, we systematically explore and compare phrases. More precisely, the detection of paraphrasing patterns is based on the matching of stemmed lexical n-grams. We first extract n-grams (Sect. 4.2.1) then match similar n-grams (Sect. 4.2.2) and generalize matching pairs of n-grams to identify lexico-syntactic patterns (Sect. 4.2.3). Then we filter and classify these patterns (Sect. 4.2.4) to detect relevant paraphrasing patterns.

4.2.1 N-Gram Extraction

We started by extracting n-grams from each part of the corpus using the Ngram Statistics Package [1]. We considered n-grams whose lengths range from 2 to 6 tokens and then filtered them according to linguistic criteria so that they would correspond to potential syntactic chunks. These criteria included removing all n-grams containing punctuation tokens, ending with a preposition or a determiner, and composed of less than two content words (adjectives, nouns, verbs and adverbs). We chose this approach over using a syntactic chunker because it is less restrictive and because chunkers for French have lower performance than those for English. We obtained two lists of n-grams, one for each part of the corpus (lay and specialized).

4.2.2 Lexical Matching

We then stemmed all content words of the n-grams with the Lingua::Stem Perl package.[12] We matched the n-grams of the specialized list to the n-grams of the lay list based on string similarity using a bag-of-content-words approach. We restricted the comparison to exact matches between stems. That is to say we extracted pairs of n-grams with identical stems of content words, without taking their order into account. Additionally we discarded pairs of n-grams with identical lemmas, because we are interested in variation of content words and not mere variation in function words (such as *treatment of a patient/treatment of the patient*).

4.2.3 Pattern Identification

The next step was to generalize the obtained pairs of n-grams by replacing tokens with their parts-of-speech and by co-indexing equivalent content words. For instance the following pair where an n-gram of the specialized list (on the left side of the pattern) is associated with an n-gram of the lay list (on the right side) is thus generalized:

---

[12] http://search.cpan.org/~snowhare/Lingua-Stem-0.83

$$\textit{treatment of the patient} \rightarrow \textit{treat the patient}$$
$$N_1\, p\, d\, N_2 \rightarrow V_1\, d\, N_2$$

Patterns are kept 'oriented', in the sense that the left side of a pattern always represents an n-gram found in the specialized part of the corpus, and the right side always represents an n-gram found in the lay part of the corpus. This means that we can encounter cases where two patterns are the reverse of one another.

4.2.4 Pattern Filtering and Classification

We then sorted all patterns according to their number of occurrences, i.e., the number of observed distinct paraphrases they are based on. We only selected patterns that are based on at least 10 distinct paraphrases, discarding the others as non-significant. Then we manually reviewed those patterns, counting the number of correct paraphrases gathered by the pattern. This provide a 'precision' measure for each of the patterns, allowing to get rid of patterns that have a zero precision. We also classified the patterns following a four-class typology based on the type of modification between the two parts of the pattern: *morphosemantic modification* (such as derivation), *simple inversion*, *verbal inflection* (including passive-active modification) and *zero variation*, which includes patterns where only word endings are changing (thus showing mainly inflection variation).

Finally we also established a distinction between patterns for which a reverse corresponding pattern had been found (e.g., N A → N Prep N and N Prep N → N A were both found in the corpus), meaning that they appear in both directions (specialized → lay and lay → specialized); and patterns that were unidirectional (e.g., only N A → V Adv was found in the corpus), i.e., that appeared only in one direction (e.g., specialized → lay). For pairs of reverse patterns, we measured the difference between the numbers of paraphrases they each are based on. This distinction gives an insight about the 'orientation' of the paraphrase pattern, and highlights clear differences between specialized and lay language.

5 Results of the Two Methods

5.1 Pre-defined Lexical Patterns: Results

This section describes the results of the extraction of paraphrases using the predefined lexical patterns presented in Sect. 4.1. Each paraphrase gathered by the patterns was manually evaluated, providing a precision measure for each pattern.

Table 2 shows the number of paraphrases and correct paraphrases acquired by the three types of patterns. A special distinction is made for the Adjective–Noun pattern (adj–noun) depending on the type of adjective: relational, prefixed or deverbal. As

Table 2 Number of paraphrases, correct paraphrases and precision for each pre-defined lexical pattern

Pattern	Paraphrases	Correct	Precision
Noun–verb pattern	214	154	0.72
Neo-classical pattern	45	30	0.67
Adjective–noun:	780	599	0.77
Relational adjective–noun	688	547	0.80
Prefixed relational adjective–noun	38	16	0.42
Deverbal adjective–noun	54	36	0.67

we can see from Table 2, the precision is quite good for the noun–verb and adj–noun patterns, and slightly lower for the neo-classical pattern, although conclusions are hard to draw for the latter because of its small number of paraphrases. For the adj–noun patterns, the overall precision is high, but precision for the prefixed-adjective pattern is low, and the number of paraphrases for these adjectives is also very small.

These results are encouraging for the use of pre-defined lexical patterns, although these patterns lack exhaustiveness and are manually designed. In the following, we present the result of the second method, that is more data-driven and allows to discover patterns according to their actual frequency and therefore relevance for these corpora.

5.2 *N-Gram-Based Patterns: Results*

This section describes the results of the n-gram-based method (see Sect. 4.2), and the methodology we applied to extract statistics and data.

5.2.1 N-Gram Extraction and Lexical Matching

N-gram extraction and grouping by lexical matching were performed and we obtained a total of 119 patterns covering a total of 4,976 paraphrases. These first extractions were carefully evaluated as explained in the following section.

5.2.2 Pattern Identification: Ranking and Classification

Each pattern was carefully evaluated by two annotators.[13] We evaluated each pattern in terms of precision by counting the number of 'valid' paraphrases that the pattern can extract. This allowed us to get rid of invalid paraphrasing patterns (those with a precision equal to 0—this was the case for 10 patterns). We ended up with 109 pat-

[13] The first two authors of this chapter.

Table 3 Four categories of paraphrase patterns

Type of paraphrase pattern	Number	Percentage (%)	Mean precision
Morphosemantic (M)	43	39.55	0.65
Simple inversion (I)	32	29.36	0.63
Verbal inflection (V)	25	22.95	0.76
Zero variation (Z)	9	8.26	0.55
Total	109	100	0.66

terns, which gathered 4,800 distinct paraphrases. Most of the patterns were bigram patterns, with a few trigram patterns (we count these in terms of their number of content words, not tokens).

Then, we ranked the patterns according to the number of correct paraphrases that were extracted with them, and we manually classified the patterns according to the four-class typology defined in Sect. 4.2.4. Table 3 presents the results of the classification, including the mean precision of each class, and Table 4 presents the 20 top-occurring patterns.[14]

The most frequent category is that of the *morphosemantic* paraphrasing patterns, followed by *simple variation* and *verbal inflection*. Moreover, morphosemantic variation is very frequent in the top-ranked patterns, where it accounts for half of the cases (see rows with an 'M' in the last column in Table 4). Morphosemantic patterns involve a change of lexical category in one or more members of the pattern. This change, because it has been captured through a comparison of stems, is due to a derivational process. In most of the cases, the pattern implies a verb > noun derivation, and in a smaller proportion, a noun > adj derivation. Further details on these paraphrasing patterns are provided in Sect. 6.1.

The *simple inversion* patterns concern cases where the two members of the pattern are inverted on the other side of the pattern. They yield reasonably good results. Errors are mainly due to the preposition that can be found between the two inverted items, which can lead to a change in meaning. Consequently, this kind of pattern should be used with appropriate constraints on prepositions.

The *verbal inflection* category mainly contains paraphrases where the verb is simply inflected, but also more complex variations, where a past participle used as an adjective is involved (provoking an inversion, as in the pattern ranked #10 in Table 4), or in cases of passive–active variation.

The *zero variation* patterns display no specific variation of any kind. Besides, their precision is not very high. They mainly help to catch inflectional variants of the same noun phrase, as in the pattern ranked #3 in Table 4 (covering inflectional variants such as *infirmier-infirmière*, EN: *nurse*, gender = m–f).

The mean precision of all the patterns is 0.66 (with a median at 0.69), which is rather high for automatically discovered patterns. Moreover, 29 patterns have a precision between 0.9 and 1.0 (among which 15 with a precision of 1.0).

---

[14] Examples of occurrences of each pattern are provided in Table 10, Appendix.

Table 4 Top 20 most frequent patterns

Rank	Correct paraphrases	Precision	Pattern (specialized/lay)	Categories
1	331	0.88	$N_1\ p\ N_2\ /\ V_1\ N_2$	M
2	183	0.94	$V_1\ N_2\ /\ N_1\ p\ N_2$	M
3	177	0.66	$N_1\ p\ N_2\ /\ N_1\ p\ N_2$	Z
4	140	0.82	$N_1\ p\ N_2\ /\ V_1\ p\ N_2$	M
5	121	0.72	$N_1\ A_2\ /\ N_1\ p\ N_2$	M
6	116	0.92	$N_1\ Vpper_2\ /\ V_2\ N_1$	V
7	96	0.64	$N_2\ p\ N_1\ /\ N_1\ A_2$	M
8	92	0.81	$N_1\ p\ N_2\ /\ N_2\ Vpper_1$	M
9	88	0.98	$V\ ADV_2\ Vpper_1\ /\ V_1\ A_2$	V
10	85	1.00	$V_2\ N_1\ /\ N_1\ Vpper_2$	V
11	83	0.99	$V_1\ ADV_2\ /\ V\ ADV_2\ Vpper_1$	V
12	76	0.65	$N_1\ p\ N_2\ /\ N_1\ A_2$	M
13	67	0.61	$V_1\ p\ N_2\ /\ N_1\ p\ N_2$	M
14	65	0.90	$V_2\ N_1\ /\ N_1\ V\ Vpper_2$	V
15	62	0.91	$V_2\ ADV_1\ /\ ADV_1\ V_2$	I
16	60	0.67	$N_2\ Vpper_1\ /\ N_1\ p\ N_2$	M
17	54	0.82	$N_1\ Vpper_2\ /\ V_2\ p\ N_1$	V
18	52	0.95	$N_1\ A_2\ /\ V_1\ ADV_2$	M
19	51	0.81	$ADV_1\ V_2\ /\ V_2\ ADV_1$	I
20	51	0.74	$N_1\ V\ Vpper_2\ /\ V_2\ N_1$	V

Categories are abbreviated as follows: *M* Morphosemantic, *Z* Zero variation, *I* Inversion, *V* Verbal inflection. The patterns are presented following standard abbreviations, i.e., *N* Noun, *V* Verb, V_{pper} Past participle of verb, *p* Preposition, *A* Adjective, *ADV* Adverb. Co-indexing (e.g., $N_i \ldots / \ldots V_i$) indicates which elements are matched across the two sides of a pattern. Recall that determiners have been purposely removed from the patterns

5.2.3 Distinction Between Paired Patterns and Unidirectional Patterns

Amongst all the patterns that have been extracted, we distinguished patterns that apply in both directions (specialized → lay and lay → specialized) from patterns that apply only in one direction. As shown in Table 5, a vast majority of patterns (more than two thirds) apply in both directions.

Paired patterns: Amongst the patterns that work in both directions, some favor one direction over the other. To observe this, we computed the difference between the

Table 5 Direction of patterns

Pattern direction	Number	Percentage (%)
Both	76	69.7
Only one	30	27.5
None	3	2.8
Total	109	100.0

Table 6 Patterns with both directions and a significant difference

Pattern (specialized/lay)	Difference (%)
$N_1\ PROrel\ V_2/V_2\ N_1$	−180
$N_1\ V_2/V_{pper_2}\ p\ N_1$	−177
$ADV_1\ N_2/N_2\ A_1$	−143
$N_1\ A_2/N_2\ p\ N_1$	−140
$N_1\ V_{pper_2}/V\ V_{pper_2}\ p\ N_1$	−90
$N_1\ p\ N_2/N_2\ p\ V_1$	−67
$N_1\ V_{pper_2}/V_{pper_2}\ p\ N_1$	−64
$N_1\ A_2/V_2\ p\ N_1$	−60
$N_1\ A_2/N_1\ p\ N_2$	+37
$N_1\ p\ N_2/N_2\ V_1$	+50
$N_1\ p\ N_2/V_1\ p\ N_2$	+52
$N_1\ p\ N_2/V\ V_{pper_1}\ N_2$	+63
$A_1\ N_2/V_2\ p\ A_1$	+67
$N_1\ p\ N_2/N_1\ p\ V_2$	+70
$N_1\ A_2/V_1\ ADV_2$	+73

number of paraphrases acquired in one and the other direction, normalized by the number of paraphrases in the source direction. Table 6 shows the patterns where this normalized difference between the specialized → lay and lay → specialized directions is over an empirical threshold of 50 % (negative percentage indicates a preference for the lay → specialized direction, and positive for the specialized → lay direction). This seems to indicate that those patterns are relevant to contrast specialized language and lay language, and can be used to acquire specialized versus lay paraphrases. Other patterns with a small difference are patterns that detect paraphrases, but with no particular relevance to contrast the two discourse types. Among relevant patterns we find several patterns with a noun–verb variation (such as $N_1\ p\ N_2/N_2\ V_1$, $N_1\ p\ N_2/\ V_1\ p\ N_2$ etc.) in the specialized → lay direction, which tends to confirm the hypothesis on which we based our pre-defined noun–verb patterns in the first method. The adj–noun pattern ($N_1\ A_2/N_1\ p\ N_2$) that we used in the first method only has a difference of +37 %. This still shows a preference for the use of the adjective in specialized discourse, and confirms common assumptions, although the difference is not as important as expected. Although this pattern displays a difference that is below the threshold, it is included in Table 6.

Table 7 Unidirectional patterns

Pattern (specialized/lay)	Precision (%)	# Correct paraphrases
$N_1\ p\ N_2\ A_3/V_1\ N_2\ A_3$	100	29
$N_1\ p\ N_2\ p\ N_3/V_1\ N_2\ p\ N_3$	100	32
$N_1\ A_2/V_1\ N_2$	68	17

Unidirectional patterns: On the other hand, some patterns are found in only one direction, which means that they might be particularly relevant in contrasting specialized versus lay languages. However among those we also find some patterns that detect a too small number of paraphrases to draw a significant conclusion, hitting a data sparseness problem. As a matter of fact, patterns with few occurrences are statistically less well represented in the corpus and have a higher chance of occurring there in only one direction. Table 7 shows example patterns[15]; we only studied those that are based on a significant number of paraphrases. The examples displayed in Table 7 also show some of the few trigrams found with the n-gram method. Again, trigrams are logically less frequent than bigrams, hence the smaller numbers.

6 Discussion: Complementarity of the Two Methods

In the previous section, we presented the results of the two extraction methods. The first method is based on a linguistic intuition of frequent phenomena that might give rise to paraphrases between specialized and lay discourse, while the second adopts a data-driven approach, by empirically gathering candidate paraphrases and selecting afterwards the potentially relevant ones. In this respect, the n-gram-based method helps to validate empirically the intuition used in the pre-defined pattern-based extraction method, and also to discover new lexical patterns. The two subsections below address these two aspects.

6.1 Pre-defined Paraphrase Patterns in the Output of the N-Gram-Based Method

In Sect. 4.1, we used pre-defined patterns to extract paraphrases, also assuming the discourse type (specialized or lay) where each member of the paraphrase patterns is to be found. The data-driven, n-gram-based method confirms these pre-defined patterns, in terms of frequency and preferences, and can bring to light new patterns (see Sect. 6.2).

By construction, the n-gram method could not discover the pre-defined patterns designed for neo-classical compounds. This can be explained by the fact that this method is based on identity of stemmed content words, which generally does not hold between neoclassical compounds and their modern-language equivalents.

In contrast, the n-gram method worked perfectly well for the other two types of pre-defined patterns. Several types of patterns were found for the verb–noun morphosemantic variation, especially in the highly-ranked patterns, as shown in Table 4. Table 8 is an excerpt of the most frequent patterns involving noun–verb variations. It clearly shows that the verb–noun variation occurs in a wide variety of

[15] Examples of occurrences of each pattern are provided in Table 11, Appendix.

Table 8 Example patterns involving a verb–noun alternation

Rank	Precision	Pattern (specialized/lay)	Examples
1	0.88	$N_1\ p\ N_2/V_1\ N_2$	Application de principes/Appliquer le principe *Application of principles/To apply the principle*
2	0.94	$V_1\ N_2/N_1\ p\ N_2$	Aggraver une pathologie/Aggravation de pathologie *To aggravate a pathology/Aggravation of pathology*
4	0.82	$N_1\ p\ N_2/V_1\ p\ N_2$	Ajout d'insuline/Ajouter de l'insuline *Addition of insulin/To add insulin*
8	0.81	$N_1\ p\ N_2/N_2\ V_{pper_1}$	Administration de produit/Produit administré *Administration of product/Administered product*
13	0.61	$V_1\ p\ N_2/N_1\ p\ N_2$	Adopter des stratégies/Adoption d' une stratégie *To adopt strategies/Adoption of a strategy*

cases, where the verb or noun is the head of the phrase, or with a change of head and order when the verb is a past participle used as an adjective (which is post-nominal in French, e.g., refer to the example at rank 8).

As regards the relational-adjective–noun pattern, we already mentioned in Sect. 5.2.3 that relational adjectives are favored in the specialized part of the corpus: the direction $N_1\ A_2 \rightarrow N_1\ p\ N_2$ is 37 % more frequent than the reverse direction.

6.2 Discovering New Paraphrase Patterns

The n-gram-based extraction method allowed us to highlight frequent morphosemantic patterns that were not considered in the pre-defined paraphrase patterns. Table 9 presents some of them, which involve qualifying adjectives (for the first two) or two morphosemantic variations, i.e., a noun–verb variation also implying a change in their modifiers: the noun is modified by an adjective, and the verb is modified by the corresponding adverb ($total \rightarrow totalement$).

As regard the adj–noun pattern involving a qualifying adjective (ranks 7 and 23 in Table 9), it has to be noticed that this pattern displays a preference in orientation: the qualifying adjective is clearly favored in the lay part, whereas the specialized part prefers the nominalization of this adjective.

Table 9 Examples of new patterns

Rank	Precision	Pattern (specialized/lay)	Examples
7	0.64	$N_2\ p\ N_1/N_1\ A_2$	Efficacité d' un soutien/Soutien efficace *Efficacy of a support/Efficient support*
23	0.33	$N_1\ A_2/N_2\ p\ N_1$	Accès facile/facilité d' accès *Easy access/Easiness of access*
18	0.95	$N_1\ A_2/V_1\ ADV_2$	Traitement adéquat/Traiter adéquatement *Appropriate treatment/To treat adequately*
53	0.82	$V_1\ ADV_2/N_1\ A_2$	Arrêter totalement/Arrêt total *To stop completely/Complete stop*

7 Conclusion

In this chapter, we presented and compared two methods for extracting paraphrases between specialized and lay discourse. The first one, based on linguistic intuitions and common hypotheses, relies on pre-defined lexical patterns and showed a rather good precision. The second one, which is more data-driven, automatically extracts paraphrasing patterns. It allowed us to confirm the frequency (and thus, the significance) of the pre-defined lexical patterns, but also unveiled new paraphrasing patterns. With this method we were also able to study the direction of the patterns (specialized $\rightarrow$ lay and/or lay $\rightarrow$ specialized) in order to see which patterns could establish a distinction between the specialized and lay discourse types. The extracted patterns confirmed the relevance of the noun–verb variation between the two discourse types: specialized language tends to prefer nominalizations where lay language uses more verbs. They also confirmed to a certain extent the preference of specialized language for phrases with relational adjectives. New relevant patterns were also discovered.

In a broader perspective, the second method further showed the importance of derivational morphology in paraphrasing: many of the most frequent patterns were based on morphological variation. Among the new patterns that were discovered, we must emphasize the important number of adverbs, found in many patterns, alternating with their corresponding adjectives. To our knowledge, this has not been studied previously in such a context.

The n-gram approach we used was based on pairing n-grams that have the same number of content words (the same n). In the future, we plan to extend this method to unbalanced pairs of n-grams (such as bigram/trigram, and vice versa). The n-gram approach can also be applied to other kinds of monolingual comparable corpora, which display other types of distinctions than specialized/lay, in order to discover new genre- and discourse-specific paraphrase variations.

Appendix

This appendix contains example patterns: Table 10 provides examples of the patterns presented in Table 4, while Table 11 shows examples of the patterns presented in Table 7.

Table 10 Examples for the top 20 most frequent patterns presented in Table 4

Rank	Pattern (specialized/lay)	Example
1	$N_1\ p\ N_2/V_1\ N_2$	Retard de cicatrisation/Retarder la cicatrisation *Delay in healing/To delay healing*
2	$V_1\ N_2/N_1\ p\ N_2$	Calculer les coûts/Calcul de coût *To compute costs/Cost computation*
3	$N_1\ p\ N_2/N_1\ p\ N_2$	Infirmière à domicile/Infirmiers à domicile *Home [female] nurse/Home [male] nurses*
4	$N_1\ p\ N_2/V_1\ p\ N_2$	Traitement de l' affection/Traiter des affections *Treatment of illness/To treat illnesses*
5	$N_1\ A_2/N_1\ p\ N_2$	Apports caloriques/Apport en calories *Caloric intakes/Intake of calories*
6	$N_1\ V_{\text{pper}_2}/V_2\ N_1$	Moyens adaptés/Adapter les moyens *Adapted means/To adapt the means*
7	$N_2\ p\ N_1/N_1\ A_2$	Complexité du problème/Problème complexe *Complexity of the problem/Complex problem*
8	$N_1\ p\ N_2/N_2\ V_{\text{pper}_1}$	Administration de produit/produit administré *Administration of product/Administered product*
9	$V\ ADV_2\ V_{\text{pper}_1}/V_1\ A_2$	A aussi permis/Permettra aussi *Has also allowed/Will also allow*
10	$V_2\ N_1/N_1\ V_{\text{pper}_2}$	Restreindre l' accès/Accès restreints *To restrict access/restricted access*
11	$V_1\ ADV_2/V\ ADV_2\ V_{\text{pper}_1}$	Fournit aussi/Être aussi fournie *Also provides/To be also provided*
12	$N_1\ p\ N_2/N_1\ A_2$	Contrôles de la glycémie/Contrôles glycémiques *Glycemia tests/Glycemic tests*
13	$V_1\ p\ N_2/N_1\ p\ N_2$	Importer des produits/Importation de produits *To import products/Importation of products*
14	$V_2\ N_1/N_1\ V\ V_{\text{pper}_2}$	Calculer les doses/Doses sont calculées *To compute doses/Doses are computed*
15	$V_2\ ADV_1/ADV_1\ V_2$	Améliore encore/Encore améliorer *Still improves/To still improve*
16	$N_2\ V_{\text{pper}_1}/N_1\ p\ N_2$	Aliments transformés/Transformation des aliments *Processed food/Food processing*
17	$N_1\ V_{\text{pper}_2}/V_2\ p\ N_1$	Expérience acquise/Acquérir de l' expérience *Acquired experience/To acquire experience*
18	$N_1\ A_2/V_1\ ADV_2$	Traitement adéquat/Traiter adéquatement *Appropriate treatment/To treat adequately*
19	$ADV_1\ V_2/V_2\ ADV_1$	Ainsi faciliter/Facilite ainsi *To thus facilitate/Thus facilitates*
20	$N_1\ V\ V_{\text{pper}_2}/V_2\ N_1$	Efficacité est renforcée/Renforcer son efficacité *Efficacy is strengthened/To strengthen its efficacy*

Table 11 Examples for the unidirectional patterns displayed in Table 7

Pattern (specialized/lay)	Example
$N_1\ p\ N_2\ A_3/V_1\ N_2\ A_3$	Augmentation du débit cardiaque/Augmenter le débit cardiaque *Increase in cardiac output/To increase the cardiac output*
$N_1\ p\ N_2\ p\ N_3/V_1\ N_2\ p\ N_3$	Analyse d' un échantillon de sang/Analyser un échantillon de sang *Analysis of a blood sample/To analyze a blood sample*
$N_1\ A_2/V_1\ N_2$	Augmentation posologique/Augmenter la posologie *Posological increase/To increase the posology*

References

1. Banerjee, S., Pedersen, T.: The design, implementation, and use of the n-gram statistics package. In: Proceedings of the Fourth International Conference on Intelligent Text Processing and Computational Linguistics, pp. 370–381, Mexico City (2003)
2. Bannard, C., Callison-Burch, C.: Paraphrasing with bilingual parallel corpora. In: Proceedings of the 43rd Annual Meeting on Association for, Computational Linguistics, pp. 597–604 (2005)
3. Barzilay, R.: Information fusion for multidocument summarization: paraphrasing and generation. PhD thesis, Columbia University (2003)
4. Barzilay, R., Lee, L.: Learning to paraphrase: an unsupervised approach using multiple-sequence alignment. In: HLT-NAACL, pp. 16–23, Edmonton, Canada (2003)
5. Barzilay, R., McKeown, K.: Extracting paraphrases from a parallel corpus. In: ACL/EACL, pp. 50–57 (2001)
6. Chiao, Y.C., Zweigenbaum, P.: Looking for candidate translational equivalents in specialized, comparable corpora. In: Proceedings of the 19th COLING, pp. 1208–1212, Taipei, Taiwan (2002)
7. Daille, B.: Identification des adjectifs relationnels en corpus. In: TALN 1999, pp. 105–114 (1999)
8. Deléger, L., Zweigenbaum, P.: Extracting lay paraphrases of specialized expressions from monolingual comparable medical corpora. In: Fung, P., Zweigenbaum, P., Rapp, R. (eds.) Proceedings of the 2nd Workshop on Building and Using Comparable Corpora: From Parallel to Nonparallel Corpora, pp. 2–10 (2009)
9. Elhadad, N., Sutaria, K.: Mining a lexicon of technical terms and lay equivalents. In: ACL BioNLP Workshop, pp. 49–56, Prague, Czech Republic (2007)
10. Fang, Z.: Scientific literacy: a systemic functional linguistics perspective. Sci. Edu. **89**(2), 335–347 (2005)
11. Fradin, B.: On the semantics of denominal adjectives. In: Sixth Mediterranean Morphology Meeting, Ithaca, Greece (2008)
12. Fung, P.: A pattern matching method for finding noun and proper noun translations from noisy parallel corpora. In: Proceedings of the 33rd Annual Meeting of the Association for Computational Linguistics, pp. 236–233, Boston, Massachusetts (1995)
13. Hathout, N., Namer, F., Dal, G.: An experimental constructional database: the MorTAL project. In: Boucher, P. (ed.) Many Morphologies, pp. 178–209. Cascadilla, Somerville (2002)
14. Ibrahim, A., Katz, B., Lin, J.: Extracting structural paraphrases from aligned monolingual corpora. In: Proceedings of the 2nd International Workshop on Paraphrasing, Association for Computational Linguistics, pp. 57–64, Sapporo, Japan (2003)
15. Jacquemin, C.: Syntagmatic and paradigmatic representations of term variation. In: Proceedings of the 37th Annual Meeting of the Association for Computational Linguistics on Computational Linguistics, pp. 341–348, College Park, Maryland (1999)
16. L'Homme, M.: Adjectifs dérivés sémantiques (ADS) dans la structuration des terminologies. In: Terminologie, Ontologie et Représentation des Connaissances, Université Jean-Moulin Lyon-3 (2004)

17. Lindberg, D.A.B., Humphreys, B.L., McCray, A.T.: The unified medical language system. Methods Inf. Med. **32**(2), 81–91 (1993)
18. Max, A.: Local rephrasing suggestions for supporting the work of writers. In: Proceedings of GoTAL, Gothenburg, Sweden (2008)
19. Max, A., Wisniewski, G.: Mining naturally-occurring corrections and paraphrases from Wikipedia's revision history. In: Calzolari, N., Choukri, K., Maegaard, B., Mariani, J., Odijk, J., Piperidis, S., Rosner, M., Tapias, D. (eds.) Proceedings of the 7th International Conference on Language Resources and Evaluation (LREC'10). European Language Resources Association (ELRA), Valletta, Malta (2010)
20. McEnery, A.M., Xiao, R.Z.: Parallel and comparable corpora: What are they up to? In: Incorporating Corpora: Translation and the Linguist—Translating Europe. Multilingual Matters, Clevedon (2007)
21. Morin, E., Daille, B., Takeuchi, K., Kageura, K.: Bilingual terminology mining: using brain, not brawn comparable corpora. In: Proceedings of ACL, Prague, Czech Republic (2007)
22. Namer, F.: Morphologie, Lexique et Traitement Automatique des Langues: l'Analyseur DériF. Lavoisier, Paris (2009)
23. Pang, B., Knight, K., Marcu, D.: Syntax-based alignment of multiple translations: extracting paraphrases and generating new sentences. In: Proceedings of HLT-NAACL 2003, pp. 102–109, Edmonton, Canada (2003)
24. Pasca, M., Dienes, P.: Aligning needles in a haystack: paraphrase acquisition across the web. In: Proceedings of IJCNLP, pp. 119–130 (2005)
25. Rapp, R.: Identifying word translations in non-parallel texts. In: Proceedings of the 33rd Annual Meeting of the Association for, Computational Linguistics, pp. 320–322 (1995)
26. Shinyama, Y., Sekine, S.: Paraphrase acquisition for information extraction. In: Proceedings of the 2nd International Workshop on Paraphrasing (IWP), pp. 65–71, Sapporo, Japan (2003)
27. Wolff, S.: Automatic coding of medical vocabulary (Chap. 7). In: Sager, N., Friedman, C., Lyman, M.S. (eds.) Medical Language Processing: Computer Management of Narrative Data, pp. 145–162. Addison-Wesley, New York (1986)

Information Network Construction and Alignment from Automatically Acquired Comparable Corpora

Heng Ji, Adam Lee and Wen-Pin Lin

Abstract In this paper we describe a novel approach to discover cross-lingual comparable corpora based on video comparison. Then we propose a new task to extract and align information networks from comparable corpora. As a case study we demonstrate the effectiveness of utilizing bi-lingual information networks, we present a weakly-supervised and language-independent approach to mine name translation pairs. Based on the fact that some certain types of expressions are written in language-independent forms, we generate seed pairs automatically. Starting from these seeds, we then apply a bootstrapping algorithm based on link comparison to mine more pairs iteratively. Results show that our approach can produce highly reliable name pairs. We also duplicate two state-of-the-art name translation mining methods and use two existing name translation gazetteers to compare with our approach. Comparisons show our approach can effectively augment the results from each of these alternative methods and resources.

Keywords Cross-lingual comparable corpora · Information network · Name mining

1 Introduction

A shrinking fraction of the world's Web pages are written in English, while about 3,000 languages are endangered [1]. Therefore the ability to access information across a range of languages, especially low-density languages, is becoming increasingly important for many applications. This need is being addressed in part by the research on Cross-Lingual Information Access (CLIA). CLIA is concerned with technologies that let users express their query in their native language, and irrespec-

H. Ji (✉) · A. Lee · W. Lin
Computer Science Department, Queens College and Graduate Center,
City University of New York, New York, NY 11367, USA
e-mail: hengji@cs.qc.cuny.edu

S. Sharoff et al. (eds.), *Building and Using Comparable Corpora*,
DOI: 10.1007/978-3-642-20128-8_13,

tive of the language in which the information is available, present the information in the user-preferred language or set of languages, in a manner that satisfies the user's information needs. The additional processing may take the form of machine translation (MT) of facts such as names and events. A key bottleneck for further research progress and improving real-world applications (especially, applications other than web search) lies in the MT performance. Traditional MT systems focus on the overall fluency and accuracy of the translation but fall short in their ability to translate certain informative words. In particular, the translation of names is fundamentally different from the translation of other lexical items. Name translation is important well beyond the relative frequency of names in a text: a correctly translated passage, but with the wrong name, may lose most of its value. We found that other tasks such as semantic role labeling [2] and question answering [3, 4] performed notably worse on machine translated texts than on texts originally written in English, and our error analysis indicated that a major cause was the low quality of name translation. Thus, it appears that better name translation can substantially improve the utility of MT and the amount of information that can be gleaned from foreign sources. In this paper we hypothesize that in order to extend CLIA to all the language pairs on the earth, or at least to some low-density languages which are lack of fundamental linguistic resources, we can start from the much more scalable task of "information" translation, or more specifically, name translation. One may argue that name pairs can be mined from the cross-lingual links between Wikipedias. However, even for the Wikipedia pages in two high-density languages (English and Chinese), only 1.68 % English pages contain links to Chinese and only 24.09 % Chinese pages contain links to English. In addition, current statistical MT methods require large and manually aligned parallel corpora as input for each language pair of interest.

Some recent work (e.g. [5]) found that MT can benefit from multi-lingual comparable corpora [6]. Recently there has also been heightened interest in discovering name pairs from comparable corpora (e.g. [7, 8]). By comparable corpora we mean texts that are about similar topics, but are not in general translations of each other. These corpora are naturally available, for example, many news agencies release multilingual news articles on the same day. There are no document-level or sentence-level alignments across languages, but important facts such as names, relations and events in one language in such corpora tend to co-occur with their counterparts in the other.

However, it is time-consuming to identify pairs of comparable texts; especially when there is lack of parallel information such as news release dates and topics. The influx of information on the Web comes from an unprecedented variety in data, including rich and complex sources such as images, videos, speech and written texts. For example, a sports news page on the Web about "*Brazil shut down the U.S. soccer team*" is composed of facts from multiple data modalities: the video (e.g. *the line up of players*), the speech in the background of the video (e.g. *the size of audience, the employee history of the coaches*), the image (*U.S. goalkeeper Tim Howard instructs his teammates*), and the texts about the detailed game procedure and comments (e.g. *Neymar is 18-years old*). The images/videos embedded in the same documents can provide additional clues for similarity computation because they are "*language-independent*". In this paper we show how a video-based comparison approach can reliably build large comparable text corpora for three languages: English, Chinese and Arabic. This provides a cross-media approach to break language barrier. Our main intuition is that if two documents include concepts (e.g. male-people, earthquake event, rescue activities, and facility-grounds) with similar frequency information, we can determine they are likely to be comparable. We represent each video as a vector of semantic concepts extracted from videos and then use standard vector space model to compute similarity.

Cross-document Information Extraction [9] and Knowledge Base Population (KBP) techniques [10]—identifying important entities, relations and events—are currently available for some non-English languages. In this paper we define "*bi-lingual information networks*" which can be extracted from comparable corpora. An information network consists of a large set of interconnected objects (e.g. entities) together with their links (e.g. relations and events). We traverse and align the information networks in two languages automatically in a bootstrapping way to discover name translation pairs. We conduct experiments on structured data (Wikipedia Infoboxes) and unstructured data (news articles) respectively. We will also compare our approach to state-of-the-art name translation mining approaches.

2 Cross-Lingual Comparable Corpora Acquisition Based on Video Comparison

2.1 Approach Overview

Figure 1 presents an example of cross-lingual multi-media comparable documents. They are both about the rescue activities for the Haiti earthquake. The background videos and images are language-independent and thus can be exploited to identify such comparable documents.

Figure 2 presents the general pipeline of discovering cross-lingual comparable documents based on background video comparison. Most document clustering systems use representations built out of the lexical and syntactic attributes. These attributes may involve string matching, agreement, syntactic distance, and document release dates. Although gains have been made with such methods, there are clearly cases where shallow information will not be sufficient to resolve clustering correctly. Therefore, we should therefore expect a successful document comparison approach to exploit world knowledge, inference, and other forms of semantic information in order to resolve hard cases. For example, if two documents include concepts referring to "*male-people*", "*earthquake event*", "*rescue activities*", and "*facility-grounds*" with similar frequency information, we can determine they are likely to be comparable. In this paper we represent each video as a vector of semantic concepts extracted from videos and then use standard vector space model to compute similarity.

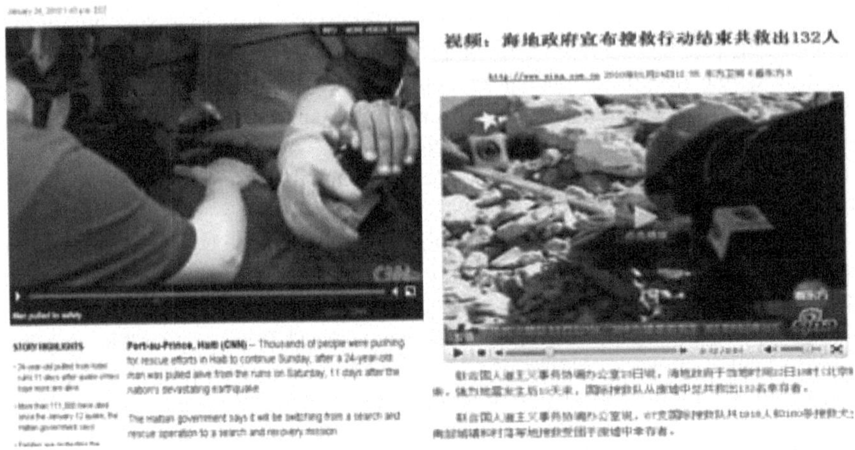

Fig. 1 An example of cross-lingual multi-media comparable documents

Fig. 2 Cross-lingual comparable text corpora acquisition based on video similarity computation

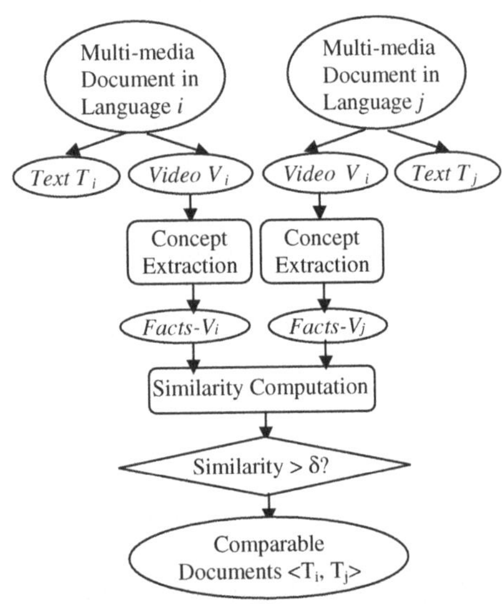

2.2 Video Concept Extraction

We apply a state-of-the-art video concept extraction system developed by IBM for the TREC Video Retrieval Evaluation (TRECVID-2005), in order to identify concepts defined in the large-scale concept ontology video for multimedia (LSCOM) program [11]. This system extracts 2,617 concepts such as "*Hospital*", "*Airplane*" and "*Female-Person*". It uses support vector machines to learn the mapping between low level features extracted from visual modality as well as from transcripts and production related meta-features. It also exploits a Correlative Multi-label Learner [12] for exploring label correlations, a Multi-Layer Multi-Instance Kernel [13] and Label Propagation through Linear Neighborhoods [14] to extract all other high-level features. For each classifier, different models are trained on a set of different modalities (e.g., the color moments, wavelet textures, and edge histograms), and the predictions made by these classifiers are combined together with a hierarchical linearly-weighted fusion strategy across different modalities and classifiers.

2.3 Video Concept Similarity Computation

Let $A = (a_1, \ldots, a_{|\sum|})$ and $B = (b_1, \ldots, b_{|\sum|})$ be the vectors for a pair of videos, we use cosine similarity to compute similarity:

$$\cos(A, B) = \frac{\sum_{i=1}^{|\sum|} a_i b_i}{\sqrt{\sum_{i=1}^{|\sum|} a_i^2}\sqrt{\sum_{i=1}^{|\sum|} b_i^2}},$$

where $|\sum|$ contains all possible concepts. We use traditional TF-IDF (Term Frequency-Inverse Document Frequency) weights for the vector elements a_i and b_i. Let C be a unique concept, V is a video consisting of a series of k shots $V = \{S_1, \ldots, S_k\}$, then:

$$tf(C, V) = \sum_{i=1}^{k} tf(C, S_i)/k$$

Let $p(C, S_i)$ denote the probability that C is extracted from S_i, we define the term frequency $tf(C, S_i)$:

$$tf(C, S_i) = \alpha^{confidence(C, S_i)}$$

where *Confidence* (C, S_i) denotes the probability of detecting a concept C in a shot S_i:

$$confidence(C, S_i) = p(C, S_i) \text{ if } p(C, S_i) > \delta, \text{ otherwise } 0.$$

Let: $df(C, S_i) = 1$ if $p(C, S_i) > \delta$, otherwise 0, assuming there are j shots in the entire corpus, we calculate *idf* as follows:

$$idf(C, V) = \log\left(j/\sum_{i=1}^{j} df(C, S_i)\right).$$

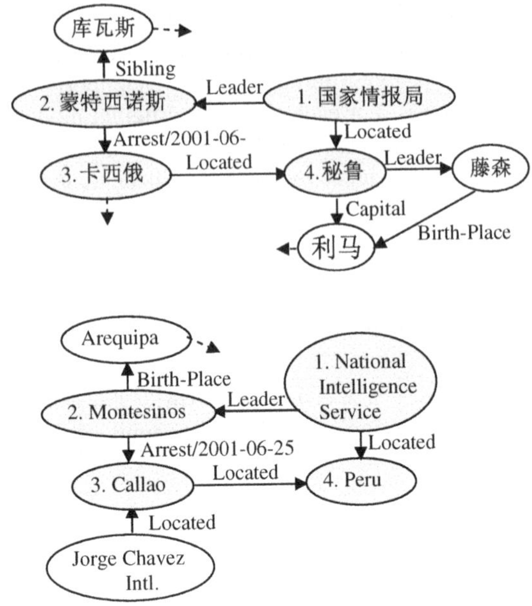

Fig. 3 An example for bilingual information networks

3 Bi-Lingual Information Network Construction

After we obtain comparable corpora from the above approach, it is important to utilize them to aid in CLIA applications. We provide a case study of constructing and aligning bi-lingual information networks from comparable corpora.

3.1 Information Network Overview

In this paper we define a new notion '*bilingual information networks*' as follows. An information network is a heterogeneous network that includes a set of "information graphs" $G = \{G_i(V_i, E_i)\}$, where V_i is the collection of entity nodes, and E_i is the collection of edges linking one entity to the other, labeled by relation or event attributes, such as "*hometown*", "*employer*", "*spouse*" etc. An example is depicted in Fig. 3.

Information networks can be constructed from different kinds of comparable corpora and therefore can provide a unifying abstraction to integrate knowledge across heterogeneous domains. We define the types of nodes and edges according to the NIST Automatic Content Extraction Program (ACE 2005) and the NIST TAC Knowledge Base Population (KBP 2010) Program [15]. ACE2005 defined seven types of entities: persons (PER), geo-political entities (GPE), locations (LOC), organizations (ORG), facilities (FAC), vehicles (VEH) and weapons (WEA), 18 types of relations (e.g., "*a town some 50 miles south of Salzburg*" indicates a "*located*" relation.); and 33 distinct types of relatively 'dynamic' events (e.g., "*Barry Diller on Wednesday quit as chief of Vivendi Universal Entertainment.*" indicates a "*personnel-start*" event). KBP 2010 defined 26 slot types for persons and 16 slot types for organizations. For example, "*Ruth D. Masters is the wife of Hyman G. Rickover*" indicates that the "*per:spouse*" of "*Hyman G. Rickover*" is "*Ruth D. Masters* ").

In this paper we consider the seven types of entities as defined in ACE program for the nodes in information networks. The edge types are derived from both of the static relations defined in ACE relation extraction and KBP, and the dynamic events defined in ACE event extraction. For each event we compose the edge type by concatenating its type and time argument (e.g. the "*Arrest/2001-06-25*" edge in Fig. 3). Some examples of edge types are presented in Table 1.

3.2 Construction from Existing Comparable Structured Knowledge Bases

Wikipedia, as a remarkable and rich online encyclopedia with a wealth of general knowledge about varied concepts, entities, events and facts in the world, may be utilized to initialize bi-lingual information networks. As of December 2010 Wikipedia contains pages from 275 languages,[1] but statistical MT techniques can only process a small portion of them (e.g. Google translate can only translate between 59 languages). Wikipedia *Infoboxes* are a highly structured form of data and are composed of a set of subject-attribute-value triples that summarize or highlight the key features of the concept or subject of each article. A large number of instance-centered

Table 1 Edge type examples

	Node pair	Edge type
Static	<PER, PER>	Spouse, parents, children, siblings
	<PER, ORG>	Member
	<PER, GPE>	Birth-place, death-place, nationality, origin
	<ORG, ORG>	Subsidiaries, parents
	<ORG, GPE>	Location, headquarter, political-affiliation
	<GPE, GPE>	Located-country, capital
Dynamic	<PER, PER>	Contact-meet, contact-phone_write, justice, sport
	<PER, ORG>	Leader, schools-attended, employee, founder, shareholder, justice
	<PER, GPE>	Resides-place, leader, conflict-attack, conflict-demonstrate, justice, movement-transport, injure
	<ORG, ORG>	Business-merge, sport, transaction
	<GPE, GPE>	Conflict-attack
	<VEH, GPE>	Movement-transport

1 http://meta.wikimedia.org/wiki/List_of_Wikipedias

knowledge-bases that have harvested this structured data are available. The most well-known are probably DBpedia [16], Freebase [17] and YAGO [18]. These existing structured comparable knowledge bases can provide a large number of seed pairs to bootstrap name translation mining. However, most of these resources only cover well-known entities. For example, while Freebase contains 116 million instances of 7,300 relations for 9 million entities, it only covers 48 % of the slot types and 5 % of the slot answers defined in the KBP task. Even for those covered slots, for 100 test queries, Freebase only contains about 5 % of the possible slot answers. Therefore, while Infoboxes are a readily available source of structured data, the content of the unstructured articles contains much richer information.

We apply the following simple rules to classify Infoboxes and slot entries:

- If an entry has *birth_date* or *death_date* slot, it's classified as a person.
- If an entry has *foundation* or *company_*(any slot type begins with company)* slot, it's classified as an organization.
- If an entry has *latitude* or *population slot*, it's classified as a geo-political name.

Because not all Wikipedia contributors follow the standard naming conventions and date/number formats for all languages, Infoboxes include some noisy instances. Fortunately the KBP task defined mapping tables which can be directly used to normalize different forms of slot types. For example, we can group "*birthdate*", "*date of birth*", "*datebirth*" and "*born*" to "*birth_date*". In addition, we also normalized all date slot values into one standard format as "*YYYY MM DD*". For example, both "1461-8-5" and "5 August, 1461" are normalized as "1461 08 05". After normalization we were able to utilize 10 edge types for GPEs, 14 edge types for organizations and 20 edge types for persons.

3.3 Automatic Construction from Unstructured Comparable Corpora Based on Information Extraction

In order to enrich bi-lingual information networks, we apply two state-of-the-art bi-lingual (English and Chinese) cross-document IE systems to extract facts from comparable corpora. These IE systems were developed for the ACE and KBP programs.

The ACE IE pipeline [9, 19–22] includes name tagging, nominal mention tagging, entity coreference resolution, time expression extraction and normalization, relation extraction and event extraction. Names are identified and classified using a Hidden Markov Model. Nominals are identified using a Maximum Entropy (MaxEnt)-based chunker and then semantically classified using statistics from the ACE training corpora. Entity coreference resolution, relation extraction and event extraction are also based on MaxEnt models, incorporating diverse lexical, syntactic, semantic and ontological knowledge. At the end an event coreference resolution component is applied to link coreferential events, based on a pair-wise MaxEnt model and a graph-cut clustering model.

In addition, we apply a state-of-the-art slot filling system [10] to identify KBP slots for every person or organization entity. This system includes a bottom-up pattern matching pipeline and a top-down question answering (QA) pipeline. In pattern matching, we extract and rank patterns based on a distant supervision approach [23] using entity-attribute pairs from Wikipedia Infoboxes and Freebase. Then we apply these patterns to extract attributes for unseen entities. We set a low threshold to include more candidate attribute answers, and then apply several filtering steps to remove wrong answers. The filtering steps include removing answers which have inappropriate entity types or involve inappropriate dependency paths to the entities. We also apply an open domain QA system, OpenEphyra [24] to retrieve more relations and events. To estimate the relevance of a query and answer pair, we use the Corrected Conditional Probability (CCP) for answer validation. Finally we exploit an effective MaxEnt based supervised re-ranking method to combine the results from these two pipelines. The re-ranking features include confidence values, dependency parsing paths, majority voting values and slot types. In the slot filling task, each slot is often dependent on other slots. For example, if the age of *X* is "*2 years old*", we can infer that there are unlikely any "employer" attributes for *X*. Similarly, we design propagation rules to enhance recall, for example, if both *X* and *Y* are children of *Z*, then we can infer*X* and*Y* are siblings. Therefore we develop a reasoning component to approach a real world acceptable answer in which all slot dependencies are satisfied. We use Markov Logic Networks (MLN) [25], a statistical relational learning language, to model these inference rules more declaratively. Markov Logic extends first order logic in that it adds a weight to each first order logic formula, allowing for violation of those formulas with some penalty.

4 Cross-Lingual Information Network Alignment

After creating the information networks from each language, we automatically align them to discover name translation pairs. In this section we start from analyzing some motivating examples, and then describe our general approach to name translation mining.

4.1 Approach Overview

Figure 4 depicts a motivating example for our approach. Based on the assumption that if two person entries had the same birth date and death date, they are likely to be the same person, we can get the seed translation pair of "*迈克尔.杰克逊 (Michael Jackson)*". We can get many seed pairs using similar language-independent clues. Then starting from these seed pairs, we can iteratively get new pairs with a large portion of overlapped slots. For example, since "*积逊五人组*" and "*The Jackson 5*"

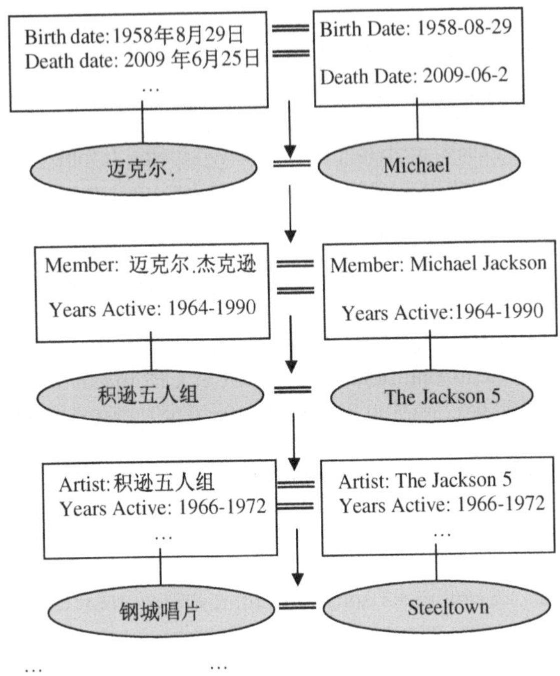

Fig. 4 A motivating example for information network alignment

share many linked nodes such as "*member*" and "*years active*", they are likely to be a translation pair. Next we can use the new pair of "*积逊五人组 (The Jackson 5)*" to mine more pairs such as "*钢城唱片*" and "*Steeltown Records*".

For the example in Fig. 3, we can start from a common name translation "*国家情报局 -National Intelligence Service (1)*", to align its *leader* as "*蒙特西诺斯 -Montesinos (2)*", align the *arrest place* of "*Montesinos*" as "*卡西俄 -Callao (3)*", and then align the *location* of "*Callao*" as "*秘鲁 -Peru (4)*".

Based on the above motivations we propose a new bootstrapping approach to mine name pairs from bi-lingual information networks. The general idea is that starting from a small seed set of name pairs, we can rely on the link attributes to align their related names. Then the new name translations are added to the seed set for the next iteration. We repeat this bootstrapping procedure until no new translations are produced. Using this approach we can discover name pairs of various types (person, organization and location) while minimizing using supervised name transliteration techniques or document-wise temporal distributions as in previous work. The following sections will focus on describing detailed algorithms for each step. The language pair that we are considering in this paper is Chinese and English.

4.2 Seed Pair Mining

Although different languages have different writing systems, a vast majority of the world's countries and languages use similar forms for representing information such as time/calendar date, number, website URL and currency [26]. In fact most languages commonly follow the ISO 8601 standard[2] so the formats of time/date are the same or very similar. We start from language-independent expressions in any two languages, and then extract those nodes which share many links, using these nodes as our seeds. Our aim for seed generation is to extract accurate pairs with high confidence so that few errors are propagated into later mining iterations. We designed the following mining rules to discover seed pairs for each entity type.

- PER

If two PER entries have the same "*birth-date*" and "*death-date*" link values, they are considered as a seed pair. The intuition is that the probability for two people to have the same birth date and the same death date is very low.

- GPE

If two GPE entries have the same "*area*" and "*population*" link values, or they have the same "*website*" link value, they are consider as a seed pair. The intuition is that the values for these three slots are language independent because they normally include numeric values and some standard units of measurement (e.g. "*km*") or URL in English.

- ORG

If two ORG entries have the same value for "Website" link, they are considered as a seed pair.

We cannot use the same strategy from person seed mining for GPEs and ORGs because they rarely include both "*foundation date*" and "*bankruptcy date*". When we chose only "*foundation date*" for ORGs, the results turned out to be quite noisy. For example, both "*瓦莱罗能源公司 (Valero Energy Corporation)*" and "*Applebee's*" were founded on "*1980/01/01*" but they are not a translation pair.

Some examples from the seed list are presented in Table 2.

4.3 Bootstrapping via Link Comparison

After we obtain the high-confidence name pairs from the approach described above, we apply them to discover more pairs in a bootstrapping framework as shown in Fig. 5.

2 http://en.wikipedia.org/wiki/ISO_8601

Table 2 Examples of mined seed name pairs

GPE	Braga Municipality	布拉加
	Goseong, Gangwon	高城郡 (韓國)
	New Delhi Municipal Council	新德里
	Regional Municipality of Niagara	尼亞加拉區
	Somerville, New Jersey	萨默维尔 (新泽西州)
ORG	Axis Communications	安讯士
	Commercial Times	工商時報
	Dow Jones & Company	道琼斯公司
	Libra Radio	非凡音廣播電台
	Lynx Aviation	邊疆航空
	Mercer (consulting firm)	美世咨询
	Rodong Sinmun	朝鲜劳动党
	Seiyu (retailer)	西友
	Wei-Chuan Food Corporation	味全
	Yctc	業強科技
PER	Georg Wenzelburger	格奧爾格·溫瑟堡
	Nelson Ikon Wu	吳訥孫

For each English node entity *e:*
For all Chinese node entities $\{c_i\}$,
(1) Compare the links of *e* and each c_i based on the seed pairs and language-independent link values, calculate the overlap rate as *overlap* (e, c_i);
(2) Find $c_{\max} = \arg\max_{c_i} overlap(e, c_i)$
(3) if *overlap* $(e, c_{max}) > \delta$ ($\delta = 50\%$ in our paper),
if the documents are unstructured and *e* and c_{max} are persons,
if the Damerau–Levenshtein edit distance between the pinyin of form of c_{max} and $e < \theta$ ($\theta = 5$ in our paper)
else add (e, c_{max}) to the seed pair list

Repeat until no new pairs are found.

Fig. 5 Bootstrapping algorithm for name pair mining

5 Experimental Results

In this section we present the experiment results for comparable corpora acquisition (Sect. 5.1) and name mining (Sect. 5.2) respectively.

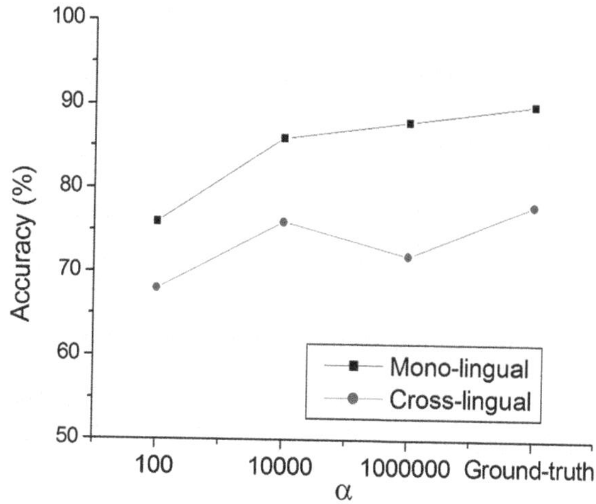

Fig. 6 Comparable corpora acquisition with IDF ($\delta = 0.6$)

5.1 Comparable Corpora Acquisition Results

We used 244 videos from TRECVID 2005 data set as our test set. This data set includes 133,918 key frames, with corresponding automatic speech recognition and translation results (for foreign languages) provided by LDC. We measured accuracy for the top 50 comparable document pairs. Due to lack of answer-keys, we asked a bi-lingual human annotator to judge results manually. The evaluation guideline generally followed the definitions in [6]. A pair of documents is judged as comparable if they share a certain amount of information (e.g. entities, events and topics). The accuracy results are presented in Fig. 6. We can see that under any conditions our approach can discover comparable documents reliably. In order to measure the impact of concept extraction errors, we also evaluated the results for using ground-truth concepts as shown in Fig. 6. Surprisingly it didn't provide much higher accuracy than automatic concept extraction, mainly because the similarity can be captured by some dominant video concepts.

5.2 Name Translation Mining Results

In this section we present the evaluation results of name translation mining from bi-lingual information networks.

5.2.1 Data

A standard Wikipedia entry ("*nodes*" in information network) includes a title, a document describing the entry, and an "*infobox*" which is a fixed-format table designed to be added to the top right-hand corner of the article to consistently present a summary

Table 3 Statistics about information networks constructed from structured data

	Language	Node types			
		PER	ORG	GPE	All
#nodes	English	326,419	81,452	357,044	1,534,809
	Chinese	10,306	5,134	8,742	62,398
#edges	English	7,321,698	2,130,222	15,323,171	37,162,712
	Chinese	191,619	108,667	283,499	1,129,717

Table 4 Number of unstructured documents

	Language	
Corpus	Chinese	English
ACE	CHSet1: XIN Oct–Dec 2000: 150 documents	ENSet1: APW Oct–Dec 2000: 150 documents ENSet2: AFP & APW Mar–June 2003: 150 documents
TDT-5	CHSet3: XIN Apr–Aug 2003: 30,000 documents	ENSet3: XIN Apr–Aug 2003: 30,000 documents ENSet4: AFP Apr–Aug 2003: 30,000 documents

of some unifying attributes (or "*edges*" in information network) about the entry. Table 3 summarizes the number of nodes and edges in the information network constructed from English and Chinese Wikipedias as of November 2010. There are in total 10,355,225 English pages and 772,826 Chinese pages.

In addition, we use some unstructured documents from the ACE (2004, 2005) training corpora and TDT-5 corpora to evaluate our approach. Table 4 shows the number of documents from different news agencies and time frames.

5.2.2 Evaluation Method

It's time consuming to evaluate the mined name pairs because sometimes the human assessor needs Web access to check the contexts of the pairs, especially when the translations are based on meanings instead of pronunciations. We implemented a baseline of mining name pairs from cross-lingual titles in Wikipedia as an incomplete answer key, and so we only need to ask two human assessors (not system developers) to do manual evaluation on our system generated pairs which are not in this answer key. A name pair is judged as correct if both of them are correctly extracted and one is the correct translation of the other. Such a semi-automatic method can speed up evaluation.

Table 5 Performance of seeds mined from structured data

Entity type	#Pairs	Accuracy (%)
PER	1,670	99.76
GPE	921	99.10
ORG	1,702	99.40
All	4,161	99.47

Table 6 Performance of bootstrapping from structured data

Entity type	#Pairs	Accuracy (%)
New pairs	1,546	91.46
All pairs	5,707	97.30

5.2.3 Structured Data Mining Results

Table 5 presents the total number of mined seed pairs and their overall accuracy. We can see that all the seeds are very reliable in all settings, with accuracy scores close to or above 99 %.

In Table 6 we summarize the number and accuracy of new pairs mined from the bootstrapping procedure under different settings. Table 6 indicates that although the accuracy scores of the new pairs are not as high as seeds, but they are all above 91 %.

5.2.4 Unstructured Data Mining Results

Table 7 shows the number and accuracy of name translation pairs discovered from unstructured data CH-Set3 and EN-Set3, using 100 common name pairs as seeds. After four iterations we discovered 968 new name translation pairs with accuracy 82.9 %. Among them there are 361 persons (accuracy 76.4 %), 384 geo-political names (accuracy 87.5 %) and 223 organization names (accuracy 85.2 %).

One major evidence exploited in the prior work is that the bilingual comparable corpora should be weakly temporally aligned. For example, Klementiev and Roth [8] used the time distribution of names to re-score name transliteration. In order to verify this observation, we investigated how well our new approach can perform on comparable corpora with different time frames. Table 8 presents the performance of two combinations: CHSet1-ENSet1 (from the same time frame) and CHSet1-ENSet2 (from different time frames) with a seed set of 10 name pairs after 5 iterations.

In addition, in order to measure the impact of news source similarity, we apply our approach to the combination of CHSet3 and ENSet4 which are from different news agencies. In total 815 name pairs are discovered after 4 iterations with overall accuracy 78.7 %, which is worse than the results from the corpora of the same news

Table 7 Overall name mining performance from unstructured data

Iteration	1	2	3	4
Number of name pairs	205	533	787	968
Accuracy (%)	91.8	88.5	85.8	82.9

Table 8 Impact of time frame similarity

Corpora	CHSet1-ENSet1	CHSet1-ENSet2
Number of name pairs	42	17
Accuracy (%)	81.0	76.5

source as shown in Table 7. Therefore we can clearly see that time and news source similarities are important to the performance of name translation pair mining.

Since in our approach we used the fully automatic IE pipeline to create the information networks, the errors from each component will be propagated into the alignment step and thus limit the performance of name translation discovery. For example, Chinese name boundary detection errors caused about 30 % of the incorrect name pairs. Some incorrect pairs were mined when one node included or entailed the other. For example, the bootstrapping algorithm mistakenly mined a pair "*List of Gaudí buildings// 安东尼·高迪 (Antoni Gaudi)*" because they share many link values. As a diagnostic analysis, we tried to discover name pairs from CHSet1 and ENSet1 but with perfect IE annotations. We obtained 63 name pairs with a much higher accuracy 90.5 %.

5.2.5 Comparison with Previous Methods and Resources

There have been some previous methods focusing on mining name translations using weakly-supervised learning. In addition there are some existing name translation gazetteers which were manually constructed. We duplicated a variety of alternative state-of-the-art name translation mining methods and mined some corresponding name pair sets for comparison. In fact we were able to implement the techniques in previous approaches but could not duplicate the same results, because we could not access the same data sets. Therefore the main purpose of this experiment is not to claim our approach outperforms these existing methods, rather to investigate whether we can mine any new information in addition to these methods from reasonable amounts of comparable documents.

(1) Name Pair Mining from Bitexts

Within each sentence pair in a parallel corpus, we ran our bi-lingual name tagger. If the types of the name tags on both sides are identical, we extract the name pairs from this sentence. Then at the corpus-wide level, we count the frequency for each name pair, and only keep the name pairs that are frequent enough. The corpora used for this approach were all DARPA GALE MT training corpora.

(2) Using patterns for Web mining

We constructed heuristic patterns such as parenthetical structure "Chinese name (English name)" [27] to extract name pairs from web data with mixed Chinese and English. We used about 1,000 web pages for this experiment.

Table 9 # Correct name pairs of adding our results to previous approaches

Method	Add our results	
	Before	After
Automatic		
(1) Bitexts	2,451	8,114
(2) Patterns for web mining	194	5,901
Manual		
(3) Bilingual gazetteer	59,886	65,539
(4) ACE2007 training data	1,541	7,221

(3) Bilingual Gazetteer

We exploited an LDC bilingual name dictionary (LDC2005T34) and a Japanese-English person name dictionary including 20126 Japanese names written in Chinese characters [28].

(4) ACE2007 Entity Translation Training Data

We also used ACE 2007 entity translation training corpus which includes 119 Chinese-English document pairs.

Table 9 shows the number of correct and unique pairs when we use our mined pairs to augment the results from each of the above approaches. It clearly shows that our name pairs can significantly augment the results from both automatic and manual methods.

5.2.6 Discussion

Besides retaining high accuracy, the final list of name pairs revealed several advantages of our approach.

Most previous name translation methods are limited to names which are phonetically transliterated (e.g. translate the Chinese name "*尤申科 (You shen ke)*" to "*Yushchenko*" in English). But many other types of names such as organizations are often rendered semantically, for example, the Chinese name "*解放之虎 (jie fang zhi hu)*" is translated into "*Liberation Tiger*" in English. Some other names involve both semantic and phonetic translations, or none of them. Our approach is able to discover all these different types, regardless of their translation sources. For example, our approach successfully mined a pair "*柏油村*" where "*Tarrytown*" is translated into "*柏油村*" neither by its pronunciation "*bai you cun*" nor its meaning "*tar village*". Name abbreviations are very challenging to translate because they need expansions based on contexts. However our approach mined a lot of abbreviations using link comparison. For example, the pair of "*Yctc/ 业强科技*" was successfully mined although its English full name "*Yeh-Chiang Technology Corp.*" didn't appear in the information networks. Huang [29] also pointed out that name translation benefited from origin-specific features. In contrast, our approach is able to discover name pairs from any origins. For example, we discovered the person name pair "*Seishi Yokomizo/ 横溝正史* " in which "*Seishi Yokomizo*" was transliterated based on Japanese pronunciation.

Furthermore, many name translations are context dependent. For example, a person name in Chinese " *亚西尔•阿拉法特* " should be translated into "*Yasser Arafat (PLO Chairman)*" or "*Yasir Arafat (Cricketer)*" based on different contexts. Our method can naturally disambiguate such entities based on slot comparison at the same time as translation mining.

More importantly, our final list includes a large portion of uncommon names, which can be valuable to address the out-of-vocabulary problem in both MT and cross-lingual information processing. Especially we found many of them are not in the name pairs mined from the cross-lingual Wikipedia title links, such as "*Axis Communications/ 安讯士* ", "*Rowan Atkinson/ 路雲·雅堅遜* ", "ELSA Technology/ 艾爾莎科技 " and "*Nelson Ikon Wu/ 吳訥孫* ". Wikipedia entries are created and updated exponentially [30] because of the increasing number of contributors, but many of whom are not multi-lingual speakers. Therefore it's valuable to align the cross-lingual entries by effective name mining.

6 Related Work

Extensive research has been done on video clustering. For example, Cheung and Zakhor [31] used meta-data extracted from textual and hyperlink information to detect similar videos on the web; Magalhaes et al. [32] described a semantic similarity metric based on key word vectors for multi-media fusion. We extend such video similarity computing approaches to a multi-lingual environment to discover cross-lingual comparable corpora.

Most of the previous name translation work combined supervised phonetic similarity based name transliteration approaches with Language Model based re-scoring (e.g. [29, 33, 34]). Our approach of extracting and aligning information network from comparable corpora is related to some prior work using comparable corpora to re-score name transliterations [7, 8] or mine new word translations [35–39]. However, most of these approaches needed large amount of seeds and suffered from information extraction errors, and thus relied on phonetic similarity or document similarity to re-score candidate name translation pairs. In this paper we extend the target names from persons to geo-political and organization names, and extract relations links among names simultaneously. And we use a bootstrapping approach to discover name translations from the bilingual information networks of comparable corpora. In this way we don't need to have a name transliteration module to serve as baseline, or compute document-wise temporal distributions.

Some recent cross-lingual information access work explored attribute mining from Wikipedia pages. For example, Bouma et al. [40] aligned attributes in Wikipedia Infoboxes based on cross-page links. Ji et al. [41] described various approaches to automatically mine name translation pairs from aligned phrases (e.g. cross-lingual

Wikipedia title links) or aligned sentences (bi-texts). Some other work mined name translations from mono-lingual documents that include foreign language texts. For example, Lin et al. [27] described a parenthesis translation mining method; You et al. [42] applied graph alignment algorithm to obtain name translation pairs based on co-occurrence statistics. This kind of data does not commonly exist for low-density languages.

7 Conclusions and Future Work

We described an effective cross-lingual comparable corpora acquisition approach based on video comparison. From the comparable corpora we further constructed bi-lingual information networks. We also described a simple, cheap and effective self-boosting approach to mine name translation pairs from information networks. This method is implemented in a weakly-supervised fashion, without using any manually created seed set, training data, transliteration or pre-knowledge about the language pair. The underlying motivation is that some certain expressions are written in language-independent forms among a large majority of languages. Therefore our approach can be applied to any language pairs including low-density languages as long as they share a small set of such expressions. Experiments on English-Chinese pair on both structured and unstructured comparable corpora showed that this approach is able to mine thousands of name pairs with high accuracy. In addition the resulting name pairs can be used to significantly augment the results from existing approaches. The mined name pairs are made publicly available.

In the future we plan to import more efficient graph mining and alignment algorithms which have been widely used for protein-protein interaction detection [43]. In the short term, our approach will provide a framework for many byproducts and directly benefit other natural language processing tasks. For example, the aligned sub-graphs with names, relations and events can be used to improve information redundancy in cross-lingual question answering; the outlier (mis-aligned) sub-graphs can be used to detect the novel or local information described in one language but not in the other; the links among names can be exploited for entity disambiguation.

Acknowledgments This work was supported by the U.S. Army Research Laboratory under Cooperative Agreement Number W911NF-09-2-0053, the U.S. NSF CAREER Award under Grant IIS-0953149 and PSC-CUNY Research Program. The views and conclusions contained in this document are those of the authors and should not be interpreted as representing the official policies, either expressed or implied, of the Army Research Laboratory or the U.S. Government. The U.S. Government is authorized to reproduce and distribute reprints for Government purposes notwithstanding any copyright notation hereon.

References

1. Krauss, M.E.: Keynote-mass Language Extinction and Documentation: The Race Over Time. The Vanishing Languages of the Pacific Rim. Oxford University Press, Oxford (2007)
2. Parton, K., McKeown, K.R., Coyne, R., Diab, M.T., Grishman, R., Hakkani-Tur, D., Harper, M., Ji, H., Ma, W.-Y., Meyers, A., Stolbach, S., Sun, A., Tur, G., Xu, W., Yaman, S.: Who, What, When, Where, Why? Comparing multiple approaches to the cross-lingual 5W task. In: Proceedings of ACL-IJCNLP (2009)
3. Hakkani-Tur, D., Ji, H., Grishman, R.: Using information extraction to improve cross-lingual document retrieval. In: Proceedings of RANLP 2007 Workshop on Multi-Source Multi-lingual Information Extraction and Summarization (2007)
4. Ji, H., Grishman, R., Wang, W.: Phonetic name matching for cross-lingual spoken sentence retrieval. In: Proceedings of IEEE-ACL SLT08 (2008)
5. Munteanu, D.S., Marcu, D.: Improving machine translation performance by exploiting non-parallel corpora. Comput. Linguist. **31**(4), 477–504 (2005)
6. Cheung, P., Fung, P.: Sentence alignment in parallel, comparable, and quasi-comparable corpora. In: Proceedings of LREC (2004)
7. Sproat, R., Tao, T., Zhai, C.: Named entity transliteration with comparable corpora. In: Proceedings of ACL2006 (2006)
8. Klementiev, A., Roth, D.: Named entity transliteration and discovery from multilingual comparable corpora. In: Proceedings of HLT-NAACL (2006)
9. Ji, H., Grishman, R., Chen, Z., Gupta, P.: Cross-document event extraction, ranking and tracking. In: Proceedings of Recent Advances in Natural Language Processing (RANLP 2009), pp. 166–172 (2009)
10. Chen, Z., Tamang, S., Lee, A., Li, X., Lin, W., Artiles, J., Snover, M., Passantino, M., Ji, H.: CUNY-BLENDER TAC-KBP2010 entity linking and slot filling system description. In: Proceedings of Text Analysis Conference (TAC2010) (2010)
11. Naphade, M.R., Kennedy, L., Kender, J.R., Chang, S.-F., Smith, J.R., Over, P., Hauptmann, A.: A light scale concept ontology for multimedia understanding for TRECVID. Technical report, IBM (2005)
12. Qi, G., Hua, X., Rui, Y., Tang, J., Mei, T., Zhang, H.: Correlative multi-label video annotation. In: Proceedings of ACM Multimedia (2007)
13. Gu, Z., Mei, T., Hua, X., Tang, J., Wu, X.: Multi-layer multi-instance kernel for video concept detection. In: Proceedings of ACM Multimedia (2007)
14. Wang, F., Zhang, C.: Label propagation through linear neighborhoods. In: Proceedings of ICML2006 (2006)
15. Ji, H., Grishman, R., Dang, H.T., Griffitt, K., Ellis, J.: An overview of the TAC2010 knowledge base population track. In: Proceedings of Text Analysis Conference (TAC2010) (2010)
16. Auer, S., Bizer, C., Kobilarov, G., Lehmann, J., Ives, Z.: DBpedia: a nucleus for a web of open data. In: Proceedings of the 6th International Semantic Web Conference (2007)
17. Bollacker, K., Cook, R., Tufts, P.: Freebase: a shared database of structured general human knowledge. In: Proceedings of the National Conference on, Artificial Intelligence, vol. 2 (2007)
18. Suchanek, F.M., Kasneci, G., Weikum, G.: Yago: a core of semantic knowledge. In: Proceedings of the 16th International World Wide Web Conference (2007)
19. Ji, H., Grishman, R.: Refining event extraction through cross-document inference. In: Proceedings of the 46th Annual Meeting of the Association for Computational Linguistics: Human Language Technologies (ACL-HLT 2008) (2008)
20. Grishman, R., Westbrook, D., Meyers, A.: NYU's Chinese ACE 2005 EDR system description. In: Proceedings of NIST Automatic Content Extraction, Workshop (ACE2005) (2005)
21. Chen, Z., Ji, H.: Language specific issue and feature exploration in Chinese event extraction. In: Proceedings of HLT-NAACL (2009)
22. Ji, H., Meyers, A., Grishman, R.: NYU's Chinese ACE 2005 EDR system description. In: ACE05 PI/Evaluation, Workshop (2005)

23. Mintz, M., Bills, S., Snow, R., Jurafsky, D.: Distant supervision for relation extraction without labeled data. In: Proceedings of Joint Conference of the 47th Annual Meeting of the Association for Computational Linguistics and the 4th International Joint Conference on Natural Language Processing of the Asian Federation of Natural Language Processing (ACL-IJCNLP 2009) (2009)
24. Schlaefer, N., Ko, J., Betteridge, J., Sautter, G., Pathak, M., Nyberg, E.: Semantic extensions of the Ephyra QA system for TREC2007. In: Proceedings of Text Retrieval Conference (TREC2007) (2007)
25. Richardson, M., Domingos, P.: Markov logic networks. Mach. Learn. **62**, 107–136 (2006)
26. IBM: IBM Globalization library. ftp://ftp.software.ibm.com/software/globalization/locales (2010)
27. Lin, D., Zhao, S., Durme, B.V., Pasca, M.: Mining parenthetical translations from the web by word alignment. In: Proceedings of ACL2008 (2008)
28. Kurohashi, S., Nakamura, T., Matsumoto, Y., Nagao, M.: Improvements of Japanese morphological analyzer JUMAN. In: Proceedings of the International Workshop on Sharable Natural Language, Resources, pp. 22–28 (1994)
29. Huang, F.: Cluster-specific name transliteration. In: Proceedings of HLT-EMNLP (2005)
30. Almeida, R.B., Mosafari, B., Cho, J.: On the evolution of Wikipedia. In: Proceedings of International Conference on Weblogs and Social Media (2007)
31. Cheung, S.-C., Zakhor, A.: Efficient video similarity measurement and search. In: Proceedings of IEEE International Conference on Image Processing (2000)
32. Magalhaes, J., Ciravegna, F., Ruger, S.: Exploring multimedia in a keyword space. In: Proceedings of ACM Multimedia (2008)
33. Al-Onaizan, Y., Knight, K.: Translating named entities using monolingual and bilingual resources. In: Proceedings of ACL (2002)
34. Huang, F., Vogel, S., Waibel, A.: Improving named entity translation combining phonetic and semantic similarities. In: Proceedings of HLT/NAACL (2004)
35. Udupa, R., Saravanan, K., Kumaran, A., Jagarlamudi, J.: MINT: a method for effective and scalable mining of named entity transliterations from large comparable corpora. In: Proceedings of EACL2009 (2009)
36. Fung, P., Yee, L.Y.: An IR approach for translating new words from nonparallel, comparable texts. In: Proceedings of COLING-ACL (1998)
37. Hassan, A., Fahmy, H., Hassan, H.: Improving named entity translation by exploiting comparable and parallel corpora. In: Proceedings of RANLP (2007)
38. Rapp, R.: Automatic identification of word translations from unrelated English and German corpora. In: Proceedings of ACL (1999)
39. Shao, L., Ng, H.T.: Mining new word translations from comparable corpora. In: Proceedings of COLING2004 (2004)
40. Bouma, G., Duarte, S., Islam, Z.: Cross-lingual alignment and complettion of Wikipedia templates. In: Proceedings of the 3rd International Workshop on Cross Lingual Information Access: Addressing the Information Need of Multilingual Societies (2009)
41. Ji, H., Grishman, R., Freitag, D., Blume, M., Wang, J., Khadivi, S., Zens, R., Ney, H.: Name translation for distillation. Handbook of Natural Language Processing and Machine Translation: DARPA Global Autonomous Language Exploitation. Springer, Berlin (2009)
42. You, G., Hwang, S., Song, Y., Jiang, L., Nie, Z.: Mining name translations from entity graph mapping. In: Proceedings of EMNLP2010 (2010)
43. Kelley, B.P., Sharan, R., Karp, R.M., Sittler, T., Root, D.E., Stockwell, B.R., Ideker, T.: Conserved pathways within bacteria and yeast as revealed by global protein network alignment. In: Proceedings of the National Academy of Sciences of the United States of America (2003)

Bilingual Terminology Mining from Language for Special Purposes Comparable Corpora

Emmanuel Morin, Béatrice Daille and Emmanuel Prochasson

Abstract In this paper we study the problem of compiling bilingual lexicon from language for special purposes (LSP) comparable corpora. We first define what would be the comparability for specialized comparable corpus and stress the distinction between expert and non-experts documents. We then turn to the contextual information method that concentrates on bilingual lexicon extraction and show its limits: the context vectors do not discriminate very much due to the small amount of data available and, the translation of the context vectors is more difficult due to the lack of specialized translations in the dictionary. For each problem, we propose a solution that relies on the LSP linguistic properties. For the first limit, we propose to strengthen the representativeness of the lexical contexts based on domain-specific vocabulary, called anchor points, notably the neoclassical compounds and the transliterations. For the second limit, we propose to use the parallel sentences present in the comparable corpus in order to build a specialized bilingual lexicon directly correlated to the specialized vocabulary of the comparable corpus. Our experiments illustrate that these two strategies are well-founded and show that candidate translations are of better quality.

E. Morin · B. Daille (✉) · E. Prochasson
Université de Nantes, LINA UMR CNRS 6241, 2 rue de la Houssinière44322, Nantes Cedex 0392208, BP , France
e-mail: beatrice.daille@univ-nantes.fr

E. Morin
e-mail: emmanuel.morin@univ-nantes.fr

E. Prochasson
e-mail: emmanuel.prochasson@univ-nantes.fr

S. Sharoff et al. (eds.), *Building and Using Comparable Corpora*,
DOI: 10.1007/978-3-642-20128-8_14, © Springer-Verlag Berlin Heidelberg 2013

1 Introduction

The automatic compilation of bilingual dictionaries from comparable corpora has received considerable attention in recent years. Parallel corpora that are translated resources are rare with regards to numerous pairs of languages, at least for those that do not include English, and for *language for special purposes (LSP)* corpora (especially from scientific domains). Comparable corpora that are sets of texts in different languages seem to be the right solution to solve textual scarcity: as monolingual productions, they are authentic texts which have not been translated, and the babel web ensures the availability of large amounts of multilingual documents. Comparable corpora have been developed for translation purposes either to train the translators or to encourage the learning of a second language [38]. They are useful for investigating usage and style, for searching for unknown equivalents or choosing between multiple options [2]. For LSP, [2, 22] stress their usefulness in choosing between terminological equivalents and term variants in the target languages and to understand specialised concepts.

The main work in the compilation of specialized dictionaries from comparable corpora is based on lexical context analysis and relies on the simple observation that a word and its translation tend to appear in the same lexical contexts [12, 31]. Using this approach, good results are obtained for lexical alignment of single-words with a precision of about 80 % for the top 20 proposed candidates using large comparable corpora (176 million bytes of English-Chinese data in [12] and 298 million-word English-German corpus in [32]). Results drop to 60 % for single-word terms using specialized small size language corpora (0.1 million-word English-German corpus in [7] and 1.5 million-word French-Japanese corpus in [23]).

The contextual information method that concentrates on bilingual lexicon extraction presents two limits in the context of specialized comparable corpora. Firstly, the context vectors do not discriminate very much due to the small amount of data available. For this first limitation, we propose to strengthen the representativeness of the lexical contexts based on domain-specific vocabulary called anchor points. Secondly, the translation of the context vectors is more difficult due to the lack of specialized translations in the dictionary. For this other limitation, we propose to use the parallel sentences present in the comparable corpus in order to build a specialized bilingual lexicon directly correlated to the specialized vocabulary of the comparable corpus. The overall architecture of the application is presented in Fig. 1.

The remainder of this paper is organized as follows: Section 2 investigates what is comparability for LSP and how it is measured. Section 3 presents the contextual information method dedicated to bilingual lexical alignment from comparable corpora. Section 4 describes how the representativeness of the lexical contexts is strengthened based on domain-specific vocabulary (i.e. anchor points) notably the neoclassical compounds and the transliterations. Section 5 presents another strategy to enhance the contextual information method by using a small specialized bilingual dictionary directly induced from parallel sentences present in the comparable corpora. Finally, Sect. 6 presents our conclusions.

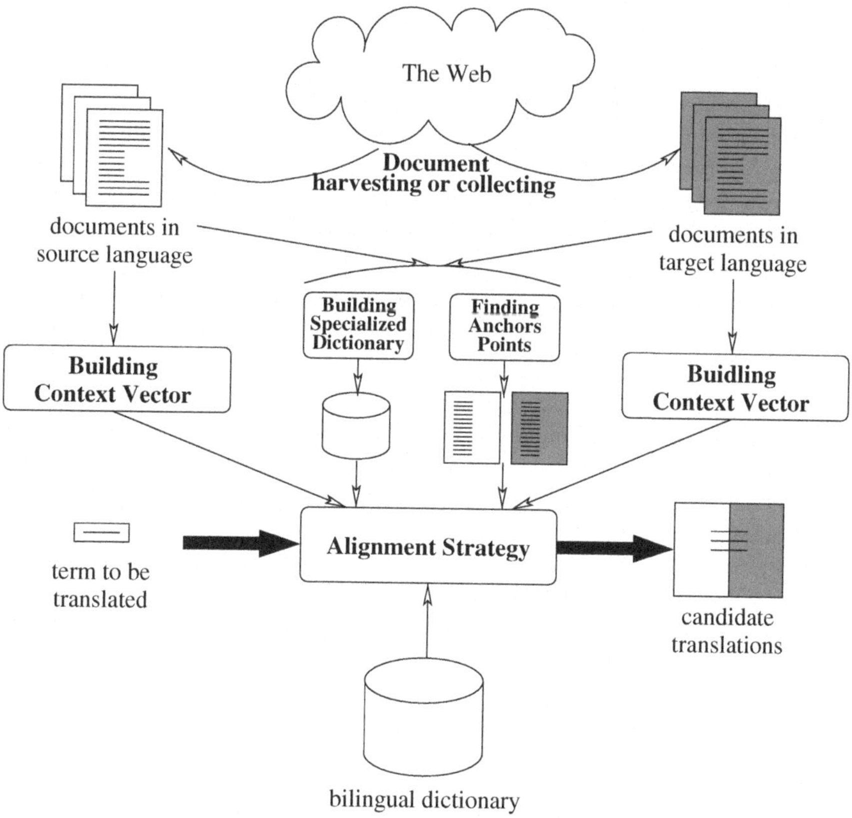

Fig. 1 Overall architecture of the application

2 Comparability

Dealing with multilingual terminology extraction requires comparable corpora that collect specialized texts. The comparability of the corpus should be defined and measured to guarantee the quality of the terminology that will be extracted. In this section, we first define what would be the comparability for LSP comparable corpus and stress the distinction between expert and non expert documents.

2.1 Definition

The EAGLES[1] gives us the following definition of a comparable corpora:

[1] http://www.ilc.cnr.it/EAGLES

« *a comparable corpus is one which selects similar texts in more than one language or variety.* »

In 1996, it stated that there was no agreement on the nature of the similarity, because there are very few examples of comparable corpora. The notion of comparability has been claimed to depend on similar objectives, but even within this requirement disagreements remain. As an example, for [2, p. 12], comparable corpora are multilingual corpora that do not contain translated texts.

« *The texts in a comparable corpora were originally written in a language A, B, C, etc., but they all have the same communicative function. In other words, they are all on the same subject, all the same type of text (e.g. instructional manual, technical report, etc.), all from the same time frame, etc.*

The objective here it to identify reliable translations. The same objective underlines within [8] for whom comparable corpora of languages A and B should be able to provide translations of the A into the B vocabularies, and even from B to A vocabularies. As a consequence, the best comparable corpora are parallel corpora that could provide a translation of each word. So an identical objective leads to opposite characteristics. In practice, this feature of translated/non-translated texts is never checked nor is the native language production.

Contrary to its finality, comparable corpora could be defined with regards to its compilation.

«*A comparable corpora can be defined as a corpus containing components that are collected using the same sampling frame and similar balance and representativeness.*» [22, p. 20]

The comparability could be ensured using various characteristics that could be automatically measured. For general language, they often share a domain or topic within a limited period. They could also share a common genre such as comparable corpora made of multilingual news reporting agencies that are provided by LDC.[2] For LSP, domains and sub-domains are requested but are not sufficient. Pearson [28] highlights the communicative settings: expert-expert, expert-initiate, relative expert-uninitiate, teacher-pupil that encode different technical levels. As stressed by [14], it is important in the medical domain to distinguish between document types that are written either for medical experts or for non-expert users; the reading can impact on users' health or communication with professionals.

2.2 Distinguishing Between Science and Popular Science

The comparability of the comparable corpora is thus a guarantee of a monolingual comparability of monolingual texts. For LSP, these features are the domain, the sub-domain, and the type of discourse: scientific discourse or popular science discourse. Scientific discourse documents are written by specialists and are addressed

[2] For English-French, AFP news: LDC2007T07 (English) and LDC2006T17 (French)

to specialists. The author and the reader share the same level of knowledge. Popular science discourse documents are addressed to the general public and can be written by both specialists and people that do not have any formal training in the field (general public). The topic of a document can be recognised by the keywords used in its web search either manually or using a crawler. The document discourse is harder to identify as its relies on several dimensions: structural, lexical and stylistic. When the documents are extracted from the web, the format and the structure of the document are useful clues: scientific articles come mostly in PDF format; resource names in the URL could be used to recognise advertising, etc. A high terminological density as well as a predominance of MWU structures should guarantee a specialist discourse. In [13], we built a classification tool that is able to distinguish, for French and Japanese languages, between expert and non-expert documents. Classifier learning is a process which observes features from a set of texts that were manually classified, and computes a model of what would be the characteristics of an expert/non-expert document. Two well-known classifiers have been tested: SVMlight [18] and C4.5 [30]. The features are:

Structural The texts are html files that were queried with keywords using search engines. Meta-information such as title and html tags are examined.

Modal The relationship between authors and readers in LSP characterises the author's attitude toward the content of his assertion. In popular sciences, the author speaks directly to the reader, although in the case of scientific documents, most sentences are impersonal [25]. Modal markers such as pronouns, verbs and polite forms are considered.

Lexical Several lexical criteria such as bibliographies and bibliographical quotations are specific to scientific documents as well as morphological units such as Greek-Latin roots.

The classifier has been trained on a corpus of the medical domain that deals with diabetes and nutrition, and evaluated on a corpus still in the medical domain but on the topic of breast cancer. The results show that French and Japanese documents are well classified, whatever the classifier, with a precision higher than 75 % and a recall higher than 75 % too.

3 Contextual Information Method

The main work in the compilation of a specialized lexicon from comparable corpora is based on lexical context analysis and relies on the simple observation that a word and its translation tend to appear in the same lexical contexts. The basis of this observation consists in identifying "First-order affinities" [15, p. 279] for each source and target language: "*First-order affinities describe what other words are likely to be found in the immediate vicinity of a given word*". These affinities can be represented by vectors, and each vector element represents a word which occurs within the window of the word to be translated (for instance a seven-word window approximates syntactic dependencies). A translation is obtained by comparing the source context vector to each translation candidate vector after having translated each element of the source vector with a general dictionary.

This method, known as the *contextual information method*, consists of the following four steps [12, 31]:

1. We collect all the words in the context of each word i and count their occurrence frequency in a window of n words around i. For each word i of the source and the target languages, we obtain a context vector v_i which groups the set of co-occurrence words j associated with the number of times that j and i occur together occ^i_j. In order to identify specific words in the lexical context and to reduce word-frequency effects, we normalize context vectors using an association score such as Mutual Information [10] or Log-likelihood [9], as shown in Eqs. 1 and 2 and in contingency Table 1 (where occ stands for 'occurrence frequency').
2. Using a bilingual dictionary, we translate the words of the source context vector. If the bilingual dictionary provides several translations for a word, we consider all of them but weight the different translations according to their frequency in the target language.
3. For a word to be translated, we compute the similarity between the translated context vector and all target vectors through vector distance measures such as Cosine [34] or weighted Jaccard [16] (see Eqs. 3 and 4 where $assoc^i_j$ stands for 'association score').
4. The candidate translations of a word are the target lexical units closest to the translated context vector according to vector distance.

$$MI(i, j) = log \frac{a}{(a+b)(a+c)} \tag{1}$$

$$\begin{aligned} \lambda(i, j) = {} & a\log(a) + b\log(b) + c\log(c) + d\log(d) \\ & +(a+b+c+d)\log(a+b+c+d) - (a+b)\log(a+b) \\ & -(a+c)\log(a+c) - (b+d)\log(b+d) - (c+d)\log(c+d) \end{aligned} \tag{2}$$

$$Cosine^{v_k}_{v_l} = \frac{\sum_t assoc^l_t \, assoc^k_t}{\sqrt{\sum_t {assoc^l_t}^2}\sqrt{\sum_t {assoc^k_t}^2}} \tag{3}$$

Table 1 Contingency table

	j	$\neg j$
i	$a = occ(i, j)$	$b = occ(i, \neg j)$
$\neg i$	$c = occ(\neg i, j)$	$d = occ(\neg i, \neg j)$

$$Weighted\ Jaccard_{v_l}^{v_k} = \frac{\sum_t min(assoc_t^l, assoc_t^k)}{\sum_t max(assoc_t^l, assoc_t^k)} \tag{4}$$

In a specialized domain, this approach presents two drawbacks. On the one hand, the context vectors do not discriminate very much due to the small amount of data available. In order to circumvent this, we suggest a contribution that strengthens the representativeness of the lexical contexts based on domain-specific vocabulary, called anchor points, notably the neoclassical compounds and the transliterations. On the other hand, the translation of the elements of the context vectors, which depends on the coverage of the bilingual dictionary vis-à-vis the corpus, is more difficult for a specialized domain because of the lack of specialized translations in the dictionary. In this case, we propose to make use of structural characteristics of the documents comprising the comparable corpus to extract parallel sentences with a high degree of quality in order to build a specialized bilingual lexicon directly correlated to the specialized vocabulary of the comparable corpus.

4 Anchor Points-Based Contextual Information Method

The first direction to improve the quality of the alignment of the contextual information method is to introduce depth in flat context-vectors, relying on selected terms that are more relevant than others called anchor points.

4.1 Specialised Vocabulary as Anchor Points

To be usable in the automatic process of bilingual lexicon extraction, anchor points need to have the following three properties:

1. They must be easily identified.
2. They must be relevant, regarding corpora topics.
3. They should not be ambiguous (no polysemy).

We propose the hypothesis that we can rely on those words to improve the discriminative strength of context-vector and therefore improve the quality of results obtained with the direct approach on small corpora. The first property allows us to use them in an automatic process. The second and third properties ensure that those anchor points are relevant, in other words, able to characterise the specialised terms we are trying to translate. They also ensure that no additional ambiguities are introduced. This particular vocabulary, that is able to characterise efficiently the specialised terms we are trying to translate, is composed of transliterations and scientific compounds.

We call *transliteration* a loan term, from one source language, which has been adapted to fit the target language speech sound and scripts (by extent, we call *transliteration* the relation between the source term and the target term). [29] shows that transliteration are pro-eminent in the Japanese language, and that they provide many links between Japanese and other languages, especially concerning English and French. Furthermore, they show that Japanese transliterations reflect specialised vocabulary used in a document. Finally, Japanese transliterations are easy to identify, since they are written using a set of symbols mostly dedicated to foreign terms, the *katakanas*. Japanese transliterations are for the most part adapted from English, but can be aligned with French terms, since French and English share a large common vocabulary. For example, the Japanese term インスリン /`i-n-su-ri-n` can be aligned with English *insulin* and with French *insuline*.

We also studied *scientific compounds* that are words built with specific roots [26]. These compounds are characteristic of a specialised vocabulary, especially in medical documents [20, 27]. For instance, [5] studying the automatic translation of medical vocabulary observe that biomedical terms are built on common Greek and Latin roots, and their derivations are consistent. Therefore, they seem to be relevant anchor points in the corpora we are using. Moreover, they can easily be identified from their morphology in different languages such as French and English.

4.2 Method

In order to strengthen the significance of context vectors by using anchor points, we chose to modify the contextual information method by giving more weight to the anchor points when calculating the association between the head of a vector and its elements. One way to do so is to dispatch association scores of non-highlighted terms on highlighted terms. That means we lower non-highlighted element scores and give them back to highlighted elements in order to keep a balanced overall score among context-vectors, see Eqs. 5–7. In these equations, AP is the set of anchor points used ($|AP|_l$ the number of anchor points found in the context-vector l and $|\neg AP|_l$ the number of other elements), $assoc^l_j$ is the association measure of element j in the context-vector l.

$$assoc\_weighted^l_j := assoc^l_j + \beta,\ if\ j \in AP \tag{5}$$

$$assoc\_weighted^l_j := assoc^l_j - offset_l,\ if\ j \notin AP \tag{6}$$

$$offset_l := \frac{|AP|_l}{|\neg AP|_l} \times \beta \tag{7}$$

The β parameter is used to calibrate the importance given to the highlighted elements. Thus, the overall weight (sum of all association scores for all items of a given vector) is equal before and after balancing. This modification of association measures implies that, if a pair of anchor points (source term and its translation) is to be found between two compared vectors, their similarity score will increase. On the other hand, if an anchor point is to be found in only one of two compared vectors, their similarity score will decrease. The last step of the direct approach is to compare translated source context-vectors with target context-vectors. If an anchor point is not transferred from the source language to the target language at the translation step, its discriminative power will be lost at the similarity computation step.

4.3 Experimental Test Bed

In this experiment, we explore the Web in order to compile an English/French/Ja-panese comparable corpus. Documents selected refer to *diabetes* and *nutrition* and are all of *scientific* discourse ("experts addressing experts"; [28], p. 36). Documents were manually extracted, following search engine results or using PubMed[3] for the English part. We converted those documents into text and cleaned them (manually removing non-informative parts such as *References*, which are frequent in scientific documents). We obtained 257,000 words for the French corpus, 235,000 words for the Japanese corpus and 250,000 words for the English corpus.

The French-Japanese bilingual dictionary required for the translation phase is composed of four dictionaries which are freely available from the Web,[4] and of a French-Japanese Scientific Dictionary (1989). It contains 173,156 entries, of which 114,461 are single word terms (SWT), with an average of 2.1 translations per entry. We used the JMDict for English/Japanese[5] which is freely available under a Creative Commons (By-SA) licence. We completed it with lists of technical terms from different domains: a list of technical terms compiled by the Japanese Ministry of Education and the National Institute of Informatics (Tokyo)[6] and the *Dictionary of Technical Terms* [19]. Overall, it contains 589,946 entries (unique words) with an average of 2.3 translations per entry and only 49,208 SWT.

To evaluate the quality of our method, we built lists of known translations. We selected the most frequen French words ($N_{occ} > 50$) for which a Japanese translation was available. Among those translations, we selected the most frequent Japanese words ($N_{occ} > 50$) in order to obtain a 98 element test list. We proceeded in the same way with English/Japanese corpus and obtained a 99 element test list. This protocol for building an evaluation term list is quite similar to the one presented in [4]. They used $N_{occ} > 100$ the source language, and $N_{occ} > 60$ the target language, to compile a test set of 95 words in an English/French comparable corpus.

---

[3] http://www.ncbi.nlm.nih.gov/PubMed/

[4] http://kanji.free.fr; http://quebec-japon.com/lexique/index.php?a=index\&d=25; http://dico.fj.free.fr/index.php; http://quebec-japon.com/lexique/index.php?a=index\&d=3

[5] http://www.csse.monash.edu.au/~jwb/j_jmdict.html

[6] http://sciterm.nii.ac.jp/cgi-bin/reference.cgi

We adopted a tool to perform automatic transliterations between English and Japanese [36]. This tool, based on the Markov chains, gives good results for English/Japanese; it generates a set of potential transliterations for a given katakana or English input. Output is then compared with the target vocabulary in the comparable corpora. Although direct French–Japanese transliterations are quite rare, a lot of English–Japanese transliterations can also be aligned with French vocabulary, due to cognate relations between French and English. We first used a specialised French–Japanese transliteration detection tool but obtained poor results (especially concerning false-positive alignments). We eventually decided to fall back on the tool used for English/Japanese. Before processing, we withdrew every diacritic specific to the French language. Using this tool, we obtained 589 pairs of transliteration for English/Japanese and 526 for French/Japanese. In order to have valuable anchor points (that is, able to be transferred from one language to another at the translation step of the direct approach), we added detected transliteration to bilingual resources.

We extracted scientific compounds using a list of 606 medical suffixes and prefixes used in English.[7] The process is quite simple: we compile regular expressions for every suffix and prefix and have them matched to the bilingual dictionaries used. The words extracted are kept with their Japanese translation. Such pairs are then used as anchor points in the alignment process. This list, dedicated to the English language can easily be adapted to French (in accordance with the [6] observation). We draw our inspiration from this work to write some simple conversion rules. For example, the `-y` suffix in English (as in *psychology*) corresponds to the—i.e. suffix in French (as in *psychologie*). After adapting rules to the French language, we performed the same extraction process as with English using the French dictionary, with the converted list of prefixes/suffixes. Some suffixes and prefixes are very productive (especially the `a-` prefix) and corresponding extracted terms are not necessarily built from this root. All suffixes and prefixes generating more than 1,000 pairs on bilingual resources were therefore withdrawn. They are however quite rare, only 12 have been discarded for English, and 17 for French. We obtained 17,210 scientific compounds in English (60,341 translation pairs, linguistic resources often give more than one translation for a given word) and 8,254 in French (24,240 translation pairs). The difference comes from the nature of linguistic resources for English and French. When projected onto our corpora vocabulary, we obtained 604 scientific compounds for English (1,197 pairs of translation) and 819 for French (822 translation pair). Unlike transliterations, scientific compounds cannot be matched in Japanese using morphological or phonetic clues. That is why they are extracted directly from bilingual resources. This also ensures that extracted scientific compound pairs are translated.

[7] http://www.medo.jp/a.htm

4.4 Experimental Results

In order to evaluate the influence of anchor points, three kinds of experiments were carried out on English/Japanese and French/Japanese alignments:

(a) direct approach (controlled experiment)
(b) taking into account automatically detected transliterations
(c) taking into account automatically extracted scientific compounds

All experiments were run on the same set of context-vectors (before the normalisation process, which is experiment-dependent), and comparisons were made between results obtained with equivalent parameters (same window size for building context-vectors, same similarity measure and equivalent association measures). We used the Cosine measure for similarity and the Log-Likelihood for the association measure. The *term frequency limit* is set to three for all experiments (it means that a word must appear three times or more in the neighbourhood of a term to be a part of its context-vector). The term lists used for evaluation are the same for all experiments. The window size for building a context vector is set to 25 words before and after the word to characterise.

The results shown here are the best that we obtained with the controlled experiments, compared with other experiments with the same set of parameters. Top_X indicates the number of correct translations found in the X first candidates output by the alignment process. Table 2 shows Top_1 and Top_{10} results for (a), (b) and (c) experiments (improvement between brackets). The results for a controlled experiment (exp. a) are quite similar to those obtained by [4]). In the case of English, the improvement when using anchor points is important: it reaches 18.2 % when using transliteration (exp. b—Top_1) and scientific compounds (exp. c—Top_1). The improvement is not that important in the case of French/Japanese alignments. It is null for Top_1 when using transliterations and reaches 10 % when using scientific compounds. This can be easily explained by the lowest quality of automatically extracted anchor points, especially concerning transliterations between Japanese and French. We think that it is not relevant to combine information brought by transliterations with information brought by scientific compounds. Indeed, those classes are barely related and are taken into account for specific, independent reasons. However, we still ran the experiment using both classes as anchor points and observed that the improvement is almost the same as when using scientific compounds alone. We showed that using anchor points can lead to an improvement of the direct approach method, for $Top_1 1$ and $Top_1 0$. Figure 2 displays the evolution of results, between the controlled experiment

Table 2 Alignment results for French/Japanese and English/Japanese ($\beta = 8$)

Experiment	a	b	c
En/Jp (Top_1)	17.1 %	20.2 % [18.2 %]	20.2 % [18.2 %]
En/Jp (Top_{10})	36.3 %	39.3 % [8.2 %]	40.4 % [11.2 %]
Fr/Jp (Top_1)	20.4 %	20.4 % [0.0 %]	22.4 % [10.0 %]
Fr/Jp (Top_{10})	36.7 %	37.8 % [2.8 %]	38.8 % [5.6 %]

and the experiment using scientific compounds in French/Japanese alignments. This figure shows all correct translations found in both experiments depending on their rank (from Top_1 to Top_{100}—ordinate) and their similarity score (on the abscissa).

In Fig. 2, the hollow triangle indicates translations that were found in the controlled experiment and can not be found with anchor points. On the contrary, the black triangle indicates translations found with anchor points that were not found in the controlled experiment. Each thin arrow displays the evolution of a translation found in both experiments. The beginning is the position of the translation in the controlled experiment, the end indicates its position when using anchor points. Finally, this thick arrows display the sum of all evolution, for each band indicated by horizontal dotted lines (Top_1 to Top_{10}, Top_{10} to Top_{20}, Top_{20} to Top_{50} and Top_{50} to Top_{100}). The results are interesting: they show that there are almost as many missing translations as new translations introduced in experiments. Moreover, the arrows show that there is an average improvement of correct translation ranking. This is especially the case for initially badly ranked translations (Top_{50} to Top_{100}). Their rank is highly improved, as indicated by the sum of evolutions for this band. This observation is available for other bands, although it is less impressive. However, initially well-ranked translations are less likely to be improved, but they are not penalised (even though their similarity scores decrease).

These observations complete the previous results: they show that correct translation candidates are re-ordered to better ranks when using anchor points, even though Top_1 and Top_{10} improvement are not that impressive. We ran a *t-test* [17] on those results. We settle as null hypothesis that using anchor points does not lead to an improvement in ranking correct translation candidates. Results of the *t-test* ($t = 1,8694$; $p = 0,0333$) allow us to reject the null hypothesis with 95 % confidence (the *Wilcoxon* test returned a *p-value* of 0.032). Those statistical tests also enable us to reject the null hypothesis in the cases of English/Japanese alignments,

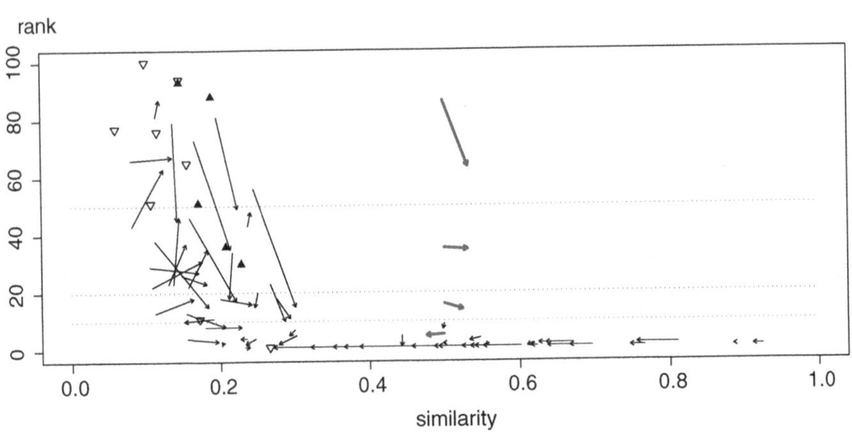

Fig. 2 Rank and similarity score of correct translations for French/Japanese alignments, with and without anchor points Scientific compounds—$\beta = 8$

using transliterations and scientific compounds, but not in the case of French/Japanese alignments using transliterations. This is once again probably due to the bad quality of automatically detected transliterations.

5 Contextual Information Method Enhanced with Parallel Sentences

Another direction to improve the quality of the alignment of the contextual information method is to improve the ratio of transfer of elements of the vector context of the word to be translated. Our assumption is that this problem can be partially circumvented by combining a general bilingual dictionary with a specialized bilingual dictionary and that it is possible to find in the comparable corpus some parallel sentences to build this specialized bilingual dictionary.

5.1 Method

Parallel sentence extraction from comparable corpora has been studied by a number of researchers [1, 3, 11, 21, 24, 33, 37, among others] and several systems have been developed such as BITS (Bilingual Internet Test Search) [21], PTMiner (Parallel Text Miner) [3], and STRAND (Structural Translation Recognition for Acquiring Natural Data) [33]. Their work relies on the observation that a collection of texts in different languages, composed independently and based on sharing common features such as content, domain, genre, register, sampling period, etc. contains probably some sentences with a source text-target text relationship. Based on this observation, dynamic programming [37], similarity measures such as Cosine [11] or word and translation error ratios [1], or maximum entropy classifiers [24] are used for discovering parallel sentences.

Although our purpose is similar to these works, the amount of data required by these techniques makes them ineffective when applied to specialized comparable corpora used to discover parallel sentences. In addition, our focus is not to propose a new technique for this task but to study how parallel sentences extracted from a comparable corpus can improve the quality of the candidate translations. For these reasons, we propose to make use of the structural characteristics of the documents comprising the comparable corpus to automatically extract parallel sentences. In fact, specialized comparable corpora are generally constructed via the consultation of specialized Web portals. For instance, [4] use CISMeF[8] for building the French part of their comparable corpora and CliniWeb[9] for the English part, and [7] use

[8] http://www.chu-rouen.fr/cismef/

[9] http://www.ohsu.edu/cliniweb/

documents extracted from MEDLINE[10] to build a German/English comparable corpus. Consequently, the documents collected through these portals are often scientific papers. Moreover, when the language of these papers is not English, the paper usually comprises an abstract, keywords and title in the native language and their translations in the English language. These characteristics of a scientific paper is useful for the efficient extraction of parallel sentences or word translations from the documents forming a specialized comparable corpus for which one part will inevitably be in English.

Here, we use the *Uplug*[11] collection of tools for alignment [35] to extract translations from our parallel sentences. The output of such a tool is a list of aligned *parts of sentences*, that has to be post-process and filtered in our case. We clean the alignment with a simple yet efficient method in order to obtain only word translations. We associate every source word from a source sequence with every target word from the target sequence. As an example, *uplug* efficiently aligns the English word *breast cancer* with the French word *cancer du sein*. We obtain the following lexical alignment:

- cancer (fr) → (en) breast, cancer
- du (fr) → (en) breast, cancer
- sein (fr) → (en) breast, cancer

With more occurrences of the French word *cancer*, we are able to align it with the English words {breast, cancer, cancer, cancer, the, of, breast, cancer}. We can then filter such a list by counting the translation candidates. In the previous example, we obtain: cancer (fr) → breast/2, the /1, of/1, cancer/4. The English word *cancer* is here the best match for the French word *cancer*. In many cases, only one alignment is obtained. For example, there is only one occurrence of the French word *chromosome*, aligned with the English word *chromosome*. In order to filter translation candidates, we keep 1:1 candidates if their frequencies are comparable in the original corpus. We keep the most frequent translation candidates (in the previous example, *cancer*) if their frequencies in the corpus are also comparable. This in-corpus frequency constraint is useful for discarding candidates that appear in many alignments (such as functional words). The criterion for frequency acceptability is: $min(f_1, f_2)/max(f_1, f_2) > 2/3$ with f_1 and f_2 the frequency of words to be aligned in the parallel corpus.

5.2 Experimental Test Bed

In this study, the documents comprising the French/English specialized comparable corpus were taken from the medical domain within the sub-domain of 'breast cancer'. These documents had been automatically selected from the Elsevier

[10] http://www.ncbi.nlm.nih.gov/PubMed

[11] http://stp.lingfil.uu.se/~joerg/Uplug/

website[12] among the articles published between 2001 and 2008 for which the title or the keywords of the articles contain the multi-word term 'cancer du sein' in French and 'breast cancer' in English. We thus collected 130 documents in French and 118 in English and about 530,000 words for each language. The documents comprising the French/English specialized comparable corpus had been normalised through the following linguistic pre-processing steps: tokenisation, part-of-speech tagging, and lemmatisation. Next, the function words were removed and the words occurring less than twice in the French and the English parts were discarded. Finally, the comparable corpus comprised about 7,400 distinct words in French and 8,200 in English. Since the 130 French documents previously collected are scientific papers, each document contains a French abstract which is accompanied by its English translation. We exploit this structural characteristic of the French documents in order to build a small specialized parallel corpus directly correlated to the sub-domain of 'breast cancer' involved in the comparable corpus.

From this small specialized parallel corpus, we used the *Uplug* toolkit as explained in the previous section to build a French/English specialized bilingual lexicon from the parallel corpus. This lexicon, called breast cancer dictionary (BC dictionary) in the remainder of this article, is composed of 549 French/English single words.

In this study, we used four types of bilingual dictionary: (i) the Wiktionary[13] free-content multilingual dictionary, (ii) the ELRA-M0033[14] professional French /English bilingual dictionary, (iii) the MeSH[15] metha-thesaurus, and (iv) the BC dictionary. Table 3 shows the main features of the dictionaries, namely: the number of distinct French single words in the dictionary (# SWs dico.), the number of distinct French single words in the dictionary after projection on the French part of the comparable corpus (# SWs corpus), and the number of translations per entry in the dictionary (# TPE). For instance, 42 % of the French context vectors could be translated with the Wiktionary (3,099/7,400).

In order to build the terminology reference list required to evaluate the performance of the alignment programs, we selected 400 French/English SWTs from the UMLS[16] meta-thesaurus and the *Grand dictionnaire terminologique*.[17] We kept only

Table 3 Main features of the French/English dictionaries

Name	# SWs dictionaries	# SWs corpus	# TPE
Wiktionary	20,317	3,099	1.8
ELRA	50,330	4,567	2.8
MeSH	18,972	833	1.6
BC	549	549	1.0

[12] http://www.elsevier.com

[13] http://www.wiktionary.org/

[14] http://www.elra.info/

[15] http://www.ncbi.nlm.nih.gov/mesh

[16] http://www.nlm.nih.gov/research/umls

[17] http://www.granddictionnaire.com/

the French/English pair of SWTs which occur more than five times in each part of the comparable corpus. As a result of filtering, 122 French/English SWTs were extracted.

5.3 Experimental Results

In order to evaluate the influence of the parallel corpus-based bilingual lexicon induced from the comparable corpus on the quality of comparable corpus-based bilingual terminology extraction, four experiments were carried out. For each experiment, we changed the bilingual dictionary required for the translation phase of the contextual information method:

1. The first experiment uses only the Wiktionary. Since the coverage of the Wiktionary from the comparable corpus is small (see Table 3), the results obtained with this dictionary yield a lower boundary.
2. The second experiment uses the Wiktionary added to the BC dictionary. This experiment attempts to verify the hypothesis of this study.
3. The third experiment uses the Wiktionary added to the MeSH thesaurus. This experiment attempts to determine whether a specialised dictionary (in this case the MeSH) would be more suitable than a specialized bilingual dictionary (in this case the BC dictionary) directly extracted from the corpus.
4. The last experiment uses only the ELRA dictionary. Since the coverage of the ELRA dictionary from the comparable corpus is the best (see Table 3), the results obtained with this one yield a higher boundary.

Table 4 shows the coverage of the four bilingual lexical resources involved in the previous experiments in the comparable corpus. The first column indicates the number of single words belonging to a dictionary found in the comparable corpus (# SWs corpus). The other column indicates the coverage of each dictionary in the ELRA dictionary (Coverage ELRA). Here, 98.9 % of the single words belonging to the Wiktionary are included in the ERLA dictionary whereas less than 95 % of the single words belonging to the Wiktionary+BC and Wiktionary+MeSH dictionaries are included in the ERLA dictionary. Moreover, the MeSH and BC dictionaries are two rather distinct specialized resources since they have only 117 single words in common.

In the experiments reported here, the size of the context window n was set to 3 (i.e. a seven-word window), the association measure was the Mutual Information and

Table 4 Coverage of the bilingual lexical resources in the comparable corpus

Name	# SWs corpus	Coverage ELRA (%)
Wiktionary	3,099	98.8
Wiktionary + BC	3,326	94.8
Wiktionary + MeSH	3,465	94.9
ELRA	4,567	100

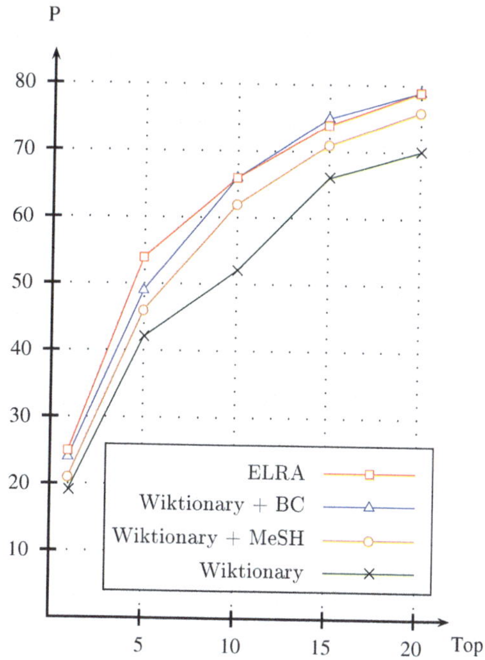

Fig. 3 Precision of translations found according to the rank

the distance measure the Cosine. Other combinations of parameters were assessed but the previous parameters turned out to give the best performance.

Figure 3 summarises the results obtained for the four experiments for the terms belonging to the reference list according to the French to English direction. As one could expect, the precision of the result obtained with the ELRA dictionary is the best and the precision obtained with the Wiktionary is the lowest. For instance, the ELRA dictionary improves the precision of the Wiktionary by about 14 points for the Top 10, and 9 points for the top 20. These results confirm that the coverage of the dictionary is an important factor in the quality of the results obtained. Now, when you add the BC dictionary to the Wiktionary, the results obtained are also much better than those obtained with the Wiktionary alone and very similar to those obtained with the ELRA dictionary alone (without taking into account the top 5). This result suggests that a standard general language dictionary enriched with a small specialized dictionary can replace a large general language dictionary.

Furthermore, this combination is more interesting than the combination of the MeSH dictionary with the Wiktionary. Since the BC dictionary is induced from the corpus, this dictionary is directly correlated to the theme of breast cancer involved in the corpus. Consequently the BC dictionary is more suitable than the MeSH dictionary (i) even if the MeSH dictionary specializes in the medical domain and (ii) even if more words in the comparable corpus are found in the MeSH dictionary than in the BC dictionary.

This last observation should make us relativize the claim: the greater the number of context vector elements that are translated, the more discriminating the context vector will be for selecting translations in the target language. We must also take into account the specificity of the context vector elements in accordance with the theme of the documents making up the corpus studied in order to improve bilingual lexicon extraction from specialized comparable corpora.

6 Conclusion

In this paper, we have studied the problem of compiling bilingual lexicon from specialized comparable corpora. First, we explained that the comparability of the comparable corpora is guaranteed by the monolingual comparability of the monolingual texts. Then, we turned to the contextual information method that concentrates on bilingual lexicon extraction and showed its limits for a specialized domain. On the one hand, the context vectors do not discriminate very much due to the small amount of data available. In order to circumvent this, we have proposed to strengthen the representativeness of the lexical contexts based on domain-specific vocabulary, called anchor points, notably the neoclassical compounds and the transliterations. On the other hand, the translation of the elements of the context vectors, which depends on the coverage of the bilingual dictionary vis-à-vis the corpus, is more difficult for a specialized domain because of the lack of specialized translations in the dictionary. In this case, we have proposed to use the parallel sentences present in the comparable corpus in order to build a specialized bilingual lexicon directly correlated to the specialized vocabulary of the comparable corpus. Our experiments illustrate that these two strategies are well-founded and show that candidate translations are of better quality. These different contributions are the first step in encouraging different approaches depending on whether you work in corpus-based specialized language or in corpus-based general language.

References

1. Abdul-Rauf, S., Schwenk, H.: On the use of comparable corpora to improve SMT performance. In: Proceedings of the 12th Conference of the European Chapter of the Association for Computational Linguistics (EACL'09), Athens, Greece, pp. 16–23 (2009)
2. Bowker, L., Pearson, J.: Working with Specialized Language: A Practical Guide to Using Corpora. Routledge, London/New York (2002)
3. Chen, J., Nie, J.Y.: Parallel web text mining for cross-language information retrieval. In: Proceedings of Recherche d'Information Assistée par Ordinateur (RIAO'00), Paris, France, pp. 62–77 (2000)
4. Chiao, Y.C., Zweigenbaum, P.: Looking for candidate translational equivalents in specialized, comparable corpora. In: Proceedings of the 19th International Conference on Computational Linguistics (COLING'02), Tapei, Taiwan, pp. 1208–1212 (2002)

5. Claveau, V.: Inférence de règles de réécriture pour la traduction de termes biomédicaux. In: Actes de la conférence Traitement automatique des langues naturelles (TALN'07) (2007)
6. Claveau, V.: Automatic translation of biomedical terms by supervised machine learning. In: Proceedings of the Language Resources and Evaluation Conference, (LREC'08), pp. 684–691 (2008)
7. Déjean, H., Gaussier, E.: Une nouvelle approche à l'extraction de lexiques bilingues à partir de corpus comparables. Lexicometrica, Alignement lexical dans les corpus multilingues, pp. 1–22 (2002)
8. Déjean, H., Sadat, F., Gaussier, E.: An approach based on Multilingual Thesauri and model combination for Bilingual Lexicon Extraction. In: Proceedings of the 19th International Conference on Computational Linguistics (COLING'02), Tapei, Taiwan, pp. 218–224 (2002)
9. Dunning, T.: Accurate methods for the statistics of surprise and coincidence. Comput. Linguist. **19**(1), 61–74 (1993)
10. Fano, R.M.: Transmission of Information: A Statistical Theory of Communications. MIT Press, Cambridge (1961)
11. Fung, P., Cheung, P.: Mining very-non-parallel corpora: parallel sentence and Lexicon Extraction via Bootstrapping and EM. In: Lin, D., Wu, D. (eds.) Proceedings of Empirical Methods on Natural Language Processing (EMNLP'04), pp. 57–63. Barcelona, Spain (2004)
12. Fung, P., McKeown, K.: Finding terminology translations from non-parallel corpora. In: Proceedings of the 5th Annual Workshop on Very Large Corpora (VLC'97), Hong Kong, pp. 192–202 (1997)
13. Goeuriot, L., Morin, E., Daille, B.: Compilation of specialized comparable corpora in French and Japanese. In: Proceedings of the 2nd Workshop on Building and Using Comparable Corpora: from Parallel to Non-Parallel Corpora (BUCC'09), Singapore, pp. 55–62 (2009)
14. Grabar, N., Krivine, S.: Application of cross-language criteria for the automatic distinction of expert and non expert online health documents. In: Bellazzi, R., Abu-Hanna, A., Hunter, J. (eds.) AIME. Lecture Notes in Computer Science, vol. 4594, pp. 252–256. Springer, Berlin (2007)
15. Grefenstette, G.: Corpus-derived first, second and third-order word affinities. In: Proceedings of the 6th Congress of the European Association for Lexicography (EURALEX'94), Amsterdam, The Netherlands, pp. 279–290 (1994)
16. Grefenstette, G.: Explorations in Automatic Thesaurus Discovery. Kluwer Academic Publisher, Boston (1994)
17. Harris, M.B.: Basic Statistics for Behavioral Science Research, 2nd edn. Allyn & Bacon, Boston(1998)
18. Joachims, T.: Learning to Classify Text Using Support Vector Machines. Kluwer Academic Publishers, Boston (2002)
19. Kotani, T., Kori, A.: Dictionary of Technical Terms. Kenkyusha, Tokyo (1990)
20. Lovis, C., Baud, R., Michel, P.A., Scherrer, J.R., Rassinoux, A.M.: Building medical dictionaries for patient encoding systems: a methodology. Lect. Notes Comput. Sci. **1211**, 373–380 (1997)
21. Ma, X., Liberman, M.Y.: Bits: a method for bilingual text search over the web. In: Proceedings of Machine Translation Summit VII, Kent Ridge Digital Labs, National University of Singapore (1999)
22. McEnery, A. and Xiao, Z.: Parallel and comparable corpora: What is happening? In: Anderman, G., Rogers, M. (eds.) Incorporating Corpora: The Linguist and the Translator. Multilingual Matters, Clevedon (2007)
23. Morin, E., Daille, B., Takeuchi, K., Kageura, K.: Bilingual terminology mining—using brain, not brawn comparable corpora. In: Proceedings of the 45th Annual Meeting of the Association for Computational Linguistics (ACL'07), Prague, Czech Republic, pp. 664–671 (2007)
24. Munteanu, D.S., Marcu, D.: Improving machine translation performance by exploiting non-parallel corpora. Comput. Linguist. **31**(4), 477–504 (2005)
25. Nakao, Y., Goeuriot, L., Daille, B.: Multilingual modalities for specialied languages. Terminology **16**(1), 51–76 (2010)

26. Namer, F.: Morphosémantique pour l'appariement de termes dans le vocabulaire médical: approche multilingue. In: Actes de la conférence Traitement Automatique des Langues Naturelles (TALN'05), pp. 63–72 (2005)
27. Namer, F., Zweigenbaum, P.: Acquiring meaning for french medical terminology: contribution of morphosemantics. In: Fieschi, M., Coiera, E., Li, Y.C.J. (eds.) Studies in Health Technology and Informatics, vol. 107, pp. 535–539 (2004)
28. Pearson, J.: Terms in Context. John Benjamins, Amsterdam (1998)
29. Prochasson, E., Kageura, K., Morin, E., Aizawa, A.: Looking for transliterations in a trilingual English, French and Japanese specialised comparable corpus. In: Proceedings of the 1st Workshop on Building and Using Comparable Corpora, Language Resources and Evaluation Conference (LREC'08), pp. 83–86 (2008)
30. Quinlan, J.R.: C4.5: Programs for Machine Learning. Morgan Kaufmann, San Francisco (1993)
31. Rapp, R.: Identify word translations in non-parallel texts. In: Proceedings of the 35th Annual Meeting of the Association for Computational Linguistics (ACL'95), Boston, MA, USA, pp. 320–322 (1995)
32. Rapp, R.: Automatic identification of word translations from unrelated English and German corpora. In: Proceedings of the 37th Annual Meeting of the Association for Computational Linguistics (ACL'99), College Park, MD, USA, pp. 519–526 (1999)
33. Resnik, P., Smith, N.A.: The Web as a parallel corpus. Comput. Linguist. **29**(3), 349–380 (2003)
34. Salton, G., Lesk, M.E.: Computer evaluation of indexing and text processing. J. Assoc. Comput. Mach. **15**(1), 8–36 (1968)
35. Tiedemann, J.: Recycling translations—extraction of Lexical data from parallel corpora and their application in natural language processing. Ph.D. thesis, Studia Linguistica Upsaliensia 1 (2003)
36. Tsuji, K., Sato, S., Kageura, K.: Evaluating the effectiveness of transliteration and search engines in bilingual proper name identifications. In: Proceedings of the 11th Annual Meeting of the Association for Natural Language Processing, pp. 352–355 (2005)
37. Yang, C.C., Li, K.W.: Automatic construction of English/Chinese parallel corpora. J. Am. Soc. Inf. Sci. Technol. **54**(8), 730–742 (2003)
38. Zanettin, F.: Bilingula corpora and the training of translators. META **43**(4), 616–630 (1998)

The Place of Comparable Corpora in Providing Terminological Reference Information to Online Translators: A Strategic Framework

Kyo Kageura and Takeshi Abekawa

Abstract This paper examines the status of comparable corpora as potential terminological resources with special reference to the applicational framework of helping online translators. In the past 15 years, we have witnessed great advances in bilingual term extraction technologies based both on parallel and comparable corpora. The use of comparable corpora is widely held to be especially important, because not many parallel corpora are available in many language pairs. However, human language practitioners, including online translators, do not make much use of terminological resources constructed using automatic methods; there seems to be a gap between what can be provided through corpus-based automatic extraction methods and what translators actually require. Against this backdrop, this paper first clarifies online translators' requirements for terminology resources. Based on this clarification, the paper examines what should be taken into account in the use of comparable corpora for bilingual term extraction if the resultant terminology resources are to be really used by translators. The discussion in this paper is deductive rather than empirical, based on the authors' experience in talking with online translators in the course of developing the integrated translation-hosting and translation-aid site Minna no Hon'yaku (MNH: translation of/by/for all) since 2005 (the site has been open to the public since April 2009).

Keywords Online translator · Terminological resources · Comparable corpora · Web

---

K. Kageura (✉)
University of Tokyo, Tokyo113-0033, Japan
e-mail: kyo@p.u-tokyo.ac.jp
http://panflute.p.u-tokyo.ac.jp/homedirkyo/

T. Abekawa
National Institute of Informatics, Tokyo101-8430, Japan

S. Sharoff et al. (eds.), *Building and Using Comparable Corpora*,
DOI: 10.1007/978-3-642-20128-8_15, © Springer-Verlag Berlin Heidelberg 2013

1 Introduction

With the rapid growth of many specialised domains in recent years, new terms are constantly being created. One of the lexical classes in which translators' reference needs are greatest is new technical terms, which are often not covered by existing terminological dictionaries [29]. While the need for up-to-date terminological reference resources is high, it is often said that manual compilation and updating of terminological resources cannot keep up with the pace of terminological development in many domains. This recognition has led to a proliferation of research into automatic extraction of bilingual terms from textual corpora, especially since the mid-1990s. While parallel corpora are used in some studies [21, 34, 62], many studies use comparable corpora as basic sources for bilingual term extraction [12, 14, 19, 20, 22, 35, 39, 40, 42, 43, 47, 50, 55], because they are available in a wider range of language pairs, text types, registers, domains, etc. on a much larger scale than parallel corpora. The techniques for bilingual term extraction from comparable corpora have steadily advanced and their performance has generally improved, while the extraction of terms for new language pairs is being carried out. In a parallel move, many investigations into methods for constructing comparable corpora have taken place and shown great advances, especially in exploiting documents on the Web [5, 13, 24, 31, 36, 58].

However, while comparable corpora themselves have been and are used in the human translation process [6, 56, 64], there are few reported cases in which bilingual terms extracted from comparable corpora are used by human translators in practical translation situations in a manner comparable to manually constructed terminological reference resources. This also holds for online translators. Here, we define "online translators" as people translating online or electronic texts, publishing their translations online, and relying on online resources in the process of translation. Although this definition is still somewhat vague and broad, it is sufficient for our purposes in this paper. Volunteer translators include those doing translation on a voluntary basis for NGOs, a group that relies more and more on online resources and tools in the process of translation. In discussions with several online translators using the translation-aid environment Minna no Hon'yaku (MNH: http://trans-aid.jp/) [60], we also found that in general they have a negative attitude towards having automatically extracted bilingual term pairs as part of the lineup of reference resources provided by the translation-aid system. As for comparable corpora, while technologies for constructing and exploring them have improved significantly, "until now there has been no agreement on the degree of similarity that documents in comparable corpora should have, or even agreement about the criteria for measuring parallelism and comparability" [54].

Given this situation, this paper examines the extraction of bilingual terms from comparable corpora and the possibility of providing automatically extracted bilingual terms as well as the use of comparable corpora within a concrete context of application, i.e. providing terminological reference resources to online translators, based on the belief that situating the discussion within a concrete framework will contribute to clarifying the potential of comparable corpora. Note that this paper is not concerned with the technical aspects of comparability, but concerned with the framework within which the concept of comparability is recognised, defined or discussed and how the framework is related to the usefulness of terminological resources. In the following, we first summarise online translators' requirements for terminological reference resources in Sect. 2. Section 3 examines the status of comparable corpora as information sources for bilingual terms. In Sect. 4, we will postulate a framework within which automatically constructed terminological resources can be practically used—together with some methods we adopted—and discuss the place of comparable corpora within this framework.

Throughout the paper, we treat online translators as a user group, characterised by their extensive use of online information as a reference source in the process of translation, although the general discussion is also applicable to other types of translators as well. We will focus on bilingual term pairs rather than multilingual term sets. When concrete methods we adopted are referred to, we assume English and Japanese as a language pair. Until we explicitly examine the definitions of comparable corpora in Sect. 3.2, we assume that comparable corpora are "sets of texts in different languages, that are not translations of each other," [8] which are somewhat comparable. As we limit our discussion to bilingual terms, the comparable corpora we refer to are naturally assumed to be bilingual comparable corpora.

2 The Nature of Terminological Reference Resources

2.1 Some Requirements for Reference Resources

In the field of library and information science, it is commonly understood that some of the most important characteristics of reference resources or materials are their authoritative nature and the comprehensiveness of their coverage in terms of their stated objective [17].

While the concept of authority and related concepts such as credibility, reliability and quality of information sources are sometimes used in a confused way, and the concept of authority is often narrowly understood as being related to the correctness of information contained in the sources in terms of truth [10, 18, 33, 46], it is first and foremost a socially defined concept. An authoritative resource is one with a social status that causes people to turn to it as a reference resource, and its authoritativeness is independent of its correctness or truthfulness, although it is unlikely that a reference resource which contained a great deal of factually incorrect information could attain such a status. This is analogous to a commercial product that is socially accepted. A well-known example is the competition between VHS and Betamax as a standard for video; the latter was widely held to be superior technologically, but it was VHS that prevailed in the end.

At this point, it is convenient to elaborate a little on what is meant here by "social". We use the word with the theoretical standpoint of social constructionism in mind [11], although in a more general sense. Throughout the paper, the term "social" is used to introduce or emphasise an aspect or process which involves discussion, understanding, recognition or agreement among people, as opposed to features defined at the level of language.

The concept of comprehensive coverage should be understood in the same way. The comprehensiveness of coverage is not defined in terms of the exhaustivity of information covered vis-à-vis the entire body of information in the world objectively defined, but rather is sustained by the social understanding that people can stop looking for information once they check a relevant resource which is held to be comprehensive.

Thus both authority and comprehensiveness socially establish limiting conditions for the lookup of reference resources. It should also be pointed out that the concepts of authority and comprehensiveness take effect one step before the concepts of truth or exhaustivity. For instance, Google is regarded as an authoritative search engine on the Web and is also regarded as comprehensive, in the sense that everybody is supposed to check Google before making an informed decision based on online information and also in the sense that Google lookup is considered to be sufficient. Whether the information obtained represents the truth or not is a different issue. It is analogous to the status of an established textbook. Every novice researcher should refer to a well-established textbook of a domain to grasp the general picture of that domain. Whether the textbook exhaustively covers the research topics of the domain or not is a different issue. At this level of understanding, authority and comprehensiveness constitute two sides of the same coin. Finally, note also that both concepts are attributed to the level of reference resources and not to the individual pieces of information contained in them.

2.2 Nature of Terminology and Terminological Reference Lookup

Maeda [37] relevantly states that the phenomena that lexicologists deal with are necessarily very concrete and anchored to the existing state of linguistic reality [37], which is reflected in the difference between lexicology and most other linguistic studies (though there are different definitions of lexicology [23, 25]; here we assume that lexicology is the study of vocabulary in a language). For instance, while the study on word formation is concerned with well-formedness or formal validity at levels such as the syntactic or the semantic and is not concerned with the existence of the given form in the socially shared lexicon of a language, lexicology is essentially concerned with what actually exists in a language in a given state and, even when it addresses the issue of potentiality, it is concerned with the realistic possibility of existence in a society rather than some sort of formal validity. Because terminology can be defined as a specialised lexicon, the study of terminology has the same traits as lexicology [27].

Put differently, lexicology and terminology are concerned with concrete language *entities*, while most other linguistic studies are concerned with *forms*, *characteristics*, *attributes*, *content*, etc. This helps us to understand the nature of lexicological and terminological reference lookup. These concern concrete entities, while in many other types of information seeking what is being looked up are attributes or characteristics (although there are borderline cases). One can check the grammaticality of a certain construction by checking similar expressions. On the other hand, one cannot make do by looking up the word "red" if one is looking for the translation of the word "brown" but cannot find it.

From a practical point of view, therefore, our reference to lexical items as a substance tends to be qualitatively different from looking up formal features of languages. Also, from the theoretical point of view, it is often said that there is a lexicological or terminological sphere and a textual sphere in language [53, 63], although it is quite often assumed that the textual sphere is primary and the lexicological or terminological sphere is secondary. One of the reasons why comparable corpora are used in the actual translation process but terminology extracted from these corpora is not is related to the nature of terminological reference lookup as described above. Incidentally, the information seeking assumed by IR is analogous to non-lexicological information lookup. This becomes clear when we note the evaluation scheme adopted by IR research. The fact that such evaluation measures as the F-measure in IR are understood to make sense reflects the recognition that relevant documents are replaceable, or that missed documents can be compensated for by retrieved documents which deal with similar topics. This does not hold for lexicological or terminological reference lookup.

Within lexicological reference lookup, we can typically recognise qualitative differences between the lookup of general words on the one hand and, say, proper names on the other. When looking up ordinary words, translators are mostly searching for meanings and other attributes. When looking up proper names, they are searching for is socially accepted or authentic expressions in the target language which cannot be replaced by other expressions even if the latter are synonymous. As terminology is situated somewhere between ordinary vocabulary and proper names [45, 48], terminology lookup is similarly situated somewhere between the two, and thus involves the aspect of searching for socially accepted concrete expressions when terminological reference resources are looked up.

2.3 Lookup of Terminological Resources by Online Translators

Translators' expectations for and behaviour in the face of terminological reference resources are established at the intersection of their requirements for the reference resources and the nature of terminological resources briefly discussed above. When dealing with terms, translators are concerned with looking up concrete entities and searching for socially accepted concrete expressions, rather than with such attributes as meanings. Incidentally, that terminologists are held to be important in industrial translations is at least partly related to this aspect of the nature of terminology.

The points made by Sato [51] in relation to reference resources for proper names hold for terminological reference resources as well [51]:

1. Translators first and foremost need and want to know whether a standard and established target language expression exists for a given proper name, or whether there is any kind of corresponding target language expression at all.
2. If they find a(n established) target language expression, they use it.
3. Only after they have satisfied themselves that no (established) target language expression exists or can be found for the proper name will they translate it themselves.

Reference lookup is related to the first two steps, which can be divided into two phases: the phase in which the existence or non-existence of target language expressions is checked; and the phase in which the degree of establishment of the found candidates is checked.

The actual behaviour of online translators in looking up terminological reference resources within the context of online translation can be summarised as follows (cf. [6, 7, 15, 29, 41]):

1. If established terminological dictionaries widely used in the domain to which the topic of the text being translated belongs are available, then translators will check them.
2. If translators cannot find terminological dictionaries or relevant entries in terminological dictionaries, then they will look up socially shared information sources which are explicitly established as reference resources, such as Wikipedia, etc.
3. If translators cannot find terminological dictionaries or relevant entries in terminological dictionaries but relevant and/or established corpora are available, then they will look up these corpora as reference resources.
4. If translators cannot find what they are looking for in these resources, then they will use Google to search the Internet—the equivalent of checking non-reference materials in libraries to look for terms in the off-line translation process.

These four steps are first and foremost related to checking whether a corresponding term in the target language exists for a given source language term, though the validation of whether the term is established is also satisfied in the case of looking up established terminological dictionaries. Using the two concepts introduced in Sect. 2.1, i.e. authority and comprehensiveness, established terminological dictionaries satisfy the need for authority, while Google search satisfies the need for comprehensiveness. For convenience, let us call the former Type A lookup and the latter Type B lookup. Lookup of Wikipedia and related resources is situated in between these two.

Incidentally, note that, quite apart from these situations, translators and translation groups conveniently look up terms automatically or semi-automatically extracted or retrieved from parallel corpora constituting of their past translations. We do not take into account this case in our discussion, because this class of information seeking belongs more to TM lookup and is qualitatively different from what we are concerned with in this paper.

3 Automatic Term Extraction and Comparable Corpora in Terminological Reference Lookup

3.1 The Place of Automatic Extraction of Bilingual Terms

The brief examination above has clarified the applicational settings in which automatic bilingual term extraction could potentially be used by online translators. There are basically two possibilities. Firstly, if we can use the methods of automatic bilingual term extraction in constructing established terminological dictionaries of a domain which would be widely accepted and used, then the dictionaries will be used by translators. Secondly, if the automatic extraction methods cover the range of bilingual terms translators can find by performing a Google search, the results will be used by translators as well. These two possibilities correspond, respectively, to Type A lookup and Type B lookup defined in Sect. 2.3.

The first choice is unrealistic if we assume fully automatic construction of bilingual terminology. That fully automatic methods cannot construct established terminological dictionaries is due as much to the fact that the construction of established terminological dictionaries is a social process, not a purely technical process, as to the fact that automatic methods still cannot achieve sufficient performance for the results to be used as is as terminological reference resources. However, just as concordancing and more advanced corpus exploration methods greatly helped in constructing dictionaries in general, we can reasonably expect automatic methods of bilingual term extraction from parallel or comparable corpora to be of great assistance in the systematic compilation of terminological dictionaries, if used properly. In relation to Type A lookup, therefore, the contribution of automatic bilingual term extraction is indirect from the point of view of online translators.

The use of automatic bilingual term extraction can be seriously considered in Type B lookup as well, in finding corresponding target language terms when they are not covered by existing terminological reference resources. We are here back to the oft-stated aim of automatic bilingual term extraction from comparable corpora, i.e. to provide bilingual terms which are not contained in existing terminological reference resources, but with a practical condition: automatic methods should be able to retrieve at least the same range of bilingual terms that translators can find heuristically through their own Google searches of information on the Web. Unlike in Type A lookup, where the concept of authority is involved, Type B lookup is at least theoretically more amenable to the direct exploration of automatic methods, because we have a level playing field in relation to the concept of comprehensiveness, i.e.

the information on the Internet is essentially open. The contribution of automatic methods would thus be more direct in Type B lookup for online translators than in Type A lookup.

3.2 Comparable Corpora and Terminological Resources

With the above pints in mind, let us now turn to examine a commonly stated framework for automatic bilingual term extraction from comparable corpora. The use of comparable corpora, mainly in contrast to parallel corpora, for automatic bilingual term extraction is based on, among others, the following perceptions and assumptions:

1. There are a greater number of language pairs for which comparable corpora are available than those for which parallel corpora are available. Parallel corpora for Tetun–Xhosa are not readily available but comparable corpora can be developed.
2. More comparable corpora are available for a given language pair than parallel corpora. While Japanese–Chinese parallel corpora are available, Japanese–Chinese comparable corpora are available on a larger scale.
3. To the extent that the corpora in two languages are comparable, they should contain a corresponding range of terms, the exploration of which would theoretically lead to successful extraction of bilingual terms.

Incidentally, it is sometimes also pointed out, in addition to these three merits, that comparable corpora can avoid the "inevitable distortion introduced by the translations of a parallel corpus" [4, 16, 26].

It is these three assumptions that need closer examination from the point of view of the practical settings we have discussed so far. To do so, let us start by clarifying the definition of comparable corpora. One of the standard definitions is:

> A comparable corpus is one which selects similar texts in more than one language or variety [16].

Bowker and Peason [8] give the following definition:

> [Comparable corpora are] sets of texts in different languages, that are not translations of each other [8].

Kilgariff [32], attempting to define the concept of comparability more clearly, answers the question "what makes comparable corpora comparable?" by stating:

> They should have roughly the same text type(s), covering the same subject matter, in the same proportions [32].

A few characteristics of comparable corpora can be immediately identified in these definitions, namely:

1. Comparable corpora consist of texts and belong to the textual sphere, which is almost a truism.

2. As such, comparability is defined by the characteristics or variety of aspects attributed to texts, as is manifested by such expressions as "text types".
3. Comprehensiveness as defined socially—whatever this may mean in relation to comparable corpora—is not a determining feature of comparable corpora, which are more a sample than the population.

Now, compare these characteristics with the nature of terminological reference resources, and the behaviour and requirements of online translators when looking for bilingual terms (as discussed in Sect. 2). These are summarised below for convenience:

1. Terms as such belong to the terminological sphere as opposed to the textual sphere.
2. Terms are concrete entities rather than characteristics, so a reference to a term cannot be compensated for by a reference to one or more terms which share similar characteristics.
3. Translators look up terminological reference resources to check whether the corresponding target language term actually exists or not.
4. Translators looking for bilingual terms need to check resources that enjoy some sort of social recognition in terms of authority or comprehensiveness. They therefore check established terminological dictionaries, relevant corpora and then check Google if no established terminological dictionaries or relevant corpora are available or if established terminological dictionaries or relevant corpora do not contain the terms or information translators are looking for.

In order to construct terminological reference resources that match these requirements by means of automatic extraction of bilingual terms from comparable corpora, as an information source comparable corpora should comprehensively contain those corresponding bilingual terms that could be found through Web searches. This in turn implies that comparability should be defined by means of the terms contained in the corpora, i.e. it should be straightforwardly claimed that any bilingual textual materials that contain corresponding terms, irrespective of the degree of correspondence from other points of view, are properly comparable. However, it would be difficult to call a set of texts that satisfies this condition comparable. This exposes a limitation of automatic bilingual term extraction from comparable corpora.

3.3 Comparability as Supportive Evidence and Comparable Corpora as Sources for Terminological Information

It is thus evident that there is a gap between comparable corpora as an information source for term extraction and online translators' requirements for terminological reference resources. Let us examine this gap, bearing in mind the place of automatic extraction as clarified in Sect. 3.1.

When we assume the construction of terminological resources for Type A lookup, we could of course count on human intervention to fill this gap. As a division of labour, however, it is more natural for humans to support authority and automatic methods to help collect bilingual term pair candidates as widely as possible, which also matches the oft-stated goal of automatic term extraction from corpora (note also the merits of comparable corpora as compared to parallel corpora in term extraction). If such is the case, instead of first constructing comparable corpora and then extracting bilingual term pairs, it would be better to collect bilingual term pair candidates as comprehensively as possible and check their validity *a posteriori* by measuring the comparability of the texts in which the terms appear. Although the technical question remains as to how bilingual term pairs can be collected without resorting to parallel or comparable corpora, this scheme worthy of serious consideration. If we can find other ways to collect bilingual terms, then the importance of comparable corpora as sources for terminological information is reduced, while the importance of the concept of comparability as supportive evidence remains intact. Within this framework, we can reverse the question "do we get comparable pages by sending comparable queries?" [58] to "what degree of comparability can be observed in pages obtained by sending comparable queries consisting of bilingual term pairs?"

This point manifests itself in a clearer way when we think of terminological resources for Type B lookup. As the important condition for Type B lookup is comprehensiveness (i.e. translators need to check whether corresponding target language terms exist in the socially accepted range of reference resources), the use of comparable corpora as information sources for bilingual terms does not match Type B lookup. On the other hand, the concept of comparability will be all the more important and useful in providing evidence supporting term candidates by measuring the comparability between texts that contain candidates, because Type B lookup in itself does not give any information related to the authority or validity of bilingual term candidates.

In summary, from the point of view of providing terminological reference resources to online translators, the status of comparable corpora can be characterised as follows:

1. As an information source for bilingual term extraction, comparable corpora cannot guarantee results matching online translators' requirements for terminological reference lookup; automatic bilingual term extraction from comparable corpora should thus be regarded as one of possibly many technical choices for extracting or collecting bilingual term pairs.
2. The importance of the concept of comparability of texts remains intact or becomes even more important in providing supporting evidence *a posteriori* after bilingual term candidates are extracted or collected in some way or another.

Two questions now remain: (a) are there methods of extracting bilingual term pairs that match terminological reference lookup (especially Type B lookup) by online translators? and (b) what sort of comparability measures or features which define comparability are important for providing evidence supporting the bilingual term candidates? In the following, we will briefly sketch the strategies we adopted and are planning to adopt in the framework of MNH, which answer these questions.

4 Automatic Construction of Terminological Resources and Use of Comparable Corpora Within MNH

4.1 General Framework

MNH is an integrated translation-aid and translation-hosting environment, which has been publicly available at http://trans-aid.jp/ since April 2009 [60]. Currently, it supports English to Japanese, Japanese to English, Chinese to English, English to Chinese, Chinese to Japanese, Japanese to Chinese, and English to Catalan translations. Users can take advantage of an integrated translation aid editor QRedit, which displays the source text in one frame and allows users to input the translated text in another frame, and which provides flexible automatic lookup of a variety of reference resources and a seamless connection to Wikipedia and to Google searches [1]. Among the language pairs, it provides the richest and most comprehensive reference resources for English to Japanese translation, including one of the most widely used comprehensive English–Japanese dictionaries [49], a widely used free dictionary [9] and terminological dictionaries in several domains. The reference resources provided by MNH all enjoy authority and/or comprehensiveness from the point of view of online translation community.

4.2 Extracting Bilingual Terms

As discussed so far, in order to construct terminological reference resources that can actually be used within the MNH environment, what is necessary at the stage of bilingual term extraction is comprehensiveness in its socially-oriented sense, i.e. it should be possible to extract at least the range of bilingual term pairs that most people involved in online translation are able to find for themselves on the Web. Taking into account what online translators are doing to terms in the target language, the range of information sources, and the characteristics of English and Japanese technical terms, we have developed or adopted the following set of methods to extract bilingual terms:

1. Extracting bilingual term pairs from within monolingual texts, using seed bilingual terms [2]. On the Web, there are many pages consisting of list of bilingual terms. Also, it is a widely adopted convention in Japanese technical documents that important technical terms are followed by their corresponding English translation in brackets. In these Web pages, term pair tokens tend to occur in the same pattern within a page. For instance, if a term pair occurs in the form "(mutual information)", other term pairs tend to occur in the same pattern, i.e. the English term is indicated within the brackets immediately after the Japanese term. This holds not only for the textual patterns but also for html tag patterns. This will enable us to collect bilingual term pairs if we can detect Web pages that contain term pairs and the patterns of occurrence of the pairs. Using available term

pairs as seeds, the system can extract bilingual term pairs that occur in the same patterns as the seed pairs from the retrieved pages. In a preliminary experiment, 3.5 million Japanese and English term pairs were collected using about 200,000 seed term pairs, with about 70 % precision. The system was also shown to be effective for a wide variety of domains [3].

2. Extracting bilingual term pairs using the compositional translation method, developed by [61]. This method takes advantage of the fact that (a) around 80 % of technical terms are complex terms and (b) technical terms tend to be formed systematically. Because of these characteristics, there is a high probability that complex terms in different languages consist of the same set of constituent elements, as in "information retrieval" and "æƒ…å ±æ¤œç´¢" " ("æƒ…å ±" corresponds to "information" and "æ¤œç´¢" to "retrieval"). Therefore, generating candidates for target language terms by translating constituent elements using bilingual dictionaries and validating the candidates is an effective way of finding and validating term translations, which human translators in fact often adopt. This method simulates this behaviour. The results of the evaluation reported in [61] show that the method achieves a maximum of 88 % precision with around 30 % recall (in relation to the experimental set). Although the factors influencing recall should be properly analysed, the result implies that the method is promising if adopted properly in combination with other methods.
3. Extracting bilingual term pairs using the transliteration method, developed by [52]. In addition to complex terms, Japanese terms are characterised by a large number of transliterated elements [27], such as "ã‚³ãƒ³ãƒ”ãƒ¥ãƒ¼ã‚¿" (computer), which are written in katakana characters. The method of detecting transliterated terms by using rules of transliteration should therefore be effective. While the effectiveness of this method in detecting technical terms has yet to be tested, it is reported to be highly effective in detecting personal names [52]. In the experiment with personal names reported in [52], the transliteration rule sets learned using the method produced 93–94 % correct answers, either unique or within several candidates, depending on the evaluation set.
4. Extracting bilingual term pairs from English and Japanese parallel documents automatically detected from the Web [28]. The method first collects bilingual parallel documents as follows: (a) search relevant documents on the Web using keywords provided by users; (b) check pre-defined keywords (such as "original", "translated from" or "English") to evaluate whether the documents are translated or not; (c) traverse links given in the document to detect the source document, and (d) evaluate similarities between the documents and adopt the pairs as parallel documents if the similarities are stronger than the given threshold. The textual area is then extracted from both source and target documents, and standard bilingual term extraction methods are applied to extract bilingual term pairs from parallel corpora. An experiment using 33 keywords showed that the parallel document was extracted with 74 % precision and 35 % recall. An overall evaluation including term extraction has not yet been carried out, as the term extraction module is not connected to the parallel text extraction module.

5. Extracting bilingual term pairs from comparable corpora constructed with qualitative factors being taken into account [39, 40]. The method is in line with standard term extraction from comparable corpora using the mapping of context vectors and evaluating the similarities between the vectors.

Among the above, methods 1–3 are fully implemented as a module, and are currently at the stage of being fine-tuned so that they can be incorporated in the MNH environment. The bilingual text crawler used in method 4 is also fully implemented. The basic algorithms for the remaining part of method 4 and method 5 are at the experimental stage, but have not yet been fully implemented as a module. While only individual evaluations of the methods have been carried out, the set of bilingual term candidates that they are able to produce as a whole should attain the required level of comprehensiveness in the sense that what most online translators, or at least asymptote to it. Instead of evaluating performance in a laboratory environment, we are planning to evaluate the usefulness of these methods within the real-world environment of actual translation carried out using MNH.

4.3 Evaluating Comparability as Supportive Evidence

At the extraction stage, achieving comprehensiveness is an essential requirement. The extracted bilingual term pair candidates should be evaluated using supportive evidence, to provide authority to the results. The main source of supportive evidence we currently assume is the degree of comparability of texts obtained by throwing parallel queries consisting of bilingual term pair candidates out to the Web.

As for the comparability metrics, while many measures have been proposed and examined so far [30, 44, 57], there is still no "agreement about the criteria for measuring parallelism and comparability" [54]. Given this situation and given the use of the concept of comparability in our context, we are currently preparing to provide a variety of points of measurement at the following two levels:

- The linguistic level: this includes units of measuring comparability, i.e. document collections, individual documents, paragraphs or sentences; features and metrics for measuring comparability, etc. Many, if not most, of the features or viewpoints in terms of which the degree of comparability are measured, such as domain, register or formality, belong to this level.
- The metalinguistic/social level: this includes attributes of authors, and of URL domains, link patterns and format of the page, etc.

Identifying useful features for measuring comparability in validating bilingual term pair candidates within the context of online translation will be the next stage, which will be carried out experimentally by making the evaluation results of comparability available to users of MNH as evidence supporting the choice of terms. This will provide answers to context-dependent question of corpus comparability, i.e. what are the required measurements of comparability from the point of view of providing online translators with supportive evidence for bilingual term pair candidates? Given the absence of general agreement regarding the concept of comparability, perhaps reporting what are good measurements for comparability in more specific and concrete contexts will contribute to our understanding of the nature of—and the nature of the nature of—comparable corpora.

A final note: within the environment of MNH, comparability to the text being translated could or should also be taken into account. This raises a practical engineering problem of how to achieve real-time calculation of comparability when the text being translated is given. This is also currently being examined and will be addressed fully in the future.

5 Conclusions

After postulating the nature of the terminological reference resources required by online translators, and examining the status of comparable corpora as a source for bilingual term extraction, we have argued that comparable corpora as an information source for bilingual term extraction, if used alone, will not be able to satisfy online translators' requirements for terminological reference resources, although they should be used *a posteriori*, after bilingual term candidates have been comprehensively extracted using whatever methods are available. In the process we emphasised authority and comprehensiveness as socially oriented concepts.

We then sketched out a framework for providing automatically constructed bilingual terminological resources in an open translation-aid environment and clarified the place of comparable corpora within the framework. Technically speaking, there is not much that is new in what was discussed in this paper. Instead, what we have tried to do is to situate the use of comparable corpora for bilingual term extraction—or construction of bilingual terminology reference resources—in an actual and specific social context in which online translators use reference resources, taking MNH as a concrete platform. What we have proposed may look like a farrago of different methodologies and small ideas if observed from a technical point of view, but it corresponds, at least to some extent, to some essential traits of reference resources and the actual situation of online translators. As such, from the social point of view, it constitutes a consistent and unified whole.

In relation to corpus comparability or comparable corpora, which is the theme of this volume, what we have proposed is a small, humble point: it could be useful to explicitly frame the discussion of corpus comparability or comparable corpora explicitly in a specific social context in which comparable corpora or the products created by using comparable corpora are to be actually used as an end social product. In the next stage of our research, we are planning to explore technical aspects more fully, including what sort of *a posteriori* corpus comparability is most useful as supportive evidence for online translators' decision making in choosing target-language terms.

Acknowledgments This work is partly supported by the Japan Society for the Promotion of Sciences (JSPS) grant-in-aid (A) 21240021 "Developing an integrated translation-aid site which provides comprehensive reference sources for translators." and the 2011 NII research cooperation project "Automatic construction of practically useful English–Japanese terminological lexica using Web information resources.".

References

1. Abekawa, T., Kageura, K.: QRedit: An integrated editor system to support online volunteer translators. Proceedings of the Digital Humanit, 3–5 (2007)
2. Abekawa, T., Kageura, K.: QRpotato: a system that exhaustively collects bilingual technical term pairs from the Web. In: Proceedings of the 3rd International Universal Communication Symposium, pp. 115–119 (2009)
3. Abekawa, T., Kageura, K.: Using seed terms for crawling bilingual terminology lists on the Web. Trans. Comp. (2011) (no pagination)
4. Baker, M.: Corpus-based translation studies: The challenges that lie ahead. In: Somers, H. (ed.) Terminology, LSP and Translation: Studies in Language Engineering in Honour of Juan C. Sager, pp. 175–186. Benjamins, Amsterdam (1996)
5. Baroni, M., Kilgarriff, A.: Large linguistically-processed Web corpora for multiple languages. In: Proceedings of the 11th EACL, pp. 87–90 (2006)
6. Bernardini, S., Castagnoli, S.: Corpora for translator education and translation practice. In: Yuste, E. (ed.) Topics in Language Resources for Translation and Localisation, pp. 39–55. Benjamins, Amsterdam (2008)
7. Bowker, L.: Corpus-based applications for translator training: Exploring the possibilities. In: Granger, S., Lerot, J., Petch-Tyson, S. (eds.) Corpus-Based Approaches to Contrastive Linguistics and Translation Studies, pp. 185–206. Rodopi, Amsterdam (2003)
8. Bowker, L., Pearson, J.: Working with Specialized Language: A Practical Guide to Using Corpora. Routledge, London (2002)
9. Breen, J. W.: A www Japanese dictionary. In: Japanese Studies Centre Symposium (1999)
10. Burbles, N.C.: Paradoxes of the web: the ethical dimensions of credibility. Libr. Trends **49**(3), 441–453 (2001)
11. Burr, V.: Soc. Constr., 2nd edn. Routledge, London (2003)
12. Chiao, Y.-C., Zweigenbaum, P.: Looking for candidate translational equivalents in specialized, comparable corpora. In: Proceedings of the19th COLING, pp. 1208–1212 (2002)
13. Collier, N., Kumano, A., Hirakawa, H.: An application of local relevance feedback for building comparable corpora from news article matching. Nat. Inst. Inform. J. **5**, 9–23 (2003)
14. Daille, B., Morin, E.: French–English terminology extraction from comparable corpora. In: Proceedings of the 2nd IJCNLP, pp. 707–718 (2005)
15. Désilets, A., Brunette, L., Malançon, C., Patenaude, G.: Reliable innovation: a tecchie's travels in the land of translators. In: Proceedings of the 8th AMTA (2008)
16. EAGLES.: Synopsis and comparison of morphosyntactic phenomena encoded in lexicons and corpora: a common proposal and applications to European languages. Technical report EAG-CLWG-Morphsyn/R, ILC-CNR, Pisa (1996)
17. Editorial Committee of the Handbook of Library and Information Science (ed.): Handbook of Library and Information Science, 2nd edn. Maruzen, Tokyo (1999)
18. Fogg, B.J., Marshall. J., Laraki, O., Osipovich, A., Varma, C., Fang, N., Paul, J., Rangnekar, A., Shon, J., Swani, P. Treinen, M.: What makes web credible? a report on a large quantitative study. In: SIGCHI Conference on Human Factors in Computing Systems, pp. 61–68 (2001)
19. Fung, P., McKeown, K.: Finding terminology translations from non-parallel corpora. In: Proceedings of the 5th Annual Workshop on Very Large Corpora, pp. 192–202 (1997)

20. Fung, P.: A statistical view on bilingual lexicon extraction: from parallel corpora to nonparallel corpora. In: Proceedings of the 3rd AMTA, pp. 1–17 (1998)
21. Gaussier, É.: Flow network models for word alignment and terminology extraction from bilingual corpora. In: Proceedings of the 36th ACL and the 17th COLING, pp. 444–450 (1998)
22. Gaussier, É., Renders, J.-M., Matveeva, I., Goutte, C., Dejean, H.: A geometric view on bilingual lexicon extraction from comparable corpora. In: Proceedings of the 42nd ACL, pp. 502–509 (2004)
23. Geeraerts, D.: Lexicology. In: Asher, R.E. (ed.) The Encyclopedia of Language and Linguistics, vol. 4, pp. 2189–2192. Pergamon Press, Oxford (1994)
24. Goeuriot, L., Morin, E., Daille, B.: Compilation of specialized comparable corpora in French and Japanese. In: Proceedings of the 2nd BUCC, Workshop, pp. 55–63 (2009)
25. Halliday, M.A.K.: Lexicology. In: Halliday, M.A.K., Teubert, W., Yallop, C., Čermáková, A. (eds.) Lexicology and Corpus Linguistics: An Introduction, pp. 1–22. Continuum, London (2004)
26. Hansen, S.: The nature of translated text: an interdisciplinary methodology for the investigation of the specific properties of translations. Ph.D. thesis. Saarbrücken: Universität des Saarlandes (2002)
27. Kageura, K.: The Dynamics of Terminology. Benjamins, Amsterdam (2001)
28. Kageura, K., Abekawa, T., Sekine, S.: QRselect: a user-driven system for collecting translation document pairs from the Web. In: Proceedings of the 10th ICADL, pp. 131–140 (2007)
29. Kageura, K., Abekawa, T.: NLP meets library science: providing a set of enhanced language reference tools for online translators. In: Proceedings of the Asia–Pacific Conference on Library and Information Education and Practice (2009)
30. Kilgariff, A.: Using word frequency lists to measure corpus homogeneity and similarity between corpora. In: Proceedings of the 5th ACL Workshop on Very Large Corpora, pp. 231–245 (1996)
31. Kilgariff, A.: Comparing corpora. Int. J. Corpus Linguist. **6**(1), 1–37 (2001)
32. Kilgariff, A.: Comparable corpora within and across languages, word frequency lists and the KELLY project. In: Proceedings of the 3rd BUCC Workshop (2010)
33. Kiousis, S.: Public trust or mistrust? perceptions of media credibility in the information age. Mass Comm. Soc. **4**(4), 281–403 (2001)
34. Kwong, O.Y., Tsou, B.K., Lai, T.B.Y.: Alignment and extraction of bilingual legal terminology from context profiles. Terminology **10**(1), 81–99 (2004)
35. Laroche, A., Langlais, P.: Revisiting context-based projection methods for term-translation spotting in comparable corpora. In: Proceedings of the 23rd COLING, pp. 617–625 (2010)
36. Li, B., Gaussier, É.: Improving corpus comparability for bilingual lexicon extraction from comparable corpora. In: Proceedings of the 23rd COLING, pp. 644–652 (2010)
37. Maeda, T.: Goi souron. In: Tamamura, F. (ed.) Nihongo no Goi, Imi, pp. 1–22. Meiji Shoin, Tokyo (1989)
38. McEnery, T., Xiao, R.: Parallel and comparable corpora: What is happening? In: Anderman, G., Rogers, M. (eds.) Incorporating Corpora, pp. 18–31. Multilingual Matters, Clevedon (2007)
39. Morin, E., Daille, B., Takeuchi, K., Kageura, K.: Bilingual terminology mining: using brain, not brawn comparable corpora. In: Proceedings of the 45th ACL, pp. 664–671 (2007)
40. Morin, E., Daille, B., Takeuchi, K., Kageura, K.: Brains, not brawn: the use of "smart" comparable corpora in bilingual terminology mining. ACM Trans. Speech Lang. **7**(1), Article 1 (2010)
41. Olohan, M.: Introducing Corpora in Translation Studies. Routledge, London (2004)
42. Prochasson, E., Morin, E., Kageura, K.: Anchor points for bilingual lexicon extraction from small comparable corpora. MT Summit XII, pp. 284–291 (2009)
43. Rapp, R.: Automatic identification of word translations from unrelated English and German corpora. In: Proceedings of the 37th ACL, pp. 519–526 (1999)
44. Rayson, P., Garside, R.: Comparing corpora using frequency profiling. In: Proceedings of the Comparing Corpora, Workshop, pp. 1–6 (2000)
45. Rey, A.: Essays on Terminology. Benjamins, Amsterdam (1995)

46. Rieh, S.Y., Danielson, D.R.: Credibility: a multidisciplinary framework. Annu. Rev. Inf. Sci. Tech. **41**, 307–364 (2007)
47. Sadat, F., Yoshikawa, M., Uemura, S.: Bilingual terminology acquisition from comparable corpora and phrasal translation to cross-language information retrieval. In: Proceedings of the 41st ACL (2003)
48. Sager, J.C.: A Practical Course in Terminology Processing. Benjamins, Amsterdam (1990)
49. Sanseido: Sanseido's Grand Concise English–Japanese Dictionary. Sanseido, Tokyo (2004)
50. Saralegi, X., San Vicente, I., Gurrutxaga, A.: Automatic extraction of bilingual terms from comparable corpora in a popular science domain. In: Proceedings of the 1st BUCC, Workshop, pp. 27–32 (2008)
51. Sato, S.: Transliteration using a large-scale candidate list. Japio Year Book 2010, pp. 258–261 (2010)
52. Sato, S.: Non-productive machine transliteration. In: Proceedings of the 9th RIAO, pp. 16–19 (2010)
53. de Saussure, F.: Cours de Linguistique Générale. Payot, Paris (1916)
54. Skadina, I., Vasiljevs, A., Skadiņš, R., Gaizauskas, R., Tufiş, D., Gornostay, T.: Analysis and evaluation of comparable corpora for under resourced areas of machine translation. In: Proceedings of the 3rd BUCC Workshop. 6–14 (2010)
55. Shao, L., Ng, T.N.: Mining new word translations from comparable corpora. In: Proceedings of the 20th COLING, pp. 618–624 (2004)
56. Sharoff, S.: Translation as problem solving: uses of comparable corpora. In: Proceedings of the Workshop on Language Resources for Translation Research and Practice (2006)
57. Sharoff, S.: Classifying Web corpora into domain and genre using automatic feature identification. In: Proceeding of the 3rd Web as Corpus Workshop (2007)
58. Sharoff, S.: Analysing similarities and differences between corpora. In: Proceedings of the 7th Conference of Language Technologies (2010)
59. Talvensaari, T., Pirkola, A., Järvelin, K., Juhola, M., Laurikkala, J.: Focused web crawling in the acquisition of comparable corpora. Inf. Ret. **11**, 427–445 (2008)
60. Utiyama, M., Abekawa, T., Sumita, E., Kageura, K.: Hosting volunteer translators. MT Summit XII (2009)
61. Tonoike, M., Kida, M., Takagi, T., Sasaki, Y., Utsuro, T., Sato, S.: A comparative study on compositional translation estimation using a domain/topic-specific corpus collected from the Web. In: Proceedings of the 2nd Web as Corpus Workshop, pp. 11–18 (2006)
62. Vintar, Š.: Bilingual term recognition revisited: the bag-of-equivalents term alignment approach and its evaluation. Terminology **16**(2), 141–158 (2010)
63. Wilks, Y., Slator, B., Guthrie, L.: Electric Words: Dictionaries, Computers, and Meanings. MIT Press, Cambridge (1996)
64. Zanettin, F.: Bilingual comparable corpora and the training of translators. Meta **43**(4), 616–630 (1998)

Old Needs, New Solutions: Comparable Corpora for Language Professionals

Silvia Bernardini and Adriano Ferraresi

Abstract Use of corpora by language service providers and language professionals remains limited due to the existence of competing resources that are likely to be perceived as less demanding in terms of time and effort required to obtain and (learn to) use them (e.g. translation memory software, term bases and so forth). These resources however have limitations that could be compensated for through the integration of comparable corpora and corpus building tools in the translator's toolkit. This chapter provides an overview of the ways in which different types of comparable corpora can be used in translation teaching and practice. First, two traditional corpus typologies are presented, namely small and specialized "handmade" corpora collected by end-users themselves for a specific task, and large and general "manufactured" corpora collected by expert teams and made available to end users. We suggest that striking a middleground between these two opposites is vital for professional uptake. To this end, we show how the BootCaT toolkit can be used to construct largish and relatively specialized comparable corpora for a specific translation task, and how, varying the search parameters in very simple ways, the size and usability of the corpora thus constructed can be further increased. The process is exemplified with reference to a simulated task (the translation of a patient information leaflet from English into Italian) and its efficacy is evaluated through an end-user questionnaire.

1 Introduction

Language professionals make ample use of technology. A survey conducted in 2009 estimated that EU language service providers invest 5–10 % of the annual turnover in multilingual technology tools, with a likely market size topping one billion euros in 2015 [30]. These tools include electronic dictionaries, terminology extraction tools

S. Bernardini (✉) · A. Ferraresi
Department of Interpreting and Translation, Corso della Repubblica 136, 47121 Forlì, FC, Italy
e-mail: silvia.bernardini@unibo.it

S. Sharoff et al. (eds.), *Building and Using Comparable Corpora*,
DOI: 10.1007/978-3-642-20128-8_16, © Springer-Verlag Berlin Heidelberg 2013

and translation memory (TM) tools as well as language training software.While corpora and corpus processing software are only mentioned in passing, and not focused upon explicitly, they would clearly belong here, since they can and indeed have been used to provide (or add to) dictionary-like insights about word meaning and use [22], to extract terminology [7, 27], to produce contextually appropriate target texts through concordance browsing [8, 36] and to improve competence in a foreign language [19]. There is no doubting that many academics who teach translation and foreign languages believe in the merits of corpus work, particularly of the comparable kind.

Freelance translators also appear to be aware that they need corpora—though they do not call them such. A survey conducted in 2005-2006 among European (mainly British) translators and students of translation [24] showed that over half the respondents (a total of 1,015) collected reference texts, but in the majority of cases either read them or searched them using word processing search facilities. Very few were aware of corpus query tools, but very many (around 80 %) claimed they would be interested in a service providing specialized corpora and/or extracting terms from them.

Lastly, some commercial TM software vendors are beginning to see the point of (slightly) more sophisticated TM concordancing functionalities, which would bridge the gap between TMs and aligned parallel corpora. Both MemoQ and MultiCorpora offer concordancing in full-text context as a selling point for their software, overcoming a common problem with competitor tools, where "the concordance passage is limited to isolated sentences that exist in the TM database [so] there is often insufficient context to provide guidance on the applicability of the found result" [25]. The latest release from market leader SDL TRADOS (Studio 2011) incorporates "character-based" concordancing to retrieve related words or "fuzzy matches" (a little like a regular expression search would), and allows searches on the target text as well as the source text. Inefficient TM concordancing is also perceived as a problem by end users, judging from complaints in user forums.

And yet use of corpora by language service providers and language professionals remains limited. The lack of widespread uptake is probably due to the existence of competing resources that are likely to be perceived as less demanding in terms of time and effort required to obtain and (learn to) use them. The translator's toolkit has never been so replete with tools and resources, from termbases and dictionaries to the web itself, from TMs to aligned or unaligned parallel or "bi-" texts. Each of these have a role to play in the translation process, and indeed expert translators resort to them depending on task requirements, the ability to choose quickly and confidently among different resources being one of hallmarks of translation competence [14]. And they all have advantages and disadvantages.

Dictionaries and termbases, particularly electronic ones, provide a wealth of "digested" information (equivalents, definitions, synonyms, typical examples) sanctioned by lexicographers and terminographers; searches are quick and easy, and solutions are often found, though they may lack the reassuring added value of a "contextual match" [35]. One obvious way of overcoming this drawback is to consult (tens or hundreds of) actual texts on the web, and draw inferences based on actual use in context. This process is more time consuming, effortful and complex than dictionary lookup, though, requiring the opening and closing of multiple pages, the quick evaluation of the reliability of sources, and an acceptance of the many limits of search engines that were not designed for linguists [3]. TMs are undoubtedly valuable productivity-enhancing resources, providing authoritative equivalents in context from previously translated texts. And yet TMs are hardly the translator's panacea. First, they are only really useful for certain text types (repetitive texts with shortish sentences) and tasks (revisions, updates). Secondly, it has been suggested that they might affect translators' strategies in a negative way, by making recyclability a priority rather than a positive side effect [6]. By preserving sentence boundaries and avoiding variation, translators may increase the amount of leveraging from one task to the next, but not necessarily produce the "best possible" translation. Lastly, aligned parallel texts/corpora are not available ready-made for most specialized domains[1] and their set up is time-consuming and technically demanding for translators.

In Sect. 2 we will suggest that some of these limitations can be overcome through the integration of comparable corpora and corpus building tools in the translator's toolkit. We shall start by providing an overview of the ways in which different types of comparable corpora can be used in translation teaching and practice.[2] We shall first look at two "traditional" corpus typologies widely discussed in the literature: small and specialized "handmade" corpora collected by end-users themselves for a specific task, and large and general "manufactured" corpora collected by expert teams and made available to end users. These occupy opposite poles of a cline going from very small and fine-tuned corpora to very large and general ones. Since besides the advantages both have disadvantages that are likely to make translators shun them, we shall suggest that striking a middleground between these two opposites is vital. To this end, in Sect. 3 we shall see how the BootCaT toolkit can be used to construct largish and relatively specialized comparable corpora for a specific translation task, and how, varying the search parameters in very simple ways, the size and usability of the corpora thus constructed can be further increased. Usability is evaluated by respondents and exemplified with reference to a simulated task (translation of a patient information leaflet from English into Italian). In Sect. 4 we conclude by summarizing our argument and looking at future prospects.

[1] This may change in the future, as more tools like Linguee (http://www.linguee.com/) provide access to the aligned web, and possibly to subsections of it.

[2] As suggested in the previous section, in this chapter we are not specifically discussing aligned parallel corpora. In our view these are more akin to TMs than to comparable corpora in terms of the technical issues involved in their construction and consultation, and of the type of insights translators can obtain from them; they are therefore not directly relevant here.

2 Comparable Corpora as Alternatives?

2.1 The Handmade Solution

Compared to other resources, comparable corpora do offer a number of advantages. Unlike in the case of web search engines and TMs, consultation of corpora through dedicated software allows translators to benefit from the querying and displaying facilities which are at the very heart of corpus-related methodologies. Linguistically sophisticated queries, e.g. through the use of wildcard and regular expressions, together with the possibility to sort results according to co-text, undoubtedly give corpora an edge in terms of time and effort required to make patterns in context emerge. Once a corpus has been compiled, translators no longer need to browse through different pages, like they would when using search engines; they can specify rules to target specific linguistic forms (as well as other querying options, e.g. setting the desired number of "empty" slots in an expression, etc.), and can variably order results according to the pattern they are searching for: no commercial TM tool that we are aware of implements all these features. Lastly, most concordancers allow full-text browsing, thus offering the rich contextual information required for decision-making, that TM tools seem to have so much trouble providing [9].[3]

Other advantages which have been suggested to be offered by comparable corpora for translation purposes depend on the "type" of corpus considered. Two kinds of corpora are particularly relevant for the present discussion, namely those that position themselves at the extremes of the "size cline" which is invoked in the literature as a yardstick to categorize corpora [18]. While size is a fuzzy criterion, especially as technological advances in computer power and memory constantly push the boundaries forward, the distinction between *small* and *large* corpora also reflects a difference in the textual populations they attempt to sample [1]. Small corpora are usually collections of "specialized" texts, intended to represent a specific domain, and include texts with homogeneous content (e.g. medicine), text type or genre (e.g. textbook), or both. On the other hand, large corpora usually aim to represent a much wider population, i.e. the whole of a language or language variety (e.g. British English), and for this reason are also called "reference" or "general purpose" corpora.

In essence, small corpora are not dissimilar from the domain-specific collections of texts translators use as reference materials (cf. Sect. 1), the main difference being in the way they are consulted, i.e. by corpus processing software instead of word processing search facilities. The advantages associated with the use of small corpora have been discussed in the literature, where such corpora are also called, crucially for the purposes of the present paper, *comparable, ad hoc* corpora (e.g. [28, 35]). *Comparability* here refers to the similarity of the (target language) texts being collected, ideally both in terms of topic and text type/genre, to the source text being

[3] "More advanced" corpus querying techniques, like extraction of keywords or computation of collocational scores can of course be of great interest to translators. However, their relevance and usefulness may be hard to grasp for less corpus-savvy users, and hence they are not discussed here.

translated, and/or to "equivalent" source language texts; the *ad hoc* label puts the stress on yet another aspect usually associated with these corpora, i.e. on the fact that they are typically built manually for specific translation tasks. According to Aston [1], specialized, *ad hoc* comparable corpora: (a) may be perceived as more familiar by translators compared to other corpus resources, especially since texts are consulted and selected manually for inclusion in the corpus; (b) facilitate the process of interpretation of concordance data, since the likelihood of encountering irrelevant examples (e.g. polysemous words used in many different senses) is reduced; (c) provide assistance in producing natural-sounding translation hypotheses based on bottom-up search strategies, counteracting the risks connected with the "hypothesis formulation-validation" cycle, which may result in overlooking potential translation alternatives.

This does not mean that small, specialized corpora do not have their shortcomings: while being powerful pedagogical instruments, whose effectiveness in the translation classroom as "performance-enhancing" tools has been convincingly argued for [35], they may not be the best option for professionals. Given their small size, they may not contain enough matches to draw confident generalizations about certain usages in the specialized domain in question, especially in the case of rarer patterns above the word unit. Indeed, Varantola [35] summarizes the advantages connected with small corpora as *reassurance*, meaning that "[w]hen relevant corpus information [is] available, the users often [gain] reassurance for their strategic decisions as well as the actual lexical choices". This definition highlights precisely the potentially negative aspect of small corpora, i.e. that small corpora simply may not provide useful evidence for the translation problem in hand and "will rarely document every word in an ST or TT" [1]. Hence the tradeoff between effort (to gather a sufficient number of texts, learn to use concordancing software, etc.) and effectiveness may be perceived by translators as not favourable enough to justify the investment of their time.

2.2 The Manufactured Solution

Large, general corpora can enable translators to overcome this limitation. Munday [26] and Hoey [23] provide evidence of contrastive insights about word usage in different languages that can be gleaned from comparable reference corpora and that are crucial for decision-making in translation. Philip [29] offers a general description of this process claiming that "[c]hoice in translation is related to choice in the SL, and this can be identified by comparing a given expression against its possible alternatives along the paradigmatic axis. [...] [I]f an equivalent paradigm of choice is set up for the TL, the most suitable correspondences can be identified and used in the translated text". General-purpose corpora are also much larger and more diverse in terms of texts sampled than small corpora, making it possible to document a large number of phraseological patterns and to relate them to specific registers. As Aston [1] puts it, these corpora "can make [translators] more sensitive to issues of phraseology, register and frequency, which are poorly documented by other tools", and hence can be used "as complements to traditional dictionaries and grammars". Add to this that the process of compiling large reference corpora has been "democratized" by the growth of the web, which has made it possible for researchers worldwide to construct and make available to the research community very large corpora in a variety of languages.

Yet reference corpora present at least three major limitations for translators. First, the specialized domain under investigation may be underrepresented (or totally absent) in the selected corpus, thus making the process of finding patterns in context matching those in the source or target language text a fruitless exercise. Conversely, "appropriate instances" [1] may be difficult to identify in the corpus due to an embarrassment of riches: too many solutions for a query, from widely different text types. Translators are forced to sift through very many and/or potentially irrelevant results, e.g. in the case of polysemous words. And third, these corpora tend to come with their own search interface, often complex and not necessarily intuitive, such that getting acquainted with it may be a rather daunting task for a time-pressed and not especially motivated professional or student. Reference corpora of the web-as-corpus type [3] have the further disadvantage of carrying no guarantee of quality or representativeness, being automatically constructed. While we have argued elsewhere [16] that this disadvantage is partly compensated for by their size and up-to-dateness, it remains a weak spot from the perspective of someone in search of a quick, reliable answer.

2.3 Using Handmade and Manufactured Corpora

In this section we briefly exemplify the use of "traditional" specialized and reference corpora for a simulated practical task: the translation from English into Italian of a Patient Information Leaflet (PIL) for paracetamol tablets. This text type was selected because it is likely to be familiar to most readers while being sufficiently specialized (in terms of subject domain) and conventional (in terms of genre) to give an idea of how different corpora might help in its translation, or fail to do so. Pedagogically, it is an ideal task since it requires students to observe, and hence be sensitized, to the existence of cross-linguistic phraseological patterning in texts. Indeed, the task was also performed in class with a group of BA-level students of translation. A small specialized comparable corpus was constructed manually to help them to observe corresponding patterns in the two languages. Nine leaflets in English and nine in Italian were collected, all accompanying paracetamol medicines, making sure that they were reliable specimens of the genre. The whole process of searching the web through appropriate queries, evaluating texts, discarding dubious ones and saving good ones, took about 5 h, and provided us with a corpus of approximately 25,000 words (13,518 in English and 11,832 in Italian).

Even such a small corpus can be useful for translating a highly conventional text like this one. Taking a very straightforward example, the ST lists "active ingredients" and "other ingredients". By searching for the Italian name of one ingredient from each category in the Italian subcorpus, say "paracetamolo" ("paracetamol") and "sodio" ("sodium"), and browsing the left-hand co-text, one easily finds equivalents for "active ingredients" (i.e. "principi attivi"), and "other ingredients" ("eccipienti"). The latter is especially tricky since an inexperienced translator may be misled to try to come up with a phrase equivalent rather than a single word equivalent, and the correct translation may simply not come to mind, especially when pressed for time. Since co-textual searches like this one (using known equivalents in the target language as "anchors" to find equivalents for unknown words used in their vicinity) take some time and effort, a small carefully constructed corpus may be more practical than a larger, dirtier one, provided of course that it contains some evidence about the expression in question.

Large reference corpora also have their uses for this task, though different from the kind just discussed: co-textual searches are certainly impractical, as are searches involving genre conventions (of the kind: *how do they say "possible side effects" in Italian PILs?*). Instead, it would make sense to look up a phrase like "serious heart condition" in a reference corpus of English, to find out whether the adjective modifies the noun phrase or is an integral part of it, and then look up a comparable reference corpus of Italian to find possible equivalents. We would thus find that "serious" is the most frequent adjectival premodifier of "heart condition" in ukWaC (32 occurrences), and that among the top 20 collocates in this position there are no obvious synonyms (if we exclude "lifethreatening", 6 occurrences) and only two antonyms, i.e. "minor" and "mild", which taken together occur five times only. This would appear to suggest that the adjective is part of the phrase, or at least that it forms a restricted collocation with it [11]. In Italian, the unmarked position of the adjective in this case would be following (rather than preceding) the noun phrase, since the pre-modifying position implies a degree of subjective judgement [32]. A search for "cardiopatia" (an equivalent for "heart condition" in Italian) followed by an adjective, conducted on a comparable reference corpus of Italian (itWaC), shows that the only viable equivalent for "serious" in this context is "grave" 11 occurrences, 7th most frequent postmodifying adjective).[4]

3 The BootCaT Way

As argued and exemplified in Sect. 2, both small ad hoc corpora and large reference corpora can be of use in the translation process—to answer questions related to genre conventions and terminology respectively. However, there are cases in which the former provide too little or no evidence, and the latter too much. For instance, in the ST we find the following sentence: "The product is available in cartons of 8 or 16 capsules". Possible translations for "carton" given by an English-Italian dictionary (*Oxford Paravia*) and potentially acceptable in this context are "scatola" and "confezione". If we look up the word "capsule", the obvious Italian equivalent of "capsules", in the Italian subcorpus of our ad hoc corpus, hoping to find either "scatola"

[4] See http://wacky.sslmit.unibo.it/doku.php for information about ukWaC and itWaC.

or "confezione" in the left-hand cotexts, unfortunately we get no results, since none of the paracetamol products described in the sampled texts come in capsules. Making a similar query to a large reference corpus would be impractical, because of the enormous amount of data one would have to sift through to find contextually-appropriate evidence: itWaC has 1,902 occurrences of "capsule", coming from the most varied types of texts.

A possible solution in this case would be to combine the advantages of manual and automatic methods of corpus building. This is the idea behind BootCaT [2], a software tool that (partly) automates the process of finding reference texts on the web and collating them in a single corpus. BootCaT is a multi-stage pipeline requiring user input and allowing varying levels of control. In the first step, users provide a list of single- or multi-word terms to be used as seeds for text collection. These are then combined into "tuples" of varying length and sent as queries to a search engine, which returns a list of potentially relevant URLs.[5] At this point the user has the option of inspecting the URLs and trimming them; the actual web pages are then retrieved, converted to plain text and saved as a single file. This can then be interrogated using one of the standalone concordancers available (e.g AntConc).[6]

Using BootCat one can build a relatively large quick-and-dirty comparable corpus (typically of about 80 texts in each language, with default parameters and no manual quality checks) in less than half an hour. The end product may be of variable quality though, and since quality comes at a cost, a "clean" corpus would require much more time and effort spent on the corpus building task—e.g. browsing/selecting URLs, trying a few re-runs with different sets of keywords or tuples, manually browsing the collected corpus to discard low-quality or irrelevant texts etc.. This flexible approach to the task makes BootCaT a very useful tool for translators and translation students, which has been used in the translation and terminology classroom to build small DIY corpora of varying size and specialization [8, 15, 17], and whose potential is worth exploring further.

3.1 Beyond Topic: BootCaT for Genre-Restricted Corpora?

In the "traditional" BootCaT pipeline, the first step in corpus creation consists in selecting seeds "that are expected to be representative of the domain under investigation" [2]. In the case of a translation task, the web pages that are retrieved by the tool based on these seeds are usually expected to deal with the same topic as the ST to be translated. However, topic similarity is not the only criterion of text selection one might adopt for documentation purposes: depending on the task, it can be argued that genre is equally if not more crucial.[7]

[5] Currently BootCaT uses Bing for URL retrieval, after both Google and Yahoo! discontinued their API services.

[6] http://www.antlab.sci.waseda.ac.jp/software.html

[7] In this paper we define genre (loosely based on Swales [34]) as a recognizable set of communicative events with a shared purpose and common formal features.

As suggested by Crowston and Kwasnik [13], "[b]ecause most genres are characterized by both form and purpose, identifying the genre of a document provides information as to the document's purpose and its fit to the user's situation". This applies equally to information retrieval in general and to text collection for reference purposes in particular: the importance of genre comparability has indeed been repeatedly stressed in the literature on specialized comparable corpora for translation, as discussed in Sect. 2. Retrieving web pages belonging to a specific genre automatically is however far more complex than retrieving pages on a specific topic.

From the perspective of the translator in need of a rough-and-ready DIY genre-restricted corpus, the techniques proposed within IR are far too computationally complex and/or require extensive linguistic modelling. Available genre classification schemes also seem inadequate, since they are unlikely to target the desired genre(s), or be available for the desired language(s) [31]. Even systems designed specifically for creating multilingual comparable corpora [20] usually present the drawback of needing substantial tuning if they are to be applied to languages other than those for which they were originally created.

In Sect. 3.2 we propose a naive approach to constructing a "genre-driven" corpus using BootCaT, i.e. a method which is intended to favour the retrieval of pages belonging to the same genre, instead of topic, as the ST under analysis. Frequent multi-word sequences have been suggested to be valuable indicators of genre: Biber and Conrad, e.g., use "lexical bundles" (i.e. frequent, uninterrupted four-word sequences) to describe register variation between conversation and academic prose [5]; Gries and Mukherjee compare regional variations of Asian English in the ICE corpus using word sequences of varying length [21]. Along similar lines, n-grams have proved to be reliable as discriminating cues in computational approaches to genre classification (cf. [12] and references therein). The approach we propose consists in using, instead of topic-specific keywords, the *n* most frequent trigrams from the manual corpus as input to the BootCaT pipeline, regardless of their being intuitively salient, syntactically complete, or lexically rich.

3.2 Corpus Construction

As a starting point for constructing the BootCaT corpora we used the manually constructed bilingual comparable corpus discussed in Sect. 2.3 above. For the topic-driven corpus, keywords were obtained from both subcorpora using a sub-set of the Europarl corpus as a reference corpus in AntConc. The top 50 keywords were selected; proper nouns were removed and the remaining words were lowercased and used as seeds, without further manual trimming (43 words in English and 45 in Italian). For the genre-driven corpus no reference corpus was required. We took the 50 most frequent trigrams in the manual subcorpora, removed those containing proper nouns and numbers, and lowercased them. The final lists of seeds contain 41 English trigrams and 46 Italian ones. The seeds were then imported into BootCaT and tuples were formed.[8] To partly compensate for the use of phrases for the genre-driven corpus, which results in a higher number of words in the queries, we used longer tuples for the topic-driven corpus queries (five single words) than for the genre-driven corpus queries (three trigrams). Table 1 shows the first three queries used in the construction of the English subcorpora:

Default parameters were used (10 tuples, 10 URLs per query) and no manual filtering of results was performed. Size information about the resulting corpus are provided in Table 2. Notice that the Italian-G subcorpus is much smaller than the rest. Another round of querying/text retrieval would have increased corpus size and made it comparable to the other subcorpora; however, in this case we favoured comparability in terms of corpus construction procedure over comparability of the final output.

Interestingly, the topic- and genre-driven corpora display very little overlap regardless of the fact that the seeds were obtained from the same text collections. The English ones share as few as 4 web pages (out of 89 in English-G and 95 in English-T), while the Italian ones only have one URL in common (out of 36 in Italian-G and 76 in Italian-T), thus suggesting that different seed selection methods do yield different results. In the next section, these differences are explored further.

3.3 Evaluation and Discussion

As a way of evaluating the output of the procedures employed to build topic- and genre-driven subcorpora, we asked informants to evaluate a sample of texts in terms of their perceived usefulness for a translation task. We randomly extracted the URLs of 10 texts chosen among the non-shared ones from each of the subcorpora, corresponding to between 10.5 and 27.8 % of the total URLs in the subcorpora, a representative sample according to [33]. They were mixed and presented to respondents in random order, as lists of 20 English and 20 Italian URLs.

The respondents were five translation teacher colleagues and 26 (BA/MA) students of translation with varying experience of corpus work who accepted to participate in the experiment on a voluntary basis. We provided them with a "model" ST, i.e. the PIL discussed in Sect. 2.3, and asked them to (a) open and quickly read

Table 1 Examples of tuples used for the construction of the English subcorpora

English-G	"the side effects" "inform your doctor" "you need to" "if you take" "solution for infusion" "doctor or pharmacist" "and what it" "effects not listed" "to your doctor"
English-T	mg mixture or ingredients your influenza pain symptoms please doctor bowl syringe capsules use kg

[8] We used the frontend developed by Eros Zanchetta [37] and available here: http://bootcat.sslmit.unibo.it/.

Table 2 Size data about the BootCaT subcorpora

Subcorpus	Number of texts	Number of words
English-G	89	166,276
English-T	95	174,397
Italian-G	36	60,478
Italian-T	76	133,965

Table 3 Overall number of relevant texts split by subcorpus

Subcorpus	Relevant texts
English-G	9
English-T	4
Italian-G	7
Italian-T	8

through the web pages associated with the URLs; (b) rate them according to whether they would include the texts in a corpus they would build for translating the ST (possibilities were: "definitely yes", "probably yes", "probably no" "definitely no"); (c) optionally, provide comments as to the reasons for their decision.

Table 3 shows the absolute number of relevant texts in each subcorpus based on respondent assessment. For a text to count as relevant, more than 50 % of the participants had to judge it "definitely" or "probably" appropriate for inclusion in a comparable corpus resource for the simulated translation task. As can be appreciated, the genre-driven corpora performed comparably well, or even better in the case of English-G, compared to the ones built following the traditional, topic-driven procedure. The subcorpus with the worst results was English-T, for which only 4 texts out of 10 were deemed useful by the majority of the respondents, while its Italian counterpart scored second best (8 "good" texts out of 10).

If we take a closer look at the results, observing the percentage of positive answers for each text in the 4 subcorpora and the related overall averages (Fig. 1), we notice that besides containing the highest number of relevant texts in absolute terms, English-G also has the highest average of positive answers (70.2 %), thus possibly pointing at high overall quality of the results. The poorer performance of Italian-G can only partly be explained by the lower number of relevant texts (7 vs. 9): in fact its lower average (61.2 %) seems to be due to the presence of three clearly "wrong" texts, each obtaining less than 20 % positive answers, with the remaining 7 texts above the 60 % threshold. The scenario is reversed for the topic-driven corpora. The average for Italian-T is very close to that for English-G (68.1 %), and much higher than English-T (44.8 %). The texts in Italian-T seem to display comparable, if not higher, percentages compared to the English genre-driven corpus, but a single text (number 10) lowers the average considerably. Finally, English-T displays the lowest average, resulting from a high number of possibly irrelevant texts, all well below the 50 % threshold.

While these results are encouraging in terms of the output quality of the proposed corpus-building procedure, they still do not shed light on whether the genre-driven corpus does indeed contain a higher number of comparable texts at the genre level. The data at our disposal do not allow us to settle the issue here, but the comments provided by the users offer interesting insights into the criteria adopted for judging the relevance of the texts. These include claims about the perceived "authoritativeness" of the websites where the texts are published and the terminological richness of the texts themselves (how many "technical terms" are found), but also, crucially, observations about topic- and genre-comparability, which are mentioned either as explicit criteria (e.g. "this is a patient information leaflet, but deals with a different medicine compared to the ST"), or implicit ones (e.g. "here terminology is probably pertinent"). In this respect it is interesting to notice that comments about all the texts with more than 80 % positive answers hint at the fact that the texts are PILs (same genre as the ST), no matter whether they refer to paracetamol medicaments or not (same topic as the ST). On the other hand, the highest scoring texts (>90 %: one in English-T, two in Italian-G, one in Italian-T) are those where both genre and topic match those of the ST.

More in-depth investigation would be required to evaluate and compare the output of the genre-driven corpus construction pipeline, and to estimate the extent to which genre comparability influences decisions regarding relevance of texts. However, the results we obtained adopting a relatively straightforward pipeline, are encouraging both for English and for Italian: starting from the same, manual corpus, and using a genre-driven procedure of seed selection besides the traditional topic-driven one, the size of the resulting corpus is doubled for English, and substantially increased for Italian (cf. Table 2), with comparable levels of perceived relevance. Furthermore, using n-grams as seeds makes seed selection more straightforward, since no reference corpus is required (differently from the topic-driven pipeline). This is certainly

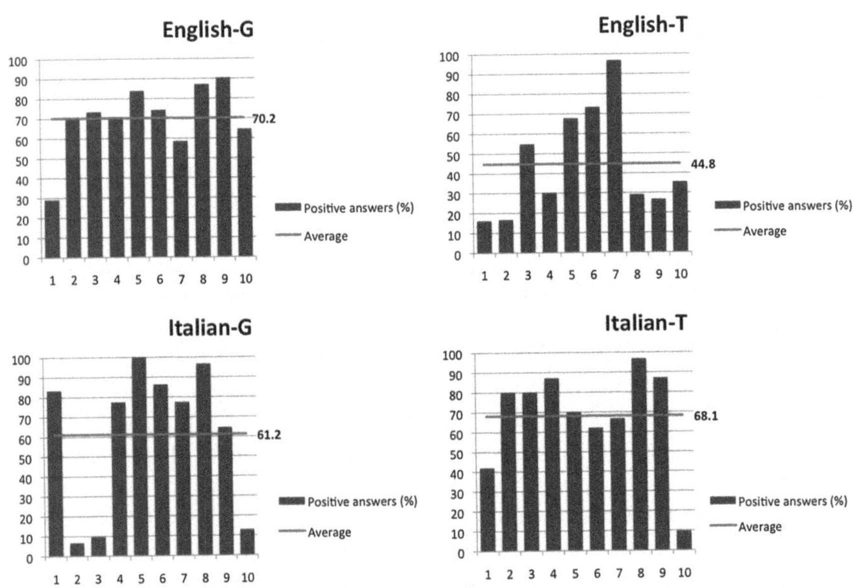

Fig. 1 Positive answers (%) for each text split by subcorpus

an advantage for translation professionals, who may not have a reference corpus available—or even understand the need for it.

Finally further analysis would be in order to shed light on the reasons for the varying quality and quantity of texts retrieved through the topic- and genre driven procedures for the two languages. Regarding quality, our tentative explanation is that the reduced relevance of the texts in the English-T subcorpus may be accounted for by the number of function words (e.g. "or", "your", cf. Table 1) that were key in the manual corpus and therefore made it to the seed list—remember that the lists were not manually edited to avoid inserting subjective biases in the procedure. As for quantity, the Italian subcorpora are consistently smaller than their English counterparts, more noticeably so in the case of the genre-driven subcorpus. In general, we got the impression that, while English PILs are available in different formats on the web, Italian ones are more likely to be in pdf, a format that is currently not supported by the BootCaT frontend we used. Specifically in the case of the Italian-G subcorpus, the n-gram procedure may work less well for languages with a rich morphology: since actual chunks of text are used as exact queries, and these include conjugated verbs, a negative effect on recall may be expected. Of course in actual practice this can be easily overcome by varying parameters of seed-selection (e.g. shorter n-grams) or increasing the number of queries submitted.

3.4 Using the BootCaT Corpus

Going back to the example in Sect. 2.3, the Italian BootCaT corpus returns 100 occurrences of the word "capsule" (English "capsules"), versus 0 in the manual corpus and 1,902 in the reference corpus. Browsing the left-end co-texts (within a span of 10 words preceding) we find both "scatola" (two occurrences) and "confezione" (two occurrences), but also the more frequent "astuccio" (six occurrences), not mentioned by the dictionary.

Moving beyond the lexical/terminological level, and considering cross-linguistic genre conventions, this comparable corpus provides evidence about the level of presupposition to be aimed at by the translator that would be difficult to obtain otherwise. The English ST is very reader-friendly, providing explanations of most terms (including rather common names for illnesses and medicines, e.g. "hypertension", "hallucinations", "vasodilators", "antidepressants"). A translator may wonder whether this tendency to explain and define terms is peculiar to this text or a convention of the genre, and whether Italian comparable texts explain domain-specific terminology to a similar degree. The first question can be answered by searching the English subcorpus, the second by searching the comparable Italian subcorpus; both are crucial for deciding how to tackle presupposition. For instance, should a rather obscure medical term like "trombocitopenia" (corresponding to English "thrombocytopenia", and meaning reduced blood platelet count) be accompanied by an explanation in Italian, as it is in the English ST, or not? The manual English subcorpus does not contain the word, suggesting that it might be quite rare in this genre. The automatic one returns

16 occurrences, 7 coming from the topic-driven corpus and 9 from the genre-driven one. None of the occurrences from the topic-driven corpus are accompanied by an explanation, while 7 (out of 9) of those from the genre-driven corpus are, suggesting that the ST is indeed following generic conventions in this case. Browsing both the manual and the automatic Italian corpus for evidence about corresponding generic conventions shows that the term is both more common than its English equivalent (9 occurrences in the manual corpus, 59 in the automatic one), and that it is virtually never accompanied by an explanation. Based on this evidence a translator should infer that the term does not require an explanation in Italian, and leave it out of the TT.

4 Conclusion

In this chapter we have presented a user's perspective on comparable corpora.We have discussed ways in which these resources can be used for reference purposes in a translation task, and highlighted positive features with respect to other, better established tools (electronic dictionaries, the web (though search engines), translation memories). Different types of comparable corpora were presented and their use exemplified, i.e. small *ad hoc* corpora, large web-derived reference corpora, and interactively constructed semi-automatic corpora, occupying a middleground between the former two. From the user's point of view, these resources are positioned along two double clines. First, in terms of usefulness/quality versus quantity: manually constructed corpora are very small, reliable and tuned to the task; as we move along the cline to semi-automatic and (web-derived) reference corpora, reliability and specialization decrease while size increases. At the same time, there is a cline in terms of time and effort required to obtain the corpora—maximum for manual corpora, minimum for reference corpora which are available in the public domain—and time and effort required for corpus searching and decision-making—minimum for the small corpora whose contents are familiar to the user and have been evaluated prior to inclusion in the corpus, and maximum for the huge reference corpora built by others for a host of different purposes; semi-automatic corpora once again occupy the middleground between these two opposites.

While we have suggested that these different corpus types do not provide alternatives but rather complementary resources, to be used in different ways and for different purposes during the translation process, we also believe that language services providers and students of translation will only engage with comparable corpora if these provide a positive tradeoff between the time and effort needed to construct and/or (learn to) consult them and their perceived usefulness. Given the reference needs of translators, the midway solution offered by semi-automatic corpora appear to be the most fruitful and likely to be taken on board by the profession.

The way forward should therefore be that of developing corpora and methods for constructing corpora that are simple, fast, and flexible, allowing (but not imposing) a degree of user control over corpus contents. We aim to pursue this objective in two main ways in future work. On the one hand, we are experimenting with other methods (besides BootCaT) for constructing specialized web-derived corpora on-the-fly. In particular, we are trying to tap the potential of Wikipedia as a repository of "virtual" comparable corpora of English and Italian [4]. All the linked entries in these languages have been downloaded, POS-tagged, lemmatized and indexed with the Corpus Workbench [10], forming a comparable corpus which is the sum of several hundred articles on the same topics. Using keywords derived from the human-generated categories accompanying the entries, users can select subsets of texts on the same topics from the two subcorpora, thus obtaining virtual comparable corpora that last the time of a search session. On the other hand, we aim to explore further the issues of corpus comparability and quality through user surveys, trying to understand how humans go about the task of selecting web texts in two or more languages for a specialized corpus/specic task. Hopefully some of these strategies can be used to improve semi-automatic corpus building methods; certainly they will help us shed some light on notions such as corpus comparability and adequacy, which have been at the basis of corpus linguistics since its early days.

Acknowledgments We would like to thank the students and colleagues who have kindly accepted to evaluate the URLs for us, Claudia Lecci for her expert insights about TM software, Federico Gaspari for fruitful lunchtime discussions on corpus construction strategies as well as the anonymous reviewer and the editors of the volume for their valuable feedback and suggestions.

References

1. Aston, G.: Corpus use and learning to translate. Textus **12**, 289–314 (1999)
2. Baroni, M., Bernardini, S.: Bootcat: Bootstrapping corpora and terms from the web. In: Proceedings of LREC 2004, pp. 1313–1316, Lisbon. ELDA (2004)
3. Baroni, M., Bernardini, S., Ferraresi, A., Zanchetta, E.: The wacky wide web: a collection of very large linguistically processed web-crawled corpora. Lang. Resour. Eval. **43**(3), 209–226 (2009)
4. Bernardini, S., Castagnoli, S., Ferraresi, A., Gaspari, F., Zanchetta, E.: Introducing comparapedia: a new resource for corpus-based translation studies. Paper Presented at the International Symposium on Using Corpora in Contrastive and Translation Studies (UCCTS 2010), Edge Hill University, Ormskirk (July 2010)
5. Biber, D., Conrad, S.: Lexical bundles in conversation and academic prose. In: Hasselgard, H., Oksefjell, S. (eds.) Out of Corpora: Studies in Honour of Stig Johansson, pp. 181–190. Rodopi, Amsterdam (1999)
6. Bowker, L.: Computer-Aided Translation Technology: A Practical Introduction. University of Ottawa Press, Ottawa (2002)
7. Bowker, L.: Examining the impact of corpora on terminographic practice in the context of translation. In: Kruger, A., Wallmach, K., Munday, J. (eds.) Corpus-Based Translation Studies, pp. 211–236. Continuum, London (2011)
8. Castagnoli, S.: Using the web as a source of LSP corpora in the terminology classroom. In: Baroni, M., Bernardini, S. (eds.) Wacky! Working Papers on the Web as Corpus, pp. 159–172. GEDIT, Bologna (2006)
9. Chama, Z.: From segment focus to context orientation. TC World, 2010. online: http://www.tcworld.info/index.php?id=167

10. Christ, O.: A modular and flexible architecture for an integrated corpus query system. In: Proceedings of COMPLEX 1994, pp. 23–32, Budapest (1994)
11. Cowie, A.P. (ed.): Phraseology: Theory, Analysis, and Applications. Oxford University Press, Oxford (2001)
12. Crossley, S.A., Louwerse, M.M.: Multi-dimensional register classication using bi-grams. Int. J. Corpus Linguist. **12**(4), 453–478 (2007)
13. Crowston, K., Kwasnik, B.H.: A framework for creating a facetted classification for genres: addressing issues of multidimensionality. Hawaii International Conference on System Sciences, 4, 2004. online: http://doi.ieeecomputersociety.org/10.1109/HICSS.2004.1265268
14. Dsilets, A., Melanon, C., Patenaude, G., Brunette, L.: How translators use tools and resources to resolve translation problems: an ethnographic study. In: Proceedings of MT Summit XII-Workshop: Beyond Translation Memories, Ottawa (2009)
15. Fantinuoli, C.: Specialized corpora from the web and term extraction for simultaneous interpreters. In: Baroni, M., Bernardini, S. (eds.) Wacky! Working Papers on the Web as Corpus, GEDIT, Bologna pp. 173–190 (2006)
16. Ferraresi, A., Bernardini, S., Picci, G., Baroni, M.: Web corpora for bilingual lexicography: a pilot study of English-French collocation extraction and translation. In: Xiao, R. (ed.) Using Corpora in Contrastive and Translation Studies, pp. 337–359. Cambridge Scholars Publishing, Newcastle (2010)
17. Gatto, M.: From Body to Web. An Introduction to the Web as Corpus. Laterza, Bari (2009)
18. Gavioli, L.: Exploring Corpora for ESP Learning. Benjamins, Amsterdam (2005)
19. Ghadessy, M., Henry, A., Roseberry, R.L. (eds.) Small Corpus Studies and ELT. Benjamins, Amsterdam (2001)
20. Goeuriot, L., Morin, M., Daille, B.: Compilation of specialized comparable corpus in French and Japanese. In: Proceedings of the ACL-IJCNLP workshop Building and Using Comparable Corpora (BUCC 2009), 2009
21. Gries, S.Th., Mukherjee, J.: Lexical gravity across varieties of English: an ICE-based study of n-grams in Asian Englishes. Int. J. Corpus Linguist. **15**(4), 520–548 (2010)
22. Heid, U.: Corpus linguistics and lexicography. In: Kytö, M., Lüdeling, A. (eds.) Corpus Linguistics: An International Handbook, pp. 131–153. Mouton de Gruyter, Berlin (2008)
23. Hoey, M.: Lexical priming and translation. In: Kruger, A., Wallmach, K., Munday, J. (eds.) Corpus-Based Translation Studies, pp. 153–168. Continuum, London (2011)
24. MeLLANGE: Corpora and e-learning questionnaire. Results summary - professionals. Internal Document (2006)
25. MultiTrans. Multitrans 4(tm): Taking the multilingual textbase approach to new heights. MultiCorpora White Paper, online: http://www.multicorpora.com/lesNVIAdmin/File/MCwhitepaper1.pdf (August 2005)
26. Munday, J.: Looming large: a cross-linguistic analysis of semantic prosodies in comparable reference corpora. In: Kruger, A., Wallmach, K., Munday, J. (eds.) Corpus-Based Translation Studies, pp. 169–186. Continuum, London (2011)
27. Pearson, J.: Terms in Context. Benjamins, Amsterdam (1998)
28. Pearson, J.: Using parallel texts in the translator training environment. In: Zanettin, F., Bernardini, S., Stewart, D. (eds.) Corpora in Translator Education, pp. 15–24. St Jerome, Manchester (2003)
29. Philip, G.: Arriving at equivalence: Making a case for comparable general reference corpora in translation studies. In: Beeby, A., Rodríguez Inés, P., Sánchez-Gijón, P. (eds.) Corpus Use and Translating, pp. 59–73. Benjamins, Amsterdam (2009)
30. Rinsche, A., Zanotti, N.P.: Study on the Size of the Language Industry in the EU. European Commission - Directorate General for Translation, Brussels (2009)
31. Santini, M.: State-of-the-art on automatic genre identification. Technical Report ITRI-04-03, ITRI, University of Brighton, UK (2004)
32. Serianni, L.: Grammatica Italiana. UTET, Torino (1991)
33. Sharoff, S.: Creating general-purpose corpora using automated search engine. In: Baroni, M., Bernardini, S. (eds.) Wacky! Working Papers on the Web as Corpus, pp. 63–98. GEDIT, Bologna

34. Swales, J.: Genre Analysis. English in Academic and Research Settings. Cambridge University Press, Cambridge (1990)
35. Varantola, K.: Translators and disposable corpora. In: Zanettin, F., Bernardini, S., Stewart, D. (eds.) Corpora in Translator Education, pp. 55–70. St Jerome, Manchester (2003)
36. Williams, I. A.: A translator's reference needs: dictionaries or parallel texts. Target **8**, 277–299 (1996)
37. Zanchetta, E.: Corpora for the masses: the BootCaT front-end. Pecha Kucha Presented at the Corpus Linguistics 2011 Conference. University of Birmingham, Birmingham (July 2011)

Exploiting the Incomparability of Comparable Corpora for Contrastive Linguistics and Translation Studies

Stella Neumann and Silvia Hansen-Schirra

Abstract For empirical research in translation studies and contrastive linguistics comparability across languages is an important and non-trivial issue, particularly, if we aim at quantifying differences and commonalities. We approach this problem by investigating the CroCo Corpus of English and German originals and translations in eight different registers concentrating on the (in)comparability of the different subcorpora. They are comparable across languages but incomparable in terms of functional, i.e. register, variation. A feature matrix is analysed to reveal differences and commonalities between the bilingually comparable texts (i.e. English and German originals) and the monolingually comparable texts (originals and translations in the same language). The study draws on deep, multilayer annotation and sophisticated querying methods to quantify the features. The interpretation is based on a statistical analysis of the query results enabling to bridge the gap between qualitative properties of the language systems involved and their quantitative instantiations in texts.

Keywords CroCo Corpus · Translation corpora · Comparable corpora · Multi-layer annotation · Alignment · Register variation

1 Motivation

Comparability of corpora is not just an issue to be considered when compiling a corpus (see for instance [13]) but also one affecting directly the analysis. Simply put, incomparability may occur each time two sets of linguistic data are compared. The

---

S. Neumann (✉)
Rheinisch-Westfälische Technische Hochschule Aachen, Kármánstr. 17/19, 52062 Aachen, Germany
e-mail: neumann@anglistik.rwth-aachen.de

S. Hansen-Schirra
Johannes Gutenberg Universität Mainz, An der Hochschule 2, 76726 Germersheim, Germany
e-mail: hansenss@uni-mainz.de

S. Sharoff et al. (eds.), *Building and Using Comparable Corpora*,
DOI: 10.1007/978-3-642-20128-8_17,

present paper focuses on this aspect of working with corpora. It pursues linguistic goals but points out some avenues for computational linguistic application.

For empirical research in translation studies and contrastive linguistics comparability across languages is an important and not trivial issue, particularly, if we aim at quantifying differences and commonalities. Even in the language pair English and German, which is in close contact, there are differences between the language systems as well as in the qualitative and quantitative instantiations in texts. The system-oriented differences have been most recently covered in a comprehensive way by [14],[15] offers an overview of quantitative and text-based contrastive differences. In addition, translations pose problems such as lexical gaps, grammatical shifts, semantic modulation, all of which result in potential divergences between source and target segments (see [7] for an overview).

Comparability of linguistic features is a long-standing issue in contrastive linguistics. König and Gast [14], for instance, address it by invoking some general semantic notions such as 'temporal relation'. Frequently, it is claimed (e.g. [20]) that underlying functions can serve as a basis of comparison. The present paper will show that this is not just an academic question having to do with traditional system-oriented studies but one that can have some serious implications for any type of cross-linguistic corpus enterprise.

Often, translations are employed for comparing linguistic units [12]. Johansson [12] also addresses the improved reliability and validity of using corpora for balancing out the interfering factors in the translation process. However, contrastive linguists frequently use corpora merely as a source of examples, thus not exploring the types of contrasts which only become visible in frequency effects.

The issue of comparability is a known problem in computational work on corpora as witnessed by some of the papers in this volume (see Chap. 2). In computational work, in particular, the use of corpora has to be efficient starting with easy accessibility of multilingual data. This is one of the reasons for resorting to comparable rather than translation corpora. This paper will show that the inclusion of translations provides additional information in various respects: Translations do seem to show some general characteristics frequently called translation properties. More specifically, they also display traits of the contrastive differences in a given language pair in the form of interference of certain features. Interference may also be specific to one translation direction. The latter point may be of particular interest to computational approaches: It entails that one direction may require clear constraints in terms of target language orientation while the opposite direction may permit more flexibility including a good deal of interference of the source language (cf. [21]).

The present paper approaches the issue of (in)comparability by investigating the register-controlled CroCo Corpus [10, 17] of English and German originals and translations in eight different registers concentrating on the (in)comparability in terms of register variation. The remainder of the paper is organized as follows. Section 2 presents the corpus design and, more specifically, the registers included in the CroCo Corpus. The analysis of register features is presented in Sect. 3.2 complemented by an introduction to the chosen methodology (Sect. 3.1). In Sect. 3.3 the findings are discussed in a more general perspective and, finally, Sect. 4 contains some concluding remarks as well as an outlook.

2 The CroCo Corpus

2.1 Corpus Design

The CroCo Corpus was specifically created in order to investigate the typical linguistic properties of translations as compared to originals in the language pair English–German. The corpus design was guided by a number of criteria including a size sufficient to allow generalizable claims, balance as well as comparability across languages (see [17]). Only those registers were considered for inclusion in the corpus which are translated in the two languages in both directions at all.

There is one paradox in compiling a corpus for the study of translation registers that continues to spark discussions: The selection of texts in the CroCo Corpus is based on the assumption that they belong to the same register, yet only an in-depth analysis of their defining features can determine whether they actually do (for a basic description of this kind of register analysis see [9]). Register analysis organizes features into those concerning referential meaning ("field of discourse"), those reflecting pragmatic aspects ("tenor of discourse") and those indicating how language is structured under the influence of the referential and pragmatic meaning ("mode of discourse"). CroCo addresses the register aspect by including a cursory and intuitive socio-cultural register analysis in the metadata. Additionally, statistical processing of the corpus findings can help identify outliers.

The core corpus is thus divided into four subcorpora: English originals (EO), German translations (GTrans), German originals (GO) and English translations (ETrans), which each contain texts from the eight registers. The corpus contains parallel parts (originals and their matching translations in the other language), monolingually comparable parts (originals and translations within one language) as well as bilingually comparable parts (originals in both languages). Following Biber's [3, 4] calculations, each register consists of at least 10 texts and 3,125 words per text (full texts and samples were collected depending on the register), yielding an overall corpus size of approximately 1 million words.

2.2 The Registers

In order to satisfy the design criterion of balance, the registers were selected with a view to covering a certain variation in terms of context of situation. The corpus meets these requirements by including texts from eight different registers which are in fact translated into both English and German: political essays (ESSAY), fictional texts (FICTION), instructional texts (INSTR), popular scientific writings (POPSCI), shareholder communication (SHARE), prepared speeches (SPEECH), tourism leaflets (TOU) and websites (WEB). In addition to being translated in both directions, these registers are assumed to foreground different registerial characteristics as determined in a preliminary analysis (cf. [17]). The registers can be roughly characterized as follows.

ESSAY contains treatises by senior politicians on a range of political topics. These texts often explain a given area of politics and aim at convincing the addressee of its importance. In terms of experiential domain under 'field of discourse', they cover various current issues in politics. The official author is typically a well known politician, although they may have been written by members of the politician's staff. The intended audience is the general public, sometimes the public in a foreign country. The mode is written to be read.

The FICTION register contains extracts from contemporary novels of which translations into the respective other language exist. One might argue that the character of literary texts as creating an imaginary world and situating the (linguistic) action within this world prohibits grouping—and analyzing—these texts as a register in its own right. However, like any other register, literary texts also reflect a given context of situation including authors/writers and readers. As Halliday and Hasan [9] argue, "we are never selecting with complete freedom from all the resources of our linguistic system". This should also apply to literary texts which, despite their originality, reflect the situation and can be said to be related to other literary texts in terms of intertextuality. The field of these texts is diverse, their audience as well. Again, all texts are in written mode.

The register INSTR consists of a variety of instruction manuals for computer software as well as hardware, for electric appliances as well as machines and tools. The overwhelming part of the texts is, of course, devoted to guiding the addressee on how to use the object, which determines the field of discourse. The tenor comprises authors who are experts in the field as well as addressees who may be experts in the field but who are not familiar with the object. Since the interest of the CroCo Project is mainly in linguistic rather than aspects of texts related to other modalities, the texts were chosen with a bias towards instruction by means of language.

POPSCI contains popular scientific articles from magazines and books aimed at the general public and their translations.[1] According to Böttger and Probst [6] popular scientific texts aim at informing a general audience in a clear and comprehensible way on findings and developments of a scientific field. The authors stress the importance of balancing out a sound scientific foundation and general comprehensibility. The field of the texts covers a wide range of topics from psychology to chemistry; the authors are typically experts addressing a lay audience. The mode is written.

SHARE contains letters from the CEOs of various companies to their shareholders. These texts inform the shareholders on the performance of the company in the last accounting period. Beyond this, they also have a persuasive character because

[1] The starting point for collecting this subcorpus was a corpus originally collected by Monika Doherty (see [8]).

their goal is to convince the shareholder of the successful or, under difficult circumstances, appropriate activities of management during the period in question. Although the managers as speakers are the experts, they address the reader in a respectful if not formal way, since the management is accountable to the shareholder as the addressee. The texts are in written mode.

SPEECH contains manuscripts of speeches held by senior government officials. They are written to be spoken; they are expected to have special properties in terms of their mode of discourse. These speeches are in the widest sense of the word political and aim at convincing the audience of the position of the speaker (and the government). The field ranges from terrorism over poverty to the financial markets. The texts are addressed at the general public, sometimes more specifically at the audience of a specific event consequently resulting in a more specialized field.

The texts in the register TOU are promotional tourism texts, typically published by the tourist information agency of a city or region. The texts inform on and promote a certain place of interest. The field of discourse thus tends to be more focused on this specific information than in other registers in the corpus. The authors of the texts are as it were experts on the place presented in the texts and address a general audience, even if some brochures target a specific group of tourists. In general, the mode of discourse is written (typically with colorful pictures)—as to whether the electronic channel of some of the texts results in a different interpretation see [19].

The eighth register is WEB and contains websites. In the selection process all kinds of websites were included that contain translations. This led to a high degree of variation between the individual websites which may even be interpreted in Santini's [18] terms as different web genres. The websites thus address topics from an artist pair's website through a website promoting the soy bean to AIDS prevention for younger people. The variation is also reflected in the goal pursued by the websites with some aiming more at informing the user while others serve clear public relations aims. This also leads to differences in the tenor as well as in the mode of discourse. While some sites are very colorful making the best of the electronic medium, others present their information in a very static, print-oriented way.

As discussed in the previous section, within-register variation cannot be kept completely constant. Register may even be comparable across languages in the relatively wide range of variation they permit. This is certainly true for WEB in the language pair English–German.

2.3 *Corpus Enrichment and Querying*

The CroCo Corpus is tokenized and annotated for part-of-speech, morphology, phrasal categories and grammatical functions. Furthermore, clauses and sentences are segmented. Words, clauses and sentences are aligned with the alignment on phrase level to be derived from the other alignment levels. A detailed description of the corpus enrichment and representation can be found in [10].

Queries for the features under investigation in this paper range from simple part-of-speech queries to complex queries combining various annotation layers. Some of the more complex queries are discussed in [22].

3 Feature Distribution Across Registers and Languages

3.1 Methodology

The subcorpora consisting of the different registers are comparable across languages but incomparable in terms of functional, i.e. register, variation. Register features can be quantified and interpreted in the following corpus combinations:

- bilingually comparable texts (English and German originals),
- monolingually comparable texts in English (English originals and English translations) and
- monolingually comparable texts in German (German originals and German translations).

A feature matrix is analyzed to reveal differences and commonalities between these corpus relations. It draws on the deep, multilayer annotation and querying methods mentioned in Sect. 2.3.

The interpretation in the perspective of (in-)comparability of the comparable corpora is based on a statistical analysis of the query results enabling to bridge the gap between qualitative properties of the language systems involved and their quantitative instantiations in texts. The analysis of the bilingually comparable texts reveals different register conventions for German and English. The monolingually comparable texts show whether translated text differs from original text in the same language and whether German is more receptive to interference effects from English than vice versa.

The feature matrix employed in this study draws on a detailed analysis of register variation [16] in which the features were grouped in various combinations to shed light on specific aspects of the abstract register parameters field, tenor and mode of discourse. Each parameter was specified in the form of—still abstract—subdimensions such as social distance for tenor of discourse. Where these subdimensions were still too abstract to be observable and thus measurable directly in texts, they were broken down into additional intermediate categories as in the case of social distance. This subdimension is reflected in different styles as, for instance, casual, neutral or formal style. Thus, different combinations of the frequencies of nouns, attributive adjectives, prepositions, forms of address, titles, salutations, imperatives, interrogatives, personal pronouns, place and time adverbs, contractions and colloquialisms as well as lexical density were compared and weighted against the respective values for reference corpora in both languages to identify differing patterns of styles as types of social distance in the registers. All in all, the feature matrix included indicators for six subdimensions of the three parameters. Even though one and the same indicator may apply to various subdimensions, their interpretation may still differ because they were analyzed in combination with other indicators in the light of the respective subdimension resulting in specific interpretations for each area.

For contrastive comparisons, Neumann [15, 16] uses relative register values, i.e. the difference between the frequency of a linguistic feature in a register-controlled corpus and the respective frequency in a reference corpus. Although this appears to be a plausible attempt to handle the influence of language-specific peculiarities, the two frequencies combined in these values still co-vary. Since the present paper also aims at providing an overview of the variability of seemingly comparable features we do not use these relative register values.

Neumann [16] employs significance testing for determining the register-specific role of the various features. However, the study still placed strong emphasis on the qualitative interpretation of the specification of each register. The comparison across languages (both in terms of contrastive and of translation-oriented comparison) was limited to two registers. This approach also made selected quantitative analyses of the data possible.

As the present study aims at giving an overview of the cross-linguistic (in) comparability of comparable corpora, this approach is somewhat softened in favor of an overview of quantitative lexico-grammatical variation across registers. As a consequence, the analyses of feature combinations will be simplified.

The 23 features included in this study are the following:

Frequencies of nouns, attributive adjectives, prepositions, infinitives, finites, past tense orientation of verbs, passive voice, modal verbs, imperatives, interrogatives, coordinating and subordinating conjunctions, pronouns, place and time adverbs, modal adverbs, various theme selections (adverbials, textual themes, verbs, subjects and objects), contractions as well as lexical density. All features are considered relative to the relevant frequency of the respective basic unit, i.e. number of tokens for parts of speech, number of sentences for sentence-based features such as finites and theme distribution for themes.

For reasons of space, the statistical analysis concentrates on the four registers of literary texts (FICTION), instruction manuals (INSTR), popular-scientific writings (POPSCI) and letters to shareholders (SHARE). These registers display a reasonable amount of variation with FICTION and INSTR representing opposite ends of the range of variation with respect to many features and POPSCI and SHARE ranging more towards the middle and displaying some similarities. For these, the relative mean values are processed in analyses of variance including a Tukey HSD posthoc test (for a discussion of this type of statistical analysis in corpus data see [16]). In order to ensure homogeneity of variance as one of the a priori assumptions of analysis of variance, only results for 10 texts per register are included in computing statistical tests.

Based on the findings of [16] we assume that there are some significant contrastive differences between arguably comparable registers. Comparing translations and originals (within one language but also across languages) we assume that the

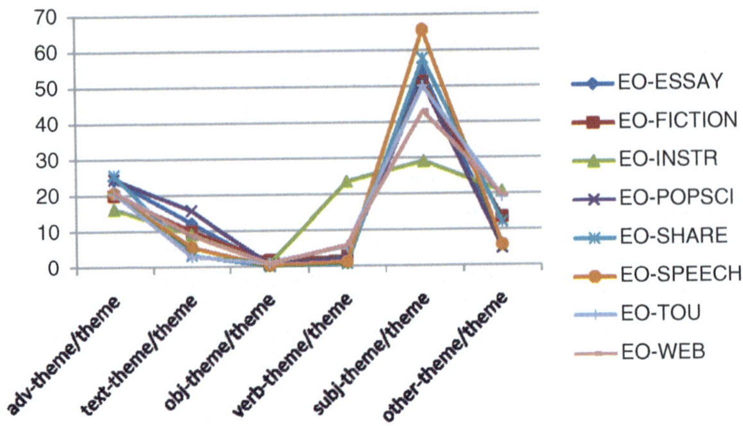

Fig. 1 Theme distribution in English originals

translations adhere to the register conventions of the target language (and audience), if there is a difference between English and German originals.

The following discussion of results proceeds in two steps. In the next subsection, we will first discuss peculiarities of the registers and corpus combinations in the light of feature frequencies. The second step in Sect. 3.3 consists in a broader overview abstracting from significant findings for individual features to tendencies of divergences between the corpus combinations on the one hand and the registers on the other.

3.2 Frequency Distributions of Register Features

We approach the first hypothesis on the basis of the distribution of themes, i.e. the distribution of syntactic functions in sentence-initial position, in English and German originals. Figure 1 shows that in English the different registers use different syntactic functions in theme position. For instance, subject theme is heavily used in speeches whereas it is rather untypical of instructions. The latter frequently use verbs (more specifically imperatives) in theme position, which is not usual for the other registers.

Comparing English and German originals, there are some predictable differences stemming from the word order differences between English and German. The relative frequencies of English themes are thus quite different from the German ones (Fig. 2), however, we can still observe some similarities in the distributions for the registers across languages. Consequently, we find the highest frequencies for subjects in speeches and the lowest in instructions. The variation for adverbials in thematic position and text-themes (and even the category "other") show similar patterns in English and German as well.

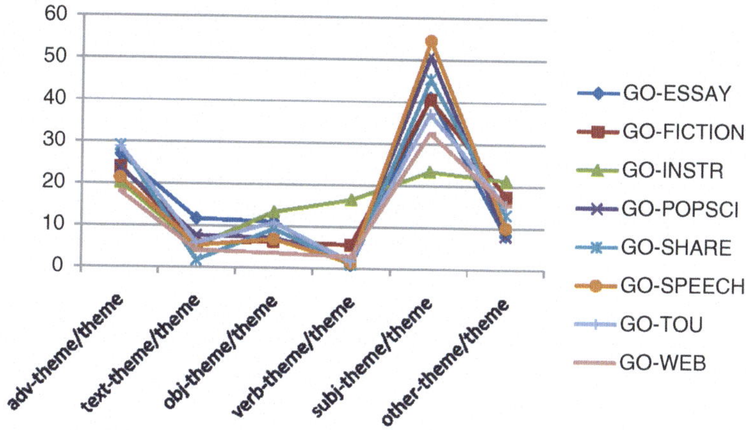

Fig. 2 Theme distribution in German originals

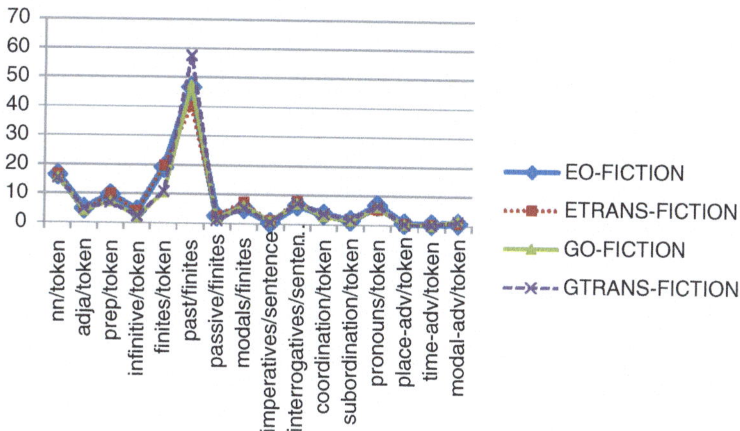

Fig. 3 Feature distribution for FICTION

If we compare translations and originals in English and German the picture is more diversified. The feature distribution[2] of translated fiction, for instance, matches quite consistently the target language originals (Fig. 3). This suggests an overwhelming tendency to adhere to target language register conventions.

However, for the majority of the CroCo registers the feature distribution varies across languages. An example is displayed in Fig. 4 on the basis of the popular-scientific register: on the one hand, the following features do not seem to be distinctive for English and German since they exhibit similar distributions in both languages: attributive adjectives, interrogatives, coordination, subordination, pronouns, place and time adverbs. On the other hand, the popular-scientific texts show contrasts with

[2] Because of limited space the default feature of certain feature bundles is not presented in the figure—e.g. imperatives and interrogatives are displayed whereas declaratives are neglected.

respect to the number of nouns, prepositions, infinitives, finites passives and modal adverbs. These features seem to be more typical of English original popular-scientific texts compared to German originals.

In English, the distribution (in terms of frequencies) of prepositions, infinitives, finite verbs and passives confirms our assumption that the translations adhere to the register conventions of the target language (i.e. normalization; [1]). We find source language interference ([21]; or shining-through, [20]) for nouns since their frequency is more similar to the German source texts than to the English originals. A hybrid translation strategy can be observed for past tense verbs since the frequency of English translations lies in between those for English and German originals.

German translations also show a clear target language orientation for prepositions, infinitives, finite verbs and passives indicating normalization. Interference effects can be found for nouns and past tense verbs. These results show that variation can also be found between translations and originals in the same language making the monolingually comparable corpora incomparable with respect to the respective features.

3.3 Contrasts in Statistical Significances

Generalizing from the frequencies of specific linguistic features of each of the four registers we will now compare the divergences between the various corpus combinations in each of the different registers. For this purpose, we discuss the number of significant divergences found for any of the features per register and corpus combination (see Table 1). This means that we abstract away from individual frequencies and only look at whether the respective comparison retrieved a significant difference. The values in Table 1 and Figs. 5, 6 are thus purely nominal. We will first discuss

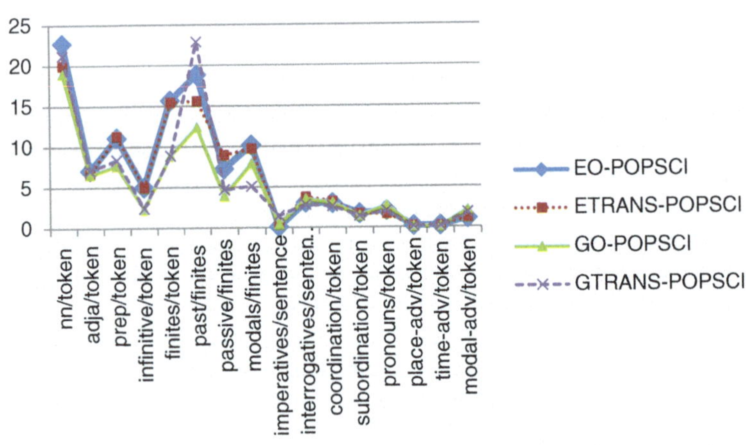

Fig. 4 Feature distribution for POPSCI

the total number of significant differences between the corpus combinations across registers and then the total number of differences between the registers across corpus combination.

In terms of corpus combinations, there is a clear difference between the monolingual combinations of originals and translations and the cross-linguistic corpus combinations with the former containing clearly fewer significant divergences than the latter. The least differences can be found between the two English subcorpora. Translators clearly follow target language preferences across registers. This seems to be true for all registers under investigation, but in particular for the literary texts in the FICTION register. Here, we can assume that the relative unrestrictedness of this register allows a high degree of flexibility on the part of the translators. To a lesser extent, this is also true for the German subcorpora. German originals and translations also display clearly fewer significant differences than the bilingual comparisons.

The most striking differences apply to GO and ETrans. Presumably, this reflects the similarity between English originals and translations and is due to the target language orientation of the English translations. These differences have a clear impact not just for modeling translation but also for language technological applications in the field of translation. In order to produce acceptable output that meets the diverging preferences of different languages, different tendencies in terms of normalization have to be considered. As far as the translation-related interpretation is concerned, the findings appear to corroborate Toury's [21] claims that the prestige of languages has an impact on the shape of translations.

The comparison of translated texts in both languages accounts for the second highest number of significant differences. This can be seen as a corollary of the differences between the other corpus combinations, in particular the contrastive differences between the originals and the behavior of the translations with respect to their aligned originals.

The comparison between the originals in both languages in the four registers (the second column in Fig. 5) shows that there are many contrastive differences in the frequencies of features. It has to be taken into account, however, that the frequencies discussed here do not include features that are functionally equivalent. A standard example is the passive in English and German which, in its central form consisting of a form of *be/werden* and the past participle, displays a clear contrastive difference in frequency, with English seemingly drawing much more on the passive voice. When

Table 1 Summary of the significant differences for the registers and corpus combinations

	FICTION	INSTR	POPSCI	SHARE	Total
EO-ETRANS	0	3	1	2	6
EO-GO	8	8	8	5	29
EO-GTRANS	8	7	10	5	30
GO-ETRANS	7	11	7	9	34
GO-GTRANS	1	4	4	1	10
GTRANS-ETRANS	6	9	11	7	33
Total	30	42	41	29	

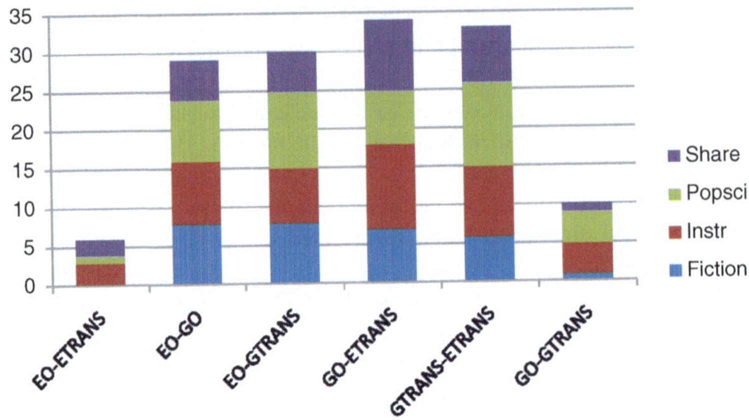

Fig. 5 Number of significant differences for the 23 features in the various corpus combinations

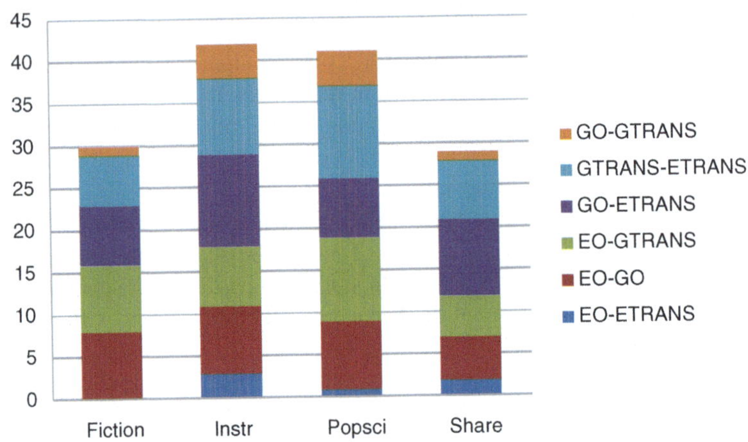

Fig. 6 Distribution of the significant differences for the 23 features per register

including the functional alternatives with, e.g. *sein* + *zu*+ infinitive in German the difference is balanced out. This highlights the limitation of too superficial an approach to cross-linguistic comparison.

The comparison of significant differences between the registers also brings to light some interesting divergences. Again, the data can be divided into two groups with INSTR and POPSCI displaying a clearly higher number of significant differences across all types of subcorpora than FICTION and SHARE. The register containing the highest number of differences is INSTR. Most of the significant differences can be found in the verbal categories and can be explained by contrastive differences between the two languages having to do with how commands are realized (see [19]). The differences between the subcorpora in POPSCI can arguably be found in an altogether different feature of cross-linguistic comparison: this register has quite diverging features in the two languages, at least as far as translations are concerned. Medium-length articles that popularize recent research findings do not occur frequently in German. Moreover, they are virtually never translated into English. As a consequence, the corpus contains somewhat different texts in the two translation directions resulting in the problems of incomparability Bernardini [2] discusses with respect to literary texts in Italian and English. The fact that the POPSCI register contains the highest number of divergences in the originals in both languages corroborates this issue. The register contributing the lowest number of significant differences is SHARE suggesting that this register is rather similar across the two languages. The reason why FICTION also scores low on divergences probably has to do with the flexibility of the register as mentioned above.

What this discussion clearly shows is that, while we can distinguish different degrees of impact of contrastive differences, these have clear implications for corpus-based quantitative comparisons even in registers which display a relatively low degree of divergences such as literary texts and letters to shareholders. The clear increase in significant differences from monolingual to cross-linguistic comparisons calls into question any approach at a simplistic quantitative exploitation of comparable corpora.

The fact that there are also differences between the monolingual corpora (EO-ETrans and GO-GTrans) signals that the impact of translation should not be underestimated.

4 Conclusions and Outlook

This paper can be read as a cautionary tale highlighting the problems a direct comparison of contrastive registers creates. There are two alternative approaches which can at least partly keep these issues in check. One is Biber's [5] inductive approach which postpones any direct cross-linguistic comparison to the last stage consisting of the interpretation of communicative functions which are derived in consecutive steps from the statistical analysis of monolingual data. This approach allows comparing data sets which diverge in corpus design as well as in features under investigation. This advantage comes at the price of limiting the explanatory space to findings that can be derived inductively from the data. A theory-driven approach to quantitative cross-linguistic comparisons that overcomes this limitation is described in [15, 16] which reflects a comprehensive overview of theory-derived features comparing only relative register values against the background of reference corpora. This approach, too, has its limitations as the relative register values co-vary and there is a certain danger of circularity with respect to hypotheses, operationalizations and their interpretation.

The incomparabilities between the different subcorpora shown in this paper provide the background against which theory-derived probabilistic distributions can be computed which in turn may facilitate automatic register classifications. These can then be exploited for various computational-linguistic tasks such as information extraction or machine translation.

In addition this paper shows that the comparison of translated versus original text in the same language provides additional information on the incomparability of corpora. It shows on the basis of a rich feature matrix and a wide range of different registers that interference effects from the source text as well as hybrid translation strategies may result in feature frequencies and distributions which are different from original texts in the same language. The same may hold true for other kinds of text production such as non-native writing which are constrained by external factors beyond situational factors and may also contribute to the incomparability of corpora.

The paper, however, cannot answer the old question of whether to rely on external or internal criteria for corpus creation and analysis. In our view both types of criteria seem to be inextricably linked. We therefore hope to have shown that sole reliance on external criteria such as some rough general classification of comparability across languages can be a source of reduced accuracy for many corpus applications.

Acknowledgments The authors would like to thank the CroCo project team in general and Erich Steiner in particular for continuous discussions on register variation and the comparability of cross-linguistic corpora. We gratefully acknowledge support by the German Research Foundation (DFG) in project no. STE 840/5-2 and HA 5457/1-2.

References

1. Baker, M.: Corpus-based translation studies: the challenges that lie ahead. In: Harold, S. (ed.) Terminology, LSP and Translation: Studies in Language Engineering in Honour of Juan C. Sager, pp. 175–186. Benjamins, Amsterdam (1996)
2. Bernardini, S.: Collocations in translated language: combining parallel, comparable and reference corpora. In: Proceedings from the Corpus Linguistics Conference Series, University of Birmingham, UK (2007). http://www.corpus.bham.ac.uk/corplingproceedings07/paper/15_Paper.pdf
3. Biber, D.: Methodological issues regarding corpus-based analyses of linguistic variation. Literary Linguist. Comput. **5**(3), 257–269 (1990)
4. Biber, D.: Representativeness in corpus design. Literary Linguist. Comput. **8**(4), 243–257 (1993)
5. Biber, D.: Dimensions of Register Variation: A Cross-Linguistic Comparison. Cambridge University Press, Cambridge (1995)
6. Böttger, C., Probst, J.: Adressatenorientierung in englischen und deutschen Texten. Arbeiten zur Mehrsprachigkeit—Folge B **23**, 1–28 (2001)
7. Čulo, O., Hansen-Schirra, S., Neumann, S., Vela, M.: Empirical studies on language contrast using the English–German comparable and parallel CroCo Corpus. In: Proceedings of the LREC 2008 Workshop on Building and Using Comparable Corpora, pp. 47–51, Marrakesh, Morrocco (2008). http://www.lrec-conf.org/proceedings/lrec2008/workshops/W12_Proceedings.pdf
8. Doherty, M.: Structural Propensities: Translating Nominal Word Groups from English into German. Benjamins, Amsterdam (2006)
9. Halliday, M.A.K., Hasan, R.: Language, Context, and Text: Aspects of Language in a Social-Semiotic Perspective. Oxford University Press, Oxford (1989)

10. Hansen-Schirra, S., Neumann, S., Steiner, E.: Cross-Linguistic Corpora for the Study of Translations: Insights from the Language Pair English–German. de Gruyter Mouton, Berlin (2012)
11. Hansen-Schirra, S., Neumann, S., Vela, M.: Multi-dimensional annotation and alignment in an English–German translation corpus. In: Proceedings of the Workshop on Multi-dimensional Markup in Natural Language Processing, pp. 35–42, Trento, Italy (2006). http://aclweb.org/anthology-new/W/W06/W06-2705.pdf
12. Johansson, S.: Contrastive linguistics and corpora. In: Granger, S., Lerot, J., Petch-Tyson, S. (eds.) Corpus-Based Approaches to Contrastive Linguistics and Translation Studies, pp. 31–44. Rodopi, Amsterdam (2003)
13. Kilgarriff, A.: Comparing corpora. Int. J. Corpus Linguist. **6**(1), 97–133 (2001)
14. König, E., Gast, V.: Understanding English–German Contrasts. Schmidt, Berlin (2007)
15. Neumann, S.: Contrasting frequency variation of grammatical features. In: Konopka, M., kubczak, f., Mair, C., Sticha, F., Waßner, U.H. (eds.) Grammatik und Korpora 2009: Dritte Internationale Konferenz, pp. 389–410. Narr, Tübingen (2011)
16. Neumann, S.: Contrastive register variation: a quantitative approach to the comparison of English and German. de gruyter mouton, Berlin (in press)
17. Neumann, S., Hansen-Schirra, S.: The CroCo project: crosslinguistic corpora for the investigation of explicitation in translations. In: Proceedings from the Corpus Linguistics Conference Series, vol. 1, Birmingham, UK (2005). http://www.corpus.bham.ac.uk/PCLC/cl-134-pap.pdf
18. Santini, M.: Web pages, text types, and linguistic features: some issues. ICAME J. **30**, 67–86 (2006)
19. Steiner, E.: Translated Texts: Properties, Variants, Evaluations. Lang, Frankfurt (2004)
20. Teich, E.: Cross-Linguistic Variation in System and Text: A Methodology for the Investigation of Translations and Comparable Texts. Mouton de Gruyter, Berlin (2003)
21. Toury, G: Descriptive Translation Studies and Beyond. Benjamins, Amsterdam (1995)
22. Vela, M., Hansen-Schirra, S., Neumann, S.: Querying multi-layer annotation and alignment in translation corpora. In: Davies, M., Rayson, P., Hunston, S., Danielsson, P. (eds.) Proceedings of the Corpus Linguistics Conference CL 2007, University of Birmingham, UK, 27–30 July 2007. http://ucrel.lancs.ac.uk/publications/CL2007/paper/97_Paper.pdf

The manufacturer's authorised representative in the EU is Springer Nature Customer Service Centre GmbH, Europaplatz 3, 69115 Heidelberg, Germany. If you have any concerns regarding our products, please contact ProductSafety@springernature.com

Printed and bound by CPI Group (UK) Ltd, Croydon, CR0 4YY
22/09/2026
02221063-0016